ADAPTIVE FILTERS

THE KLUWER INTERNATIONAL SERIES IN ENGINEERING AND COMPUTER SCIENCE

VLSI, COMPUTER ARCHITECTURE AND DIGITAL SIGNAL PROCESSING

Consulting Editor
Jonathan Allen

ADAPTIVE FILTERS:

STRUCTURES, ALGORITHMS, AND APPLICATIONS

MICHAEL L. HONIG
Bell Communications Research, Inc.
Murray Hill, New Jersey

DAVID G. MESSERSCHMITT
University of California
Berkeley, California

KLUWER ACADEMIC PUBLISHERS
Boston—The Hague—London—Lancaster

Distributors for North America:
Kluwer Academic Publishers
190 Old Derby Street
Hingham, MA 02043, U.S.A.

Distributors outside North America:
Kluwer Academic Publishers Group
Distribution Centre
P.O. Box 322
3300 AH Dordrecht
THE NETHERLANDS

Library of Congress Cataloging in Publication Data

Honig, Michael L.
Adaptive filters.

(The Kluwer international series in engineering and computer science)
Includes bibliographies and index.
1. Adaptive filters. I. Messerschmitt, David G.
II. Title. III. Series.
TK7872.F5H66 1984 621.381'7324 84-12634
ISBN 0-89838-163-0

Printed in the United States of America

Second Printing, 1985

CONTENTS

PREFACE

The subjects of this book are the *structures*, *algorithms*, and *applications* of adaptive filters. Applications are an integral part of the book because we feel that the material is more palatable, particularly for the applications engineer, when accompanied by applications examples. Applications further serve to enhance and motivate the material, and give opportunities for displaying many of the practical constraints which must be considered when designing adaptive filters. The applications, which are drawn primarily from the field of telecommunications, are introduced in chapter 2, and are then used throughout the book to illustrate the algorithms and introduce practical constraints.

Most of the material covered here can be found scattered through the signal processing literature. As is usually the case, the many contributors to this field generally use disparate notation and mathematical techniques, making the material less accessible. One of the contributions of this book is therefore to present the variety of adaptive filtering techniques in a common notational and mathematical framework so that the concepts may be assimilated more efficiently and the relationships among the different concepts can be better appreciated.

The mathematical nature of the adaptive filtering field presents a problem, particularly for the applications oriented reader. One approach would be to present the material in an intuitive and motivational way without a complete mathematical foundation, requiring the reader to accept many concepts without proof or derivation. We have chosen instead to make the book as complete and self contained as possible by including derivations of virtually all the concepts presented. At the same time, we have attempted within this framework to make the book accessible to the reader who is not interested in all the mathematical details. Techniques used for increasing the accessibility of the material include covering topics in the order of increasing mathematical sophistication, introducing the necessary mathematical machinery only when it is necessary, and placing the more esoteric mathematics in appendices and problems.

The book presumes a background in discrete-time signal processing, including Z-transforms, and a working knowledge of discrete-time random processes. In the latter case familiarity with the concepts of expectation, autocorrelation function, and power spectrum is sufficient. Readers familiar only with continuous-time random processes should find it easy to make the connection with the discrete-time case. In chapters 4 and 6, extensive use has been made of the geometric concepts of inner products and projections. An introduction to these techniques is included at the beginning of chapter 4 so that readers with no prior familiarity will find these chapters accessible.

This book is an outgrowth of a first-year graduate course in adaptive filtering, but should also be accessible to advanced undergraduates and professional readers. Problems have been included at the end of each chapter (except the first) in order to make the book suitable for use in a graduate course. The problems also offer opportunities to supplement the material with additional applications, provide extensions of the material, and present and prove less important results. Instructors can obtain solutions to many of the problems by writing the authors.

The applications presented have been drawn from our experience in telecommunications, and the treatment has been limited to the scalar signal case. There are many applications, particularly outside of telecommunications, which require vector or matrix valued signals, and the techniques in this book can be readily adapted to those cases. It is our belief that treating the most general case would obscure the notation and the underlying principles, while not enhancing the reader's understanding or adding much in the way of new concepts.

Another limitation in the scope of the book is that overlapping fields such as system identification, spectral estimation, adaptive control, etc., are mentioned, but no attempt is made to rigorously relate them to adaptive filtering. A final limitation is that hardware implementation of adaptive filters is not discussed, although the applications examples give many opportunities to point out common modifications to the algorithms necessary in many applications.

What we do hope to communicate to the reader of this book is a basic understanding of the filter structures and algorithms used for adaptive filtering, and what modifications to these algorithms have to be made due to practical constraints. The structures covered include the transversal and digital lattice filters. In the domain of adaptation algorithms, the widely used LMS stochastic gradient algorithm is covered in depth, as are the block LS algorithms often used in speech processing. An important objective is to also treat in a common notational and mathematical framework the more recently proposed recursive least squares algorithms. These latter algorithms perform well and are computationally efficient, and can be implemented economically using available integrated circuit technology. Thus, they deserve consideration in applications of adaptive filtering. Unfortunately, these algorithms are not widely accessible to applications engineers because of their mathematical complexity, and we hope that this book helps to rectify this situation.

A word about notation is in order. The notation required for the more advanced algorithms in this book is necessarily intricate, and we feel that careful choice of notation is critical to the understanding of the material without undue effort. Therefore, we have expended considerable effort in the choice of notation, and in making the notation consistent throughout the book. The time index or variable used has consistently been T, suggestive of time, to distinguish it from such auxiliary parameters as filter order and vector component. The components of a vector or the filter coefficient number has consistently been written as a subscript. All signals have been written with the time variable in parentheses, and where the order of a filter is important in the generation of this signal the order always follows the time variable with a delimiting vertical slash. Thus, for example, $f_j(T|n)$ is the j-th component of the filter coefficient vector **f** at time index T for a filter of order n. The reader can begin to appreciate the aforementioned intricacy of the notation!

The authors of this book are indebted to many friends and colleagues for their stimulating interaction and discussion of the topics in this book. Deserving of special mention is P. Chu, who first derived many of the results of chapter 5. The authors also thank D. Lin for his comments and suggestions for improving the manuscript. Other colleagues who have contributed to our knowledge on the subject include R. Aaron, O. Agazzi, J. Cioffi, D. Duttweiler, D. Falconer, A. Gersho, R. Gitlin, J. Mazo, and S. Weinstein. We are all indebted to the major contributors to adaptive filtering, who include in addition to the above B. Friedlander, L. Griffiths, T. Hoff, T. Kailath, L. Ljung, R. Lucky, J. Makhoul, J. Markel, M. Morf, J. Proakis, B. Widrow, and numerous others.

The manuscript was prepared using the UNIX typesetting system developed at Bell Laboratories (how would we have done without it?), and the book was typeset at the University of California using the same system with the patient assistance of Mrs. Robin Hoek.

Finally, this book was written during the turbulent period in which the Bell System, and in particular, American Telephone and Telegraph Company, was restructured. Despite changing computer systems, computer links by which the authors communicated, and changing management, facilities for continuing this work were always available. We are especially grateful to (in chronological order) Bell Laboratories, AT&T Information Systems, and Bell Communications Research, Inc. for use of their word processing facilities and art studios. We are also grateful to our loved ones for their patience while we slaved at our computer terminals.

ADAPTIVE FILTERS

1

INTRODUCTION

The subject of this book is the *structures* and *algorithms* for adaptive filtering. Since many of these structures and algorithms are discussed in the context of specific applications, a supplementary goal is the enumeration and discussion of *applications* of adaptive filters.

Specifically what is meant by structures and algorithms of adaptive filters? Perhaps the requisite question to ask is what is meant by a filter? The usual sense of the term filter is as depicted in figure 1-1; namely a continuous-time linear time-invariant system or a discrete-time linear shift-invariant system. In either case what is meant intuitively is a system whose output depends linearly

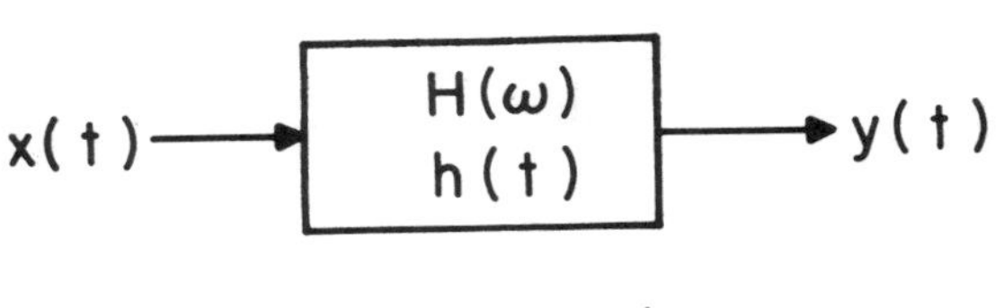

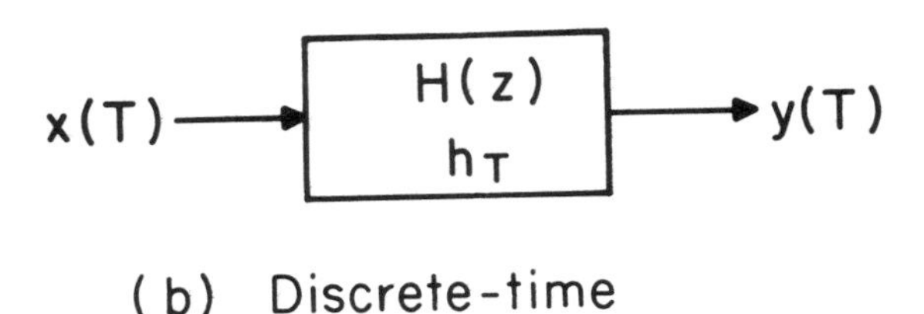

Figure 1-1. Linear time- and shift-invariant filters

on its input and whose internal structure is not varying with time. Filters are of course widely used for operations like equalizing frequency-dependent attenuation, eliminating out-of-band noise, etc. Such filters are usually specified or characterized by their impulse response (for a continuous-time filter) or unit pulse response (for a discrete-time filter), or in either case the frequency-domain equivalent transfer function (commonly termed the "frequency response").

At this point a word about notation is in order. Unless otherwise specified, all signals in this book are discrete time, and the variable T will consistently be used to denote the current sample number or time index. This choice is suggestive of "time," and is not to be confused with a continuous time variable. Further, the sample number or time index will consistently be enclosed in parenthesis, as in figure 1-1. The filter coefficient index, or the component of a vector, will consistently use a subscript, as in c_k.

An *adaptive* filter is consistent with this concept of filter in figure 1-1, with two important distinctions. First, it is a filter with a finite number of internal parameters which can be used to control the transfer function over a useful range. There are many ways to construct a filter with a transfer function which depends on a finite set of parameters, and what is meant by *structure* of the filter is the particular way of realizing the filter. Sometimes what we refer to here as "structure" is called the "realization" of the filter. Second, an adaptive filter has an *adaptation algorithm* which enables the filter transfer function to track in a useful manner some feature of its external environment. Specifically, the adaptation algorithm monitors the external influence or environment of the filter and controls the filter transfer function by varying the aforementioned filter parameters. Examples of where it is advantageous to track an external environment by adapting the internal parameters of an adaptive filter are given in chapter 2.

An adaptive filter is, ignoring possible finite precision effects, usually linear. However, by its very nature an adaptive filter is time-variant (for a continuous-time filter) or shift-variant (for a discrete-time filter), since its internal parameters are being deliberately varied with time. However, most adaptation algorithms necessarily change the filter parameters slowly, and most external influences which the filter is tracking necessarily vary slowly (of course "slowly" is a relative term; what is meant here is slowly in relation to the highest frequency component of the signal being filtered). As a result, a useful representation of an adaptive filter is a linear time- or shift-invariant filter, which is thereby characterized by a transfer function, but whose transfer function is varying slowly with time.

1.1. FILTER STRUCTURES FOR ADAPTIVE FILTERING

The most convenient structure for an adaptive filter, and the one most commonly employed, is the finite impulse response (FIR) discrete-time filter. This filter can be represented by a difference equation,

$$y(T) = \sum_{k=0}^{n} c_k x(T-k) \tag{1.1.1}$$

or equivalently by its unit-pulse response

$$h_k = c_k, \quad 0 \leqslant k \leqslant n \tag{1.1.2}$$

or transfer function (Z-transform of impulse response)

$$H(z) = \sum_{k=0}^{n} c_k z^{-k}. \tag{1.1.3}$$

The utility of this structure derives from its simplicity and generality. Its transfer function can be changed easily by controlling the $n+1$ coefficients c_k, $0 < k \leqslant n$. There is a simple relationship between the parameters of the filter, namely the filter coefficients, and the transfer function of the filter (it is in fact a linear relationship). Furthermore, any filter over the rather general class of FIR filters can be realized in this fashion.

There are many filter structures which equivalently implement the FIR filter. There are two particular filter structures which are almost universally utilized in adaptive filters. These two structures, the *transversal* and *lattice* structures, are illustrated in figure 1-2. The transversal structure is a rather direct realization of the transfer function of (1.1.3) in terms of delays and multipliers, and is considered in chapter 3. The lattice structure is equivalent, in the sense that any transfer function which can be represented by the transversal structure can also be represented within a multiplicative constant by the lattice structure. The coefficients of the lattice structure, k_j, $1 \leqslant j \leqslant n$, are commonly called the *reflection* or *PARCOR* coefficients. The reason that the lattice structure has only n parameters is that its equivalent transfer function of (1.1.3) requires that $c_0 = 1$. The origin and utility of the lattice structure is perhaps not obvious at this point, but will be developed in chapter 4.

Many other structures could potentially represent the FIR filter. For example, the transfer function of (1.1.3) could be factored into a product of second-order polynomials (each one representing an FIR filter with $n=2$) and a structure with a cascade of these FIR filters, each one realized as a transversal or lattice filter, could be employed. What criterion does one use, then, for choosing among the infinite number of possible filter structures, even if restricted to FIR filters? In the case of nonadaptive filters, the primary criterion for choice of a filter structure for a digital implementation is finite precision effects, such as the effect of round-off error internal to the filter. This is an important consideration for an adaptive filter also, but considerably more important for an adaptive filter is that there be a simple and analytically tractable relationship between the transfer function of the filter and the parameters of the filter. In the case of a transversal filter, this relationship is simple because it is linear, whereas for most other structures such as the cascade discussed earlier the relationship is nonlinear. The lattice structure also has a nonlinear relationship, but fortunately there is a convenient *recursive* representation described in chapter 4. The complicated nonlinear relation between parameter and transfer function for most other structures makes the conception of the corresponding adaptation

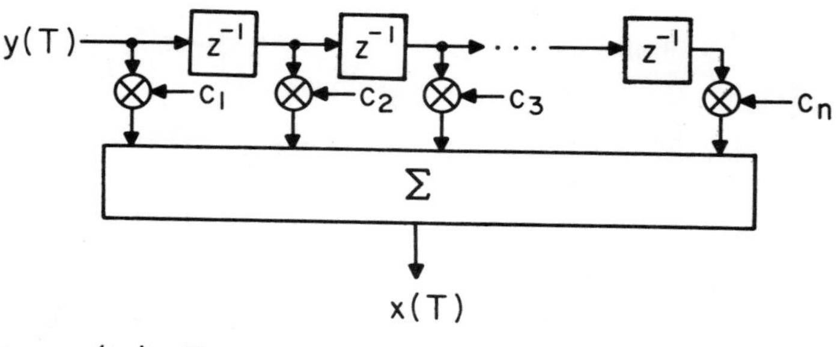

(a) Transversal filter

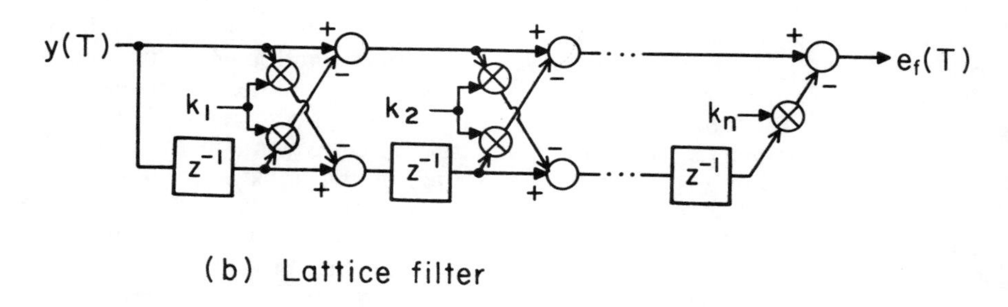

(b) Lattice filter

Figure 1-2. Transversal and Lattice Filter Structures

algorithm too difficult. This book is restricted almost exclusively to consideration of the transversal and lattice structures for adaptive filtering.

Most adaptive filters are discrete-time because of the dual advantages of digital implementations and the fact that there is no realizable continuous-time equivalent of the FIR filter. However, it is easy to conceive of a generalization of the FIR filter which can be carried over to continuous time. Specifically, observe the structure of figure 1-3a, where each of the filters $H_i(z)$ can be FIR or (more interestingly) infinite impulse response (IIR). The transfer function of this filter,

$$H(z) = \sum_{k=1}^{n} c_k H_k(z) \tag{1.1.4}$$

has a simple linear relationship between parameters and transfer function. This filter structure includes the transversal filter as a special case for

$$H_k(z) = z^{-k}. \tag{1.1.5}$$

Furthermore, by replacing $H(z)$ by $H(s)$ in figure 1-3a, a continuous-time filter structure shown in figure 1-3b results. At least one application where this continuous-time filter structure has been proposed will be seen in chapter 2.

1.2. ADAPTATION ALGORITHMS

Once a filter structure has been chosen, an adaptation algorithm must also be chosen. Several alternatives are available, and they generally exchange increased complexity for improved performance. Complexity is simple to characterize, but how is the performance of an adaptation algorithm measured? There are generally two measures of performance applied: the speed of adaptation and the accuracy of the transfer function after adaptation. There is a tradeoff between these two measures: for a particular class of adaptation algorithm, as the speed of adaptation is increased the accuracy of the transfer function after adaptation gets poorer.

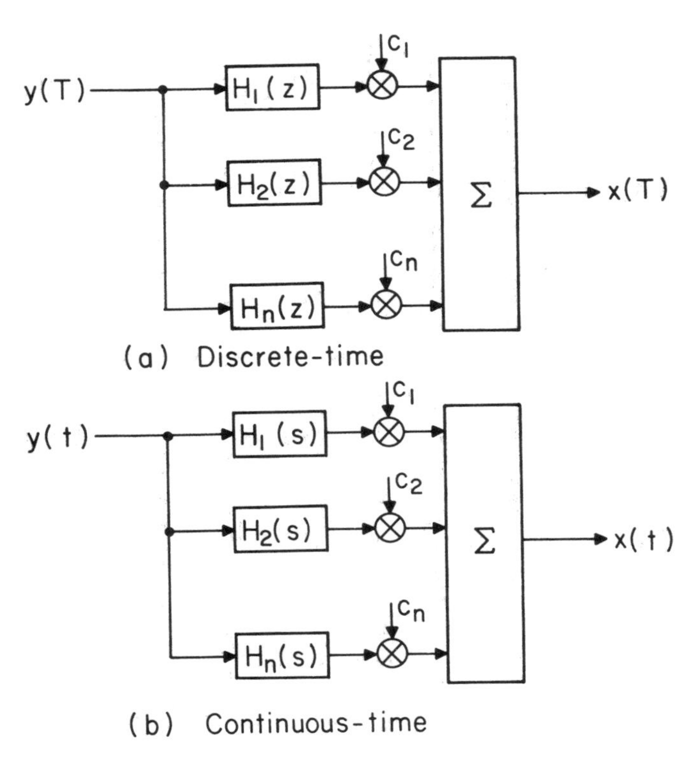

Figure 1-3. A generalized adaptive filter structure

Actually, it is inappropriate to discuss the performance of an adaptive filter outside the context of its external environment. There are usually two motivations for using an adaptive filter: either the external environment of the filter is not known in advance, so that adaptation is a convenient way of tailoring the filter to the environment, or the external environment is slowly varying so that the adaptive filter is necessary to obtain the best performance in light of this changing environment. Thus, it is common to employ adaptive filtering where the environment is not changing (usually said to be stationary), or where it is slowly varying. For the stationary case, the accuracy of the final adaptation transfer function is usually the most critical factor, although the speed at which the filter gets to this asymptotic state is sometimes important also. For the nonstationary environment, the situation is not nearly so simple, since the accuracy of the filter is also related to the speed of adaptation; in particular, it is important for the filter to adapt at least as fast as the environment is changing in order for the filter to effectively track the environment.

In practice it is almost impossible to characterize the performance of an adaptive filter for a nonstationary environment except by experimentation. Thus, for analytical purposes the stationary environment is frequently assumed. The manner in which the filter adapts to a stationary environment from an arbitrary initial condition is fortunately very informative with respect to its performance in a nonstationary environment, at least on an intuitive level. The stationary environment performance of the various structures and adaptation algorithms is considered in chapter 7.

The types of adaptation algorithms, which are considered in chapters 3 and 6, can be classified as *block* or *recursive*. In the block algorithms, the input signal is divided in time into blocks, and each block is processed independently (although there is commonly some overlap between the adjacent blocks). Within each block, the optimum parameters of the adaptive filter can be determined by a series of matrix operations. The end result is a new set of filter parameters at the end of each block. In the recursive algorithms, on the other hand, the adaptive algorithm is implemented as a continually operative set of recursive equations, so that a new set of parameters is generated at each input data sample (at least in the discrete-time case). Aside from the obvious implementation differences, the primary distinction is that the block algorithms have a finite memory and the recursive algorithms usually have infinite memory. That is, the current set of parameters for a block algorithm depends only on a finite segment of the past input signal, namely the current block, while the parameters for a recursive algorithm usually depend (at least in theory) on the complete past history of the input signal. The distinction is clouded by the fact that recursive algorithms can also have finite memory.

Block algorithms and finite memory recursive algorithms are of greatest interest where the environment is changing relatively rapidly, since in those environments infinite memory is deleterious rather than helpful. An example of an input signal with rapidly varying statistics is the speech signal, and here it is common to use block algorithms. On the other hand, in a stationary or very slowly varying environment both block and recursive algorithms can give very accurate estimates. An infinite memory recursive algorithm can be designed to have a very long memory by the choice of adaptation parameters, and a long memory can also be achieved in a block algorithm by choosing a very large block size. In both cases a long memory does not come at the expense of an additional computational overhead. (In a block algorithm the size of the vectors and matrices which must be manipulated are independent of block size, and the computation required to generate those vectors and matrices increases linearly with block size which results in a fixed computational load per input sample).

The comparison between block and recursive adaptive filtering algorithms is roughly analogous to the comparison between FIR and IIR discrete-time filters. In particular, the FIR filter has a finite memory of past input samples, as does the block adaptation algorithm, while the IIR filter has an infinite memory like the recursive adaptation algorithm. Both the FIR and IIR filters can achieve a very low bandwidth (long impulse response or longer memory), but the IIR filter requires less computational complexity for narrow bandwidth than an FIR filter with its finite memory.

The idea of the tradeoff between speed of adaptation and accuracy for adaptive filters is intimately tied to the length of memory of the algorithm. In particular, as the length of the memory increases more of the input signal is considered in the adaptation and hence the accuracy can be greater. The penalty is that as the memory increases the adaptation slows since new information is not assimilated as rapidly. In a nonstationary environment it is not useful to have a memory which exceeds in some sense the interval over which the environment can be considered approximately stationary.

The block adaptation algorithms are usually of the *least squares (LS)* type. In this class of algorithms, the difference between the desired response of the filter and the actual response is squared and summed over the block (there are actually several variations on this theme). This squared difference metric is then minimized over the choice of the filter parameters. The minimization implicitly involves the calculation of a sampled autocorrelation function followed by a matrix inversion. This type of *block LS* algorithm is considered in chapter 3.

The LS approach can be extended to the recursive case by extending the summation back to the beginning of the input signal, often with a weighting function which decays to zero in the infinite past to limit or control the memory of the algorithm. This leads to a minimization problem in which the signal vector is growing in extent, but usually a recursive solution resulting in a fixed computational load can be found if the weighting function is chosen properly. This type of *recursive LS* algorithm is illustrated in chapter 3 and considered in detail in chapter 6. There are also finite memory versions of recursive LS algorithms.

An older and simpler approach to recursive adaptation algorithms is the class of *least mean square (LMS)* or *stochastic gradient (SG)* techniques. This is an approach to adaptation in which an explicit solution of a set of linear equations is avoided by incrementally moving the parameter vector in the direction opposite to a gradient vector. This is the simplest and most widely applied adaptation algorithm, and is considered in chapters 3 and 4 in the context of the transversal and lattice filter structures respectively.

1.3. OUTLINE OF THE BOOK

This book is generally organized along historical lines, starting with the oldest and most widely applied adaptive filtering techniques and moving toward the most modern but (at present) less widely used techniques. Since the more modern techniques are generally more mathematically sophisticated, this has the additional advantage that the necessary mathematical machinery can be minimized at each stage.

Some specific applications are introduced in chapter 2, and then used to illustrate adaptive filtering concepts throughout the book. The transversal filter and lattice filter structures are then considered in chapters 3 and 4. In these chapters, gradient adaptation algorithms are emphasized, although some alternative algorithms of the LS type are introduced in chapter 3 as a harbinger of the more modern recursive LS algorithms considered in chapter 6. The important zero sensitivity and stability properties of the transversal and lattice algorithms are considered in chapter 5. Finally, in chapter 7 the stationary environment convergence performance of the various adaptation algorithms are compared.

1.4. FURTHER READING

The background assumed in this book is a working knowledge of discrete-time signals and systems, and specifically filters, convolution, the Z-transform, and second order analysis of discrete-time random processes. An excellent thorough treatment of all these topics can be found in the text by Oppenheim and Schafer [1]. A more thorough book, which also covers many applications and hardware implementation details, is Rabiner and Gold [2]. It does not, however, cover discrete-time random processes. Books on a more introductory level are those by Peled and Liu [3] and Antoniou [4]. A short review of these topics can be found in chapter 2 of Rabiner and Schafer [5].

REFERENCES

1. A.V. Oppenheim and R.W. Schafer, *Digital Signal Processing,* Prentice-Hall, Englewood Cliffs, N.J. (1975).
2. L.R. Rabiner and B. Gold, *Theory and Application of Digital Signal Processing,* Prentice-Hall, Englewood Cliffs, N.J. (1975).
3. A. Peled and B. Liu, *Digital Signal Processing,* Wiley, New York (1976).
4. A. Antoniou, *Digital Filters: Analysis and Design,* McGraw-Hill, New York (1979).
5. L.R. Rabiner and R.W. Schafer, *Digital Processing of Speech Signals,* Prentice-Hall, Englewood Cliffs, N.J. (1978).

2

APPLICATIONS

In this chapter some applications of adaptive filtering are developed. This will motivate the need for adaptive filtering, as well as better illuminate some of the performance issues. These applications also give an opportunity to explore some of the practical issues which arise in the application of adaptive filtering techniques. These applications will be further explored in later chapters to illustrate the concepts developed there.

The list of applications which follows is not comprehensive, being limited primarily to applications within the field of telecommunications. Adaptive filtering has been extensively used in the context of many other fields, including among others geophysical signal processing, biomedical signal processing, the elimination of radar clutter, and sonar processing.

The applications of adaptive filtering can be divided into several classes, as differentiated by the external connections to the filter, and illustrated in figure 2-1. In figure 2-1a, the adaptive filter has only an input signal and no explicit output signal. The filter, in the course of adapting to the external environment (the input signal), is implicitly estimating some characteristics of that environment. The internal parameters of the adaptive filter are used as the estimate. The most common use of an adaptive filter in this context is as an estimator of the spectrum of the input signal. An example use of spectral estimation is the LPC analysis of speech discussed in section 2.1.

In the second class illustrated in figure 2-1b, the filter is used to filter an input signal, $y(T)$ to yield an output signal $e_f(T)$. Very often the objective of the filter is to minimize the size of the output signal $e_f(T)$ within the constraints of the filter structure. The name of the signal "e" stands for error, and the subscript f coincides with the filter coefficients which will be used in this type of adaptive filter, f_j.

The third class of adaptive filter, illustrated in figure 2-1c, is the *joint process estimator*. In this case, there are two inputs, $y(T)$ and $d(T)$. The $y(T)$ input is filtered and the result subtracted from $d(T)$ input to yield the error $e_c(T)$. The subscript c corresponds to the coefficients for a joint process estimator, which will be c_j. The objective is usually to minimize the size of the output $e_c(T)$, in which case the objective of the adaptive filter itself is to generate an estimate of $d(T)$ based on a filtered version of $y(T)$.

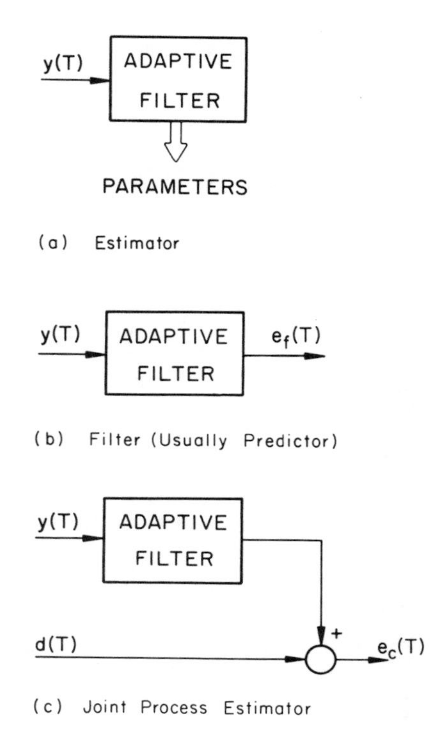

Figure 2-1. Classes of adaptive filtering.

The internal structure of the filter in the first two classes of adaptive filtering is often a *predictor*, in which a linear weighting of some finite number of past input samples is used to *estimate* or *predict* the current input sample, and the filter output $e_f(T)$ is the difference between the current input sample and its predicted value, given by [1]

$$e_f(T) = y(T) - \sum_{i=1}^{n} f_i y(T-i) \; . \tag{2.0.1}$$

A transversal predictor filter structure is shown in figure 2-2a. Except in degenerate cases, the signal $e_f(T)$ cannot be forced to zero, but it can be minimized over the choice of the transversal filter coefficients. As examples of the use of an adaptive predictor, LPC analysis will be discussed in section 2.1 and waveform coding of speech using adaptive differential pulse code modulation (ADPCM) will be described in section 2.2.

The predictor of figure 2-2a can also be used as an estimator in the manner of figure 2-1a. Since the adaptive predictor coefficients can be considered a function of the spectrum of the input signal $y(T)$, the filter coefficients can be turned into an estimate of that spectrum. This approach will be illustrated in section 2.1.

A transversal filter structure can be used to realize a joint process estimator as illustrated in figure 2-2b. In this case the error in the estimate of $d(T)$ formed as a weighted linear combination of current and past inputs $y(T)$ is given by

$$e_c(T) = d(T) - \sum_{i=1}^{n} c_i y(T - i + 1) . \qquad (2.0.2)$$

Observe that the predictor can be considered to be a special case of the joint process estimator, for the particular values

$$\begin{aligned} c_1 &= -1 \\ c_{i+1} &= f_i, \quad 1 \leqslant i \leqslant n \\ d(T) &= 0 . \end{aligned} \qquad (2.0.3)$$

Applications of joint process estimation which will be discussed in this chapter are adaptive equalization in section 2.3, noise cancellation in section 2.4, and echo cancellation in section 2.5.

2.1. SPECTRAL ESTIMATION OF SPEECH

A major application of adaptive filtering is in the digital encoding of speech [2, 3]. There are two basic approaches to speech encoding. The first is *analysis-synthesis*, in which the speech generation process is characterized by a simple model, and the input speech is used to estimate the parameters of that model. This is the analysis part of the encoding, after which these parameters are quantized yielding the digital representation of the speech. To recover the speech from this digital representation, the speech is synthesized by emulating the speech model using the quantized parameter values.

A second approach, which yields higher quality speech at the expense of a higher bit rate for the digital representation, is *waveform encoding*. In the encoder the speech waveform is quantized, and in the decoder reconstructed from the quantized representation. The major distinction is that analysis-synthesis techniques only promise to maintain the intelligibility of the speech, while the waveform encoders attempt to reconstruct the speech waveform itself, and thus the naturalness of the speech, the speaker identity, etc.

Both approaches to speech encoding often utilize adaptive filtering techniques. A fundamental reason for this is that speech is highly nonstationary, and thus requires adaptation of the encoder to fully exploit its changing characteristics. In this section one specific technique for analysis-synthesis encoding will be discussed, and in section 2.2 one technique for waveform encoding will be discussed.

A characteristic of the speech which is adequate for distinguishing among the different sounds is its spectrum. Most analysis techniques thus depend on the estimation of the short-term spectrum; that is, the spectrum over a time interval such that the speech signal can be considered stationary (usually 10-20

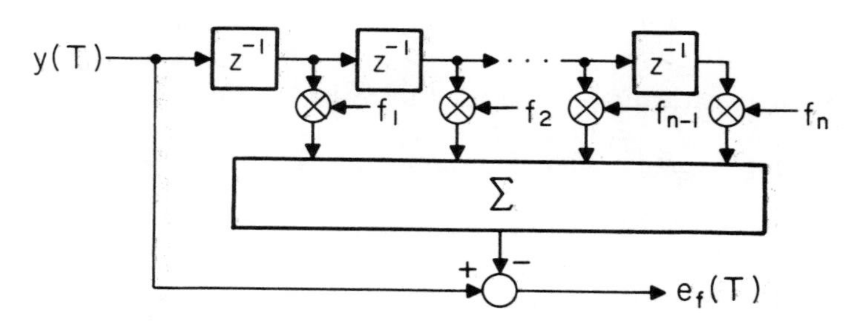

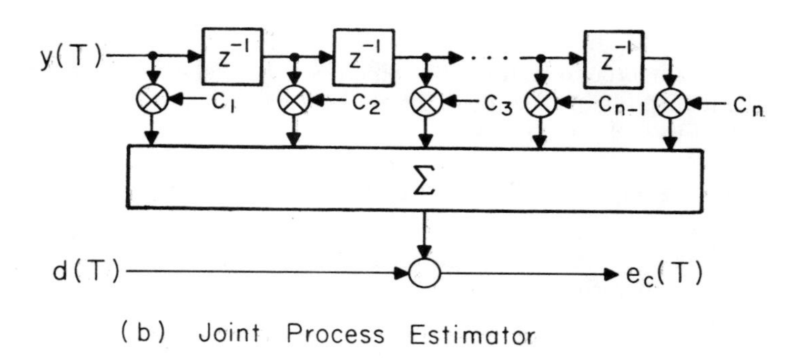

Figure 2-2. Transversal filter as predictor and joint process estimator.

msec). One obvious and reasonable way to obtain this short-term spectrum is to build a bank of bandpass filters, modulating each filter output down to be centered at d.c., and then sampling the output of each of these filters at a rate approximately equal to the filter bandwidth. Each of these filter outputs, which is complex-valued because the spectrum is not necessarily symmetric about the center frequency of the bandpass filter, can be considered an estimate of the short-term spectrum of the speech at the center of the filter band.

The output of such a bank of filters can be used to reconstruct the speech waveform with high accuracy. In order to reconstruct intelligible speech it is adequate to retain only the magnitude of the short-term spectrum and discard the phase information. This is done in the *channel vocoder*, the earliest analysis-synthesis method [4]. The channel vocoder does not require adaptive filtering, but does require realization of a bank of bandpass filters.

A more recently proposed speech analysis-synthesis technique, the *linear predictive coding (LPC) vocoder*, uses an adaptive predictor as a part of the analysis processing [5]. This avoids the realization of an expensive bank of filters. It is based on a model of speech production illustrated in figure 2-3a, which consists of either an impulse train in the case of voiced sounds (vowels), or a white noise source in the case of unvoiced sounds (fricatives such as /s/,/f/) as the input $\eta(T)$ to a linear time varying filter. The filter in figure 2-3a has a rational all-pole form with no zeros. Since the input to the filter is assumed to have a flat spectrum, the power spectrum of the output (namely of the speech) is proportional to the magnitude of the frequency response of the filter $H(z)$,

$$S_y(e^{j\omega}) = N_0 |H(e^{j\omega})|^2$$

$$H(z) = \frac{G}{1 - \sum_{i=1}^{m} a_i z^{-i}} \tag{2.1.1}$$

where $S_y(e^{j\omega})$ is the power spectrum at angle ω on the unit circle, and N_0 is the (constant) power spectrum of the excitation input. The desired spectral information is therefore contained in the coefficients G and a_j, $1 \leqslant j \leqslant m$, where m is typically on the order of 10 to 15. The analysis of the speech consists of extracting these coefficients (which are slowly varying with time) from the input speech waveform.

The filter of figure 2-3a is of course a special case of the general model of an excitation driving a filter. The choice of a filter with only poles in figure 2-3a results in a speech model waveform which is said to be *autoregressive* or *AR*. If

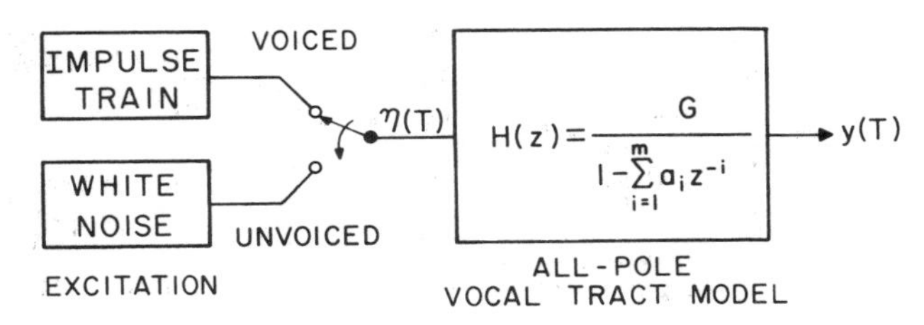

(a) Speech Generation Model

y(T) → $F(z) = 1 - \sum_{i=1}^{n} f_i z^{-i}$ → $e_f(T)$

ALL-ZERO ANALYSIS FILTER

(b) Speech Analysis Filter

Figure 2-3. LPC speech modeling and analysis.

we call the white input to the filter $\eta(T)$ and the output $y(T)$, the AR model is characterized by the difference equation,

$$y(T) = \eta(T) + \sum_{j=1}^{m} a_j y(T-j) \ . \tag{2.1.2}$$

This is only one of three possible models we could choose. The second is the *moving average* or *MA* model, in which the filter has only zeros, characterized by the difference equation

$$y(T) = \sum_{j=0}^{l} b_j \eta(T-j) \ , \tag{2.1.3}$$

where the "moving average" terminology is of obvious origin. A third model would combine poles and zeros, resulting in the *autoregressive moving average* or *ARMA* model characterized by the difference equation

$$y(T) = \sum_{j=0}^{l} b_j \eta(T-j) + \sum_{j=1}^{m} a_j y(T-j) \ . \tag{2.1.4}$$

Obviously it would be desirable to choose the most general ARMA model for speech production because of its greater generality, but it turns out to be much more difficult to estimate the model parameters for the ARMA model than for the AR model. In addition, it is fortuitous that the AR model is a fairly accurate model of speech production, particularly for the subjectively important vowels. Another oft-stated justification for the AR model is that zeros can be approximated by poles, as illustrated by the equation

$$1 - az = \frac{1}{1 + az + a^2z^2 + \cdots} \ . \tag{2.1.5}$$

If the AR model for speech is accepted, then a hint as to how the parameters of the model can be estimated can be seen from figure 2-3b where the speech (generated by an AR model) is passed through an all-zero FIR filter resulting in a signal called $e_f(T)$. If we temporarily make the presumptuous assumption that the speech signal is wide sense stationary, at least for the short term, then when the output of the all-zero filter

$$e_f(T) = y(T) - \sum_{j=1}^{n} f_j y(T-j) \tag{2.1.6}$$

is white, the magnitude of the transfer function of the all-zero filter is the inverse of the magnitude of transfer function of the model all-pole filter. Since (2.1.6) is precisely the transversal predictor filter of figure 2-2a, the problem of estimating the parameters of the AR model of speech is equivalent to adapting the parameters of a transversal filter predictor to achieve a white output signal. This white output signal can be achieved only if the predictor order n is greater than or equal to the AR model order m. The criterion for choice of the predictor coefficients is to make the prediction error $e_f(T)$ white; that is, to make the error samples uncorrelated with one another.

It will be shown in chapter 3 that an equivalent criterion is to minimize the mean-square prediction error. That is, f_j, $1 \leqslant j \leqslant n$ should be chosen to minimize

$$\epsilon \equiv E[(y(T) - \sum_{j=1}^{n} f_j y(T-j))^2] . \tag{2.1.7}$$

The f_j, $1 \leqslant j \leqslant n$, are called the "prediction coefficients" for an nth-order linear predictor. Given an input signal $y(T)$ produced by the model in figure 2-3a, the optimal prediction coefficients (the set of f_j's which minimize ϵ), turn out to be

$$f_j = \begin{cases} a_j & 1 \leqslant j \leqslant m \\ 0 & m < j \leqslant n \end{cases} . \tag{2.1.8}$$

This choice also results in a white output error signal.

The previous reasoning is based on the assumption that the speech is a wide-sense stationary random process. Perhaps more disturbing is the assumption that the statistics (ensemble averages) of the input process are known. In fact, speech is highly nonstationary, and the best that can be hoped for is that the speech be "quasi-stationary" over short intervals of time (approximately 10-20 msec). In addition, in actuality we have only the speech signal $y(T)$ and have no knowledge of the ensemble statistics (which is what it is desired to estimate!). Thus, in the practical world the use of ensemble averages must be abandoned, and the algorithms must deal only with the input sequence $y(T)$. When the spectral estimation is done on the basis of the input data sequence alone, it is called the *given data* case. The basic approach to the given data case is to substitute time averages for the unavailable ensemble averages [2].

The practical methods for estimating the prediction coefficients in the given data case fall into two categories. The first divides the speech into blocks, which are sometimes called frames, and estimates the predictor coefficients for each block using a time average. One such method, which is termed *block least squares (LS)*, finds the set of predictor coefficients which minimizes

$$\sum_{T=Mk}^{M(k+1)-1} w_{T-Mk} \left\{ y(T) - \sum_{j=1}^{n} f_j y(T-j) \right\}^2 \tag{2.1.9}$$

where M is the number of samples in the block, k is the number of the block, and w_i is an appropriate weighting function. Note that this is simply a time-average squared prediction error, averaged over successive blocks of input samples. It is also common to overlap the blocks to reduce end effects.

The second category of methods, called *recursive least squares (LS)*, avoids the use of blocks and attempts to estimate the prediction coefficients recursively. In particular, given the prediction coefficients at time $T-1$ and the new sample $y(T)$, a recursive rule is used to compute new predictor coefficients. Both the block and recursive LS algorithms will be discussed in more detail in chapter 3, and more sophisticated versions of the recursive algorithms will be developed in chapter 6.

The ability to accurately extract the "spectral" coefficients a_j in figure 2-3 has led to very low bit rate speech communication and storage systems. A particular speech sound can be represented accurately in most cases by the parameters of the model of figure 2-3a. These include a decision as to whether the sound is voiced or unvoiced; if the sound is voiced, the period of the impulse train driving the filter $H(z)$; the coefficients a_j, $1 \leqslant j \leqslant m$, which represent the spectral information; and finally, the coefficient G which represents the power or level of the speech signal. Speech can be synthesized from these three items by realizing the AR model shown in figure 2-3a and using the estimated parameters. The white input noise is generated by a noise generator.

A practical problem that arises here (and in the ADPCM coding of section 2.2) is the potential instability of the all-pole filter in the decoder. If the poles of that filter should wander outside the unit circle, the filter will be unstable for the duration of that condition and erroneous behavior can be expected. This raises two issues. First, does the filter adaptation algorithm in the encoder always result in a set of filter coefficients which correspond to a stable all-pole filter in the decoder? Second, can it be insured that even in the case of an affirmative answer to the first question, that stability can be maintained in the face of the quantization of the filter coefficients between encoder and decoder and the potential for transmission errors? These issues will be addressed in chapter 5.

2.2. WAVEFORM CODING OF SPEECH

In the last section it was shown how the number of information bits required to represent a speech sound was reduced by extracting parameters from the time waveform and quantizing these parameters. Alternatively, the time waveform can be encoded directly, a process called *waveform coding* [6]. The simplest approach, to sample and quantize the input waveform directly, is known as *pulse code modulation* or PCM [7, 8]. PCM is simple and robust, but requires a relatively high bit rate to accurately represent the speech.

A modification of this technique shown in figure 2-4, known as *differential pulse code modulation* or DPCM, employs a linear predictor in a feedback loop so that the prediction error rather than the input signal is quantized and sent to the receiver [9, 10, 11]. This is one of many waveform quantization techniques, and one that utilizes adaptive filtering in a manner similar to LPC analysis-synthesis coding. In fact, DPCM can be viewed as quantizing and transmitting the output of the LPC predictor (the so-called linear prediction residual), a signal which is normally discarded in LPC encoding!

DPCM can be understood as follows. In figure 2-4a, a predicted value of the input sample $y(T)$, called $\hat{y}(T)$, is subtracted from $y(T)$ in the encoder prior to quantization in the A/D converter. Deferring for the moment the method of deriving this predicted value, this same value $\hat{y}(T)$ is added in the receiver at the output of the D/A converter to yield a reconstructed and quantized sample $\tilde{y}(T)$. The encoder and decoder are then represented by the two relations

$$e_f(T) = y(T) - \hat{y}(T) \tag{2.2.1}$$

$$\tilde{y}(T) = \hat{y}(T) + \tilde{e}_f(T) , \tag{2.2.2}$$

where $e_f(T)$ is the prediction error in the encoder, $\tilde{e}_f(T)$ is the quantized version of the prediction error available in the decoder, and $\tilde{y}(T)$ is the reconstructed speech sample in the decoder.

If $\hat{y}(T)$ is a good prediction of $y(T)$, then the prediction error $e_f(T)$ which is actually quantized is much smaller than the input sample $y(T)$ on average. It is therefore possible to reduce the step-size of the quantizer without increasing the frequency of overloading it. Further, since the same predicted value $\hat{y}(T)$ is subtracted in the encoder and added in the decoder, it follows from (2.2.1-2) that

$$y(T) - \tilde{y}(T) = e(T) - \tilde{e}(T) , \tag{2.2.3}$$

where the quantity on the right side is precisely the quantization error. Thus, the overall error is the same as the quantization error. The scaling down of the quantizer step size results in a corresponding reduction in the quantization error and hence a reduction in the overall error between input of the encoder and output of the decoder. This is the advantage of using prediction in the ADPCM coder. Ultimately it implies that the quantizer precision can be reduced, with a corresponding reduction in bit rate, for the same encoding accuracy.

Normally, the predicted value $\hat{y}(T)$ would be generated by forming a weighted linear combination of past input samples $y(T)$. However, in this case the input samples are not available in the decoder (that would be cheating!). In order to avoid mistracking between encoder and decoder, the past quantized samples $\tilde{y}(T)$ are therefore used as the input to the linear predictor in both transmitter and receiver. This is illustrated in figure 2-4b, where the box labeled "P" is the transversal filter predictor. The predicted value generated in both the encoder and decoder is, in analogy to figure 2-2a,

$$\hat{y}(T) = \sum_{j=1}^{n} f_j \tilde{y}(T-j) . \tag{2.2.4}$$

The circuitry in the encoder which generates the prediction $\hat{y}(T)$ is simply a replication of the entire decoder.

The coefficients of the predictor can be either fixed or allowed to adapt. If they are fixed, they can be selected to minimize the mean square prediction error based upon the long term statistics of the input signal. However, due to the nonstationarity of speech, use of fixed predictor coefficients will be suboptimal. It is therefore advantageous to be able to track the optimal predictor coefficients as functions of time as in the case of analysis-synthesis encoding. This can be done by implementing identical adaptation algorithms in both the transmitter and receiver. Since these adaptation algorithms both work on the same signal, it is not necessary to transmit the prediction coefficients as in LPC (where the prediction error is not available for adaptation in the decoder). A DPCM scheme which uses an adaptive predictor is called *adaptive differential pulse code modulation* or ADPCM. In addition to using an adaptive predictor it is also common in ADPCM to use an adaptive quantization scheme (labeled "QA" in figure 2-4b), or automatic gain control, which attempts to quantize a normalized version of the input signal, $e_f(T)/\sigma(T)$, where $\sigma^2(T)$ is an estimate of the prediction error variance at time T. Adjusting the step size of the quantizer

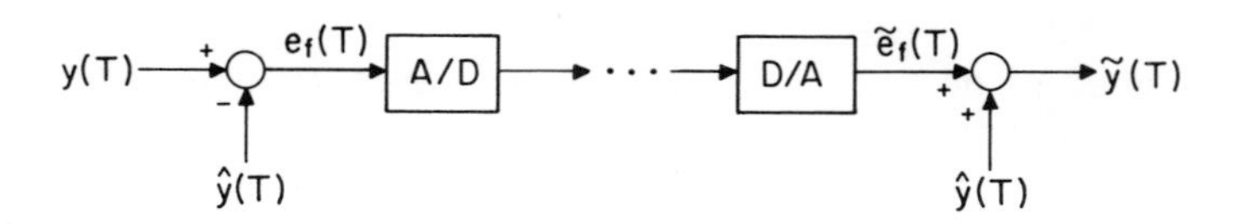

(a) Principle

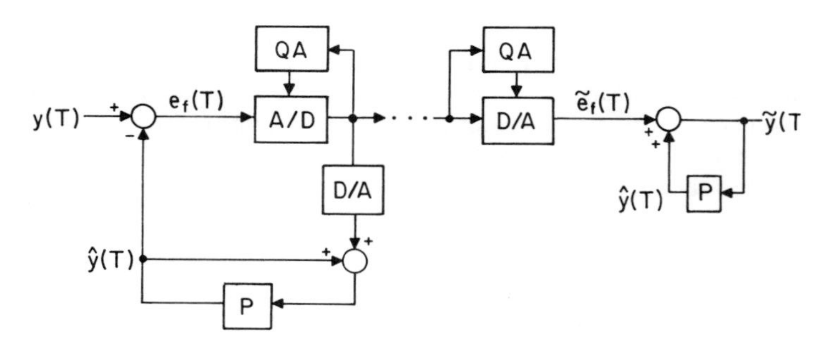

(b) Complete system

Figure 2-4. Adaptive differential pulse code modulation.

in proportion to $\sigma(T)$ is equivalent to normalizing the input signal. Either technique is advantageous due to the wide dynamic range of talker power in most practical speech communication systems. Note that there are two identical adaptive quantizers, one in the encoder and one in the decoder, with each using the same input signal. A common method of realizing this adaptive quantizer will be described in chapter 3.

The analogy between LPC and ADPCM extends to the issue of stability as well. As in LPC, the reconstruction filter in the decoder is an all-pole filter, and it must be ensured that it is stable at all times. The reason is that when the quantization is ignored, in effect the speech signal is passed through an all-zero filter in the transmitter and an all-pole filter in the receiver, where the latter filter is the causal inverse of the former. The steady state response of the cascade of filters is a system with a unity transfer function. However, when the two filters have different initial conditions, such as when the system is turned on or after a transmission error, the transient response due these different initial conditions must die away to zero, which will only occur when the all-pole filter is stable.

There is one difference between this application and LPC; namely, in ADPCM the coefficients of the filter are not quantized for transmission, but rather are derived by identical adaptation circuits in both transmitter and receiver. Hence, on first reflection the stability of the all-pole filter in the decoder would depend only on the adaptation algorithm. However, this is not the case because in most practical situations transmission errors can occur. This suggests several practical issues, including whether stability in the decoder can be ensured, and whether the encoder and decoder adaptive predictors and quantizers are able to eventually recover to the same state starting from the different internal states that will inevitably result after transmission errors. The tracking of the two adaptive predictors is considered in section 3.2.2.1, and the stability of the all-pole filter in the decoder is addressed in chapter 5.

ADPCM communication systems have been extensively simulated and shown to result in high quality reproduction of speech at moderate bit rates (24 to 32 kb/s as compared to 56 to 64 kb/s for PCM) The naturalness of the reproduced speech and the preservation of the speaker identity are in fact significantly better than that produced by linear predictive vocoders. However, this increase in quality is usually accompanied by a substantial increase in bit rate (which is typically chosen to be in the range of 2 to 10 kb/s for LPC).

2.3. ADAPTIVE EQUALIZATION

One of the first applications of adaptive filtering in telecommunications was the equalization of frequency-dependent channel attenuation in data transmission [12, 13, 14, 15]. The frequency response of most (but by no means all) data channels does not vary significantly with time. However, this response is frequently not known in advance. Hence, it is necessary for the data transmission system designer to build in the means to adapt to the channel characteristics, and perhaps also track time-varying characteristics.

A typical approach to digital data transmission is *pulse amplitude modulation* (PAM), in which the amplitudes of successive pulses in a periodic pulse train are multiplied by data symbols. These data symbols assume one of a small number (often two or four) discrete levels representing a data stream. These pulses are passed through the channel with frequency-dependent attenuation, and at the receiver the data signal is sampled synchronously with the transmitted data signals. If this sampling rate is equal to the data symbol rate, then the overall channel response can be characterized by the sampled response of the channel to a single data symbol, which we denote by h_T. If the data symbols themselves are further denoted by $d(T)$, then the sampled channel output can be represented as a convolution of the input data symbols with the channel response,

$$y(T) = \sum_j d(j) h_{T-j} + \eta(T) , \tag{2.3.1}$$

where $\eta(T)$ is an inevitable undesired noise signal. The need for adaptive filtering arises when the received pulse shape h_T is not known in advance and/or is slowly time-varying.

The signal of (2.3.1) is baseband; i.e., its spectrum is centered at d.c. It is common in data transmission over a passband channel to transmit two signals of the form of (2.3.1), each amplitude modulating carriers having a 90 degree phase relationship. In this event, the equalization and adaptive filtering problem can be referred to baseband if desired by first demodulating by quadrature carriers.

For simplicity consider only the baseband case, where the entire PAM receiver is shown in figure 2-5. The first step is to filter the input signal to eliminate out-of-band noise prior to sampling (ideally this is a matched filter but often just a low-pass filter in practice). The next task is to derive, from the data signal itself, the timing clock (at the rate of the data symbols). Because most adaptive filters are implemented in discrete-time, the signal is then sampled in accordance with this timing information at the symbol rate. The result of this sampling is the received signal given by (2.3.1). The adaptive equalizer is an adaptive filter which compensates for the frequency-dependent attenuation of the channel, and generates an estimate of the data symbol. There is a delay through the adaptive equalizer, so that the estimate is delayed by L symbols, and we denote that estimate as $\hat{d}(T-L)$, an estimate of $d(T-L)$. The slicer applies a set of thresholds to this estimate to recover the original data symbols. The thresholds are chosen to be nominally halfway between each pair of amplitudes in $d(T)$. These detected data symbols are denoted as $\tilde{d}(T-L)$. The hope is that $d(T-L)=\tilde{d}(T-L)$, and the probability that this is not the case (the probability of error) is a good indication of the quality of the data transmission.

Rewriting the received samples of (2.3.1) in the form

$$y(T)=d(T)h_0+\sum_{j\neq T} d(j)h_{T-j}+\eta(T) \tag{2.3.2}$$

clearly displays the first term as the signal needed for detection of the current data symbol, and the second term as *intersymbol interference* among symbols due to the frequency-dependent dispersion of the channel. The object of the adaptive equalizer is to remove this intersymbol interference.

A transversal filter can be used as an adaptive equalizer. This is shown in figure 2-6, where the equalizer forms a linear combination of input samples to yield the estimate $\hat{d}(T-L)$ of the data symbol. The object of the adaptation algorithm is to choose the filter coefficients so that $\hat{d}(T-L)\approx d(T-L)$. A filter so adapted can be viewed as providing a frequency response which is roughly the inverse of the frequency response of the channel (ignoring aliasing

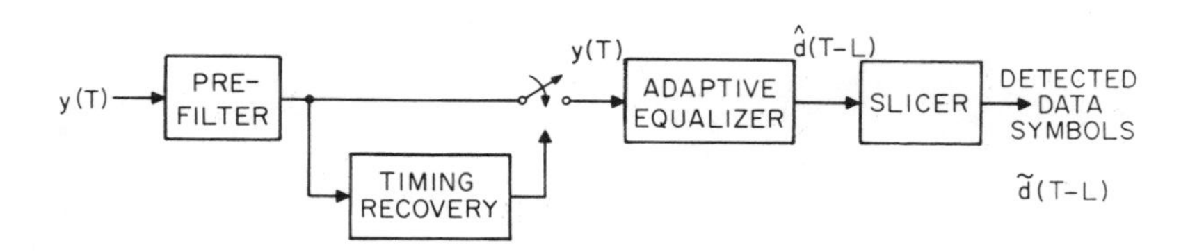

Figure 2-5. Receiver for PAM data transmission.

effects). In practice intersymbol interference can be contributed by future as well as past data symbols (this does not violate causality because of the flat group delay of the channel on the long-distance transmission), and thus the data symbol centered L samples in the past are estimated based on the current and past $2L$ input samples as per the modified transversal filter equation

$$\hat{d}(T-L) = \sum_{j=-L}^{L} c_j y(T+L+j) \ , \tag{2.3.3}$$

where the c_j, $-L \leqslant j \leqslant L$, are the $n = 2L+1$ transversal filter coefficients. We have subscripted these coefficients from $-L$ to $+L$ to be consistent with the literature in adaptive equalization; this emphasizes the fact that the equalizer is compensating for intersymbol interference from both sides of the current data symbol. For the same reason, and because of the causality constraint, the output of the transversal filter is considered as an estimate of $d(T-L)$, rather than $d(T)$, and is labeled $\hat{d}(T-L)$. This signal is connected to the slicer to make the decision on the data symbol. The estimation error is given by

$$e_c(T-L) = \hat{d}(T-L) - d(T-L) \ . \tag{2.3.4}$$

A criterion that can be used to select the filter coefficients is the minimization of the output mean square error $\epsilon = E[e_c^2(T-L)]$. This criterion is not equivalent to minimizing the probability of error, but is close enough for practical purposes.

Comparing figures 2-2b and 2-6, channel equalization is a classical joint process estimation problem (with minor variations in notation). Of course, using $d(T-L)$ as an input to the filter as in figure 2-6 would seem to be cheating, since it is precisely this data symbol which must be detected after equalization. In actuality, this is not a problem for several reasons. First, it is not $e_c(T-L)$ but rather $\hat{d}(T-L)$ which is used in the detection of the current data symbol (by applying this quantity to a slicer). Thus, $d(T-L)$ and $e_c(T-L)$ are used

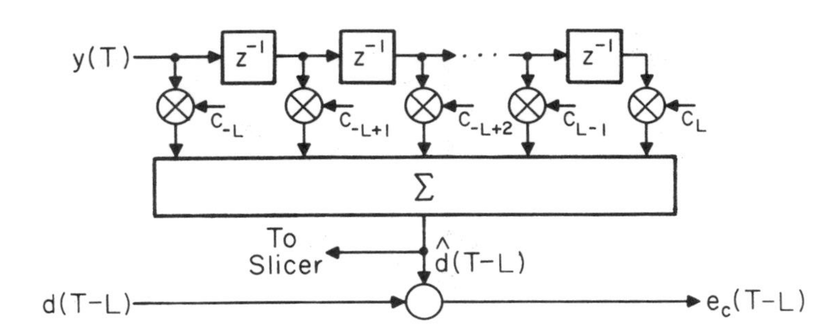

Figure 2-6. Transversal filter PAM equalizer.

only for purposes of the adaptation algorithm. Second, it is possible to envision, using $d(T-L)$ in the adaptation algorithm, if the detected data symbol is used in place of the actual data symbol. It turns out that as long as this detected data symbol is correct most of the time, the adaptation is not adversely affected. An error rate as high as 10% does not cause problems in practice [12] due to the long averaging effect of the adaptation algorithm. This use of the detected data is called *decision direction*, and is commonly applied in adaptive filtering for data transmission.

Decision directed adaptation is acceptable for tracking a slowly varying channel, but cannot be used as a means to adapt to a new channel where the error rate would initially be too large to enable convergence of the adaptation algorithm. The approach in this case is to use a *training period*, in which a deterministic data sequence known to the receiver is transmitted. The receiver's adaptive equalizer can then be trained to at least initially equalize the channel accurately, since the actual $d(T-L)$ is known at the receiver.

The time it takes for training of the equalizer is dependent on the speed of adaptation of the filter. This undesirable delay at the beginning of data transmission has resulted in an emphasis on finding adaptation algorithms which converge rapidly. The Kalman filter may be applied to yield a fast linear LS recursive estimator as discussed in section 3.5. This is currently the fastest known equalizer adaptation algorithm.

An enhancement of the adaptive equalizer concept is the *decision feedback equalizer* (DFE) [16, 17] which is illustrated in figure 2-7. The DFE can actually be viewed as an application of adaptive prediction, as illustrated in figure 2-7a. At the output of the forward equalizer, which is the transversal filter which has just been discussed, the intersymbol interference has been eliminated and thus the samples are equal to the true data symbols plus noise, $d(T-L) + \eta(T)$. The key observation is that the noise component is in general not white (that is, the noise samples are correlated). This leads to the idea of exploiting this noise correlation to reduce the noise variance at the slicer input, very much like prediction is used to reduce the variance of the speech signal at the quantizer input in DPCM. First the current detected data symbol $\tilde{d}(T-L)$ is subtracted from this signal, leaving just the noise term $\eta(T)$ assuming that there was no decision error. The resulting noise is run through a strictly causal prediction filter $F(z)$, the output of which is a prediction of the current noise sample. This prediction is subtracted from the current noise sample, and the prediction filter can be chosen to minimize the noise variance at the slicer input.

This is another instance of a decision-directed algorithm, and is fair because the detected data is passed through the causal prediction filter before it is used in the detection of future data symbols. An incorrect decision results in the incorrect cancellation of a data symbol at the predictor input, and this tends to cause additional decision errors (this phenomenon is known as *error propagation*) [18].

An equivalent form of the DFE is shown in figure 2-7b. In this case the predictor filter is combined with the forward equalizer to yield a single filter, and a

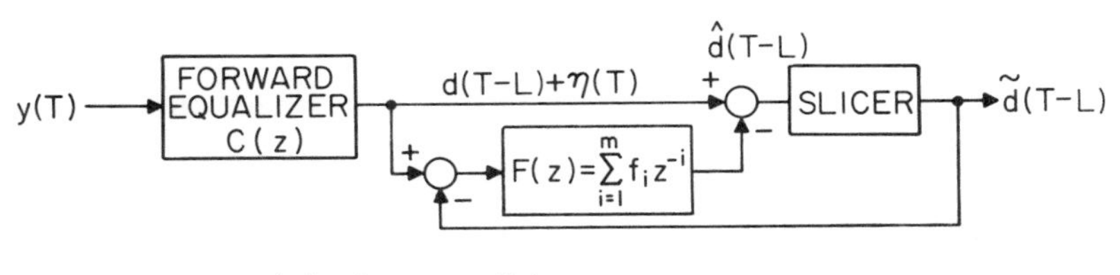

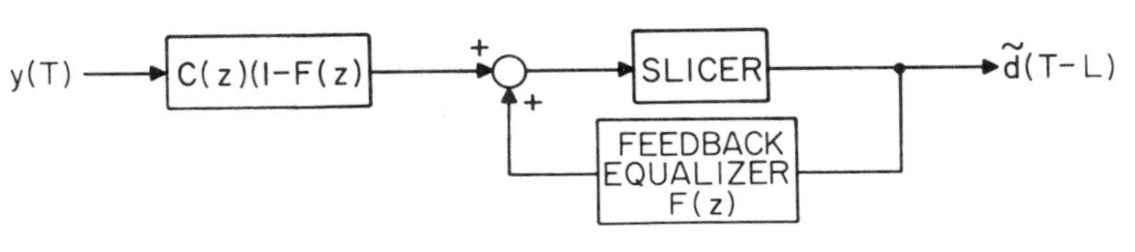

Figure 2-7. Decision feedback equalizer.

separate instance of the predictor filter is put in a feedback loop around the slicer and called the feedback filter. In this form, both the forward and feedback equalizers can be jointly adapted to yield the minimum error component at the slicer input, as will be described in chapter 3. The interpretation of this configuration is that the forward equalizer is eliminating the intersymbol interference from future data symbols (present because of the group delay of the channel), resulting in a signal at its output of the form

$$d(T-L) - \sum_{j=1}^{m} d(T-L-j)f_j + \eta(T) \, , \tag{2.3.5}$$

which has intersymbol interference only from past data symbols. The remaining intersymbol interference is canceled by passing the decisions on the data symbols through a transversal filter with the appropriate impulse response, and then subtracting the result from (2.3.5). In this interpretation, the advantage of the DFE is gained because canceling this interference with the forward equalizer alone would increase the size of the noise, while using the past decisions has no impact on the noise.

There are even more sophisticated receiver algorithms for countering intersymbol interference, including the minimum probability of error receiver [19] and the sequence estimation approach [20]. These techniques can also be combined with adaptive filtering for countering a channel with unknown or time varying impulse response.

2.4. ADAPTIVE NOISE CANCELING

Another application of joint process estimation is adaptive noise cancellation [21]. This application is illustrated in figure 2-8, where the canceler has two inputs, the so-called *primary input* and the *reference input*. Roughly speaking, the primary input consists of signal plus noise and the reference input consists of noise alone. The two noises, one in the primary input and the other in the reference input, must be correlated. The object is to use the reference input to reduce the effect of the noise in the primary input.

Assuming that the signals are discrete-time, the primary input is assumed to be of the form

$$d(T) = s(T) + \eta_1(T) , \qquad (2.4.1)$$

where $s(T)$ is the signal which is to be extracted from the noise and $\eta_1(T)$ is a noise which is assumed uncorrelated with the signal. The reference input is then

$$y(T) = \eta_2(T) , \qquad (2.4.2)$$

where $\eta_2(T)$ is a second noise which is assumed to be correlated with $\eta_1(T)$. It is this correlation which can be exploited to minimize the noise in the primary input. It is perhaps counterintuitive that the exact nature of this correlation need not be known in advance.

With this notation, noise cancellation is seen to be simply the joint process estimation problem, and the structure of figure 2-2b is applicable. The adaptive joint process estimation algorithms developed in chapters 3 and 4 will be able to exploit the unknown correlation between the two input signals to minimize the mean-square error $E[e_c^2(T)]$, where

$$e_c(T) = d(T) - \eta_3(T) , \qquad (2.4.3)$$

and $\eta_3(T)$ is the noise at the output of the adaptive filter. Taking into account

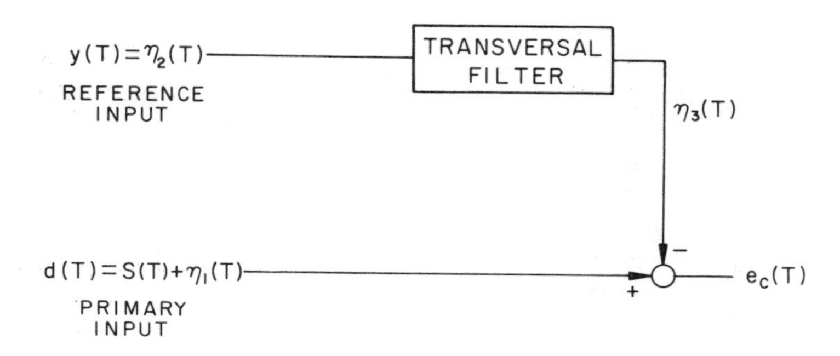

Figure 2-8. Adaptive noise canceler.

the assumption that $\eta_2(T)$ and hence $\eta_3(T)$ are uncorrelated with $s(T)$,

$$E[e_c^2(T)] = E[s^2(T)] + E[(\eta_1(T) - \eta_3(T))^2] . \qquad (2.4.4)$$

Since $E[s^2(T)]$ is a constant, minimizing (2.4.4) is equivalent to minimizing the mean-square noise component in $e_c(T)$; that is, making $\eta_3(T)$ the best replica of $\eta_1(T)$.

Applications of adaptive noise canceling include the canceling of various forms of periodic interference in electrocardiography, the canceling of periodic interference in speech signals, the canceling of broadband interference in the sidelobes of an antenna array, and the elimination of tape hum or turntable rumble during the playback of recorded broadband signals.

2.5. ECHO CANCELLATION

A specific application of noise canceling which is is becoming prevalent in telecommunications is echo cancellation [22, 23]. An echo canceler counteracts a common but undesirable phenomenon in the telephone network, namely undesired echo.

An understanding of the source of echo in the telephone network can be gained from figure 2-9a, where a simplified telephone connection is shown. This connection contains "two-wire" segments on the ends (the subscriber loops and possibly some portion of the local network), and a "four-wire" segment in the center (carrier systems for medium to long-haul transmission). The distinction is that in the two-wire portion of the connection both directions of transmission are carried on the same wire pair, while in the four-wire portion the two directions of transmission are segregated on two different transmission paths (which may or may not consist simply of "wires"). The hybrid performs the conversion from two-wire to four-wire and vice versa. The purpose of the hybrid becomes apparent in the remainder of the figure. The vital talker speech path, shown in figure 2-9b, requires that the hybrid not have any appreciable attenuation between its two-wire and either four-wire port.

There are two distinct echo mechanisms shown in figures 2-9c and 2-9d. Talker echo results in the talker hearing a delayed version of his or her own speech, while in listener echo it is the listener who hears a delayed version of the talker's speech. Both these echo mechanisms are mitigated if the hybrid has significant loss between its two four-wire ports. Achieving this large loss unfortunately depends on the knowledge of the two-wire line impedance, which varies considerably over the population of subscriber loops. The four-wire to four-wire frequency dependent loss of the hybrid cannot be depended on to be greater than about 6 dB.

The effects of echo can be controlled by adding insertion loss to the four-wire portions of the connection, since the echo signals experience this loss two or three times (for talker and listener echo respectively) while the talker speech suffers this loss only once. However, on long connections, this loss itself becomes a significant impairment, and other echo control techniques must be used. On terrestrial connections longer than about 1800 miles echo suppressors are used, in which a large attenuation is inserted in one direction or the other

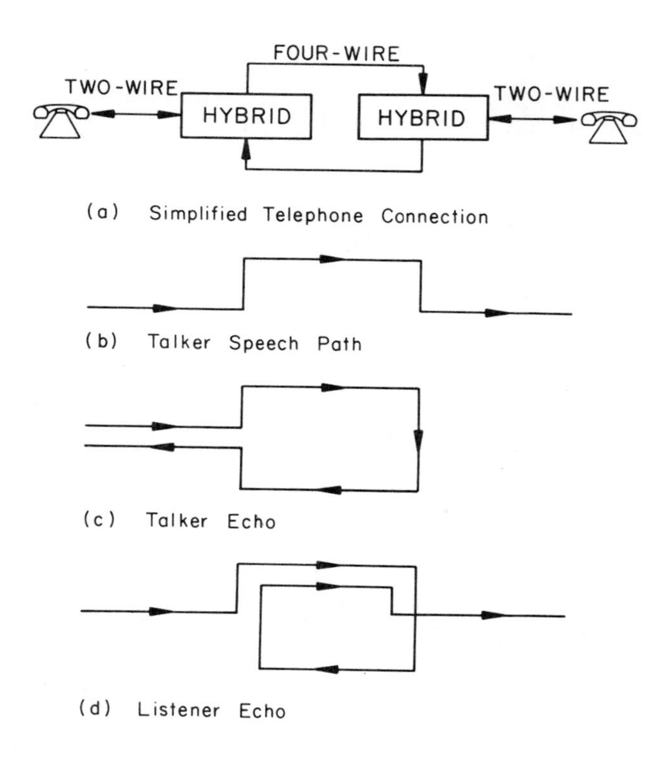

Figure 2-9. Sources of echo in the telephone network.

depending on the direction of active speech. On satellite facilities, which have very long delays on the order of 600 msec round trip, echo suppressors have significantly degraded performance during the doubletalk situation (which occurs when one party to the conversation tries to interrupt the other).

The solution on the long-delay channel is to use the echo canceler, as illustrated in figure 2-10. Shown is the canceler for mitigating the echo in only one direction of the connection. It is a four-terminal device which interfaces to both directions of transmission. Ports C and D are coupled to the near-end talker through a hybrid, while ports A and B are coupled to the far-end talker. The purpose of the echo canceler is to combat the echo representing the feed-through from port C through the hybrid to port D. This path, assuming it to be linear and time-invariant, can be represented by some equivalent transfer function $H(z)$. The canceler is a joint process estimator or noise canceler. The primary input D ($d(T)$) is a superposition of the near-end talker signal with the undesired echo signal. The reference input A ($y(T)$) consists of the far-end talker alone, and is used to cancel the echo component in the primary input. In the notation of (2.4.1-2), $s(T)$ is the near-end talker signal which the echo canceler passes through unchanged, $\eta_1(T)$ is the far-end talker signal after passing through the echo channel appearing in the primary input which it is desired to cancel, and $\eta_2(T)$ is the far-end talker signal appearing in the reference input. The two signals $\eta_1(T)$ and $\eta_2(T)$ are correlated because the former is obtained by passing the latter through the echo channel.

In order to effectively cancel this echo component, it must be assumed that the transfer function $H(z)$ of the hybrid from four-wire port to four-wire port has a finite impulse response. Then the canceler replica $\hat{d}(T)$ can be generated by a transversal filter with input $y(T)$

$$\hat{d}(T) = \sum_{j=1}^{n} c_j y(T-j+1) \ . \tag{2.5.1}$$

Since the echo transfer function $H(z)$ is not known in advance, it is necessary to adapt the filter coefficients c_j of the cancellation filter.

Of course in practice it is desirable to cancel the echos in both directions of a trunk. For this purpose two adaptive transversal filters are necessary, as shown in figure 2-11. One of the filters cancels the echo from each end of the connection. The near-end talker for one of the filters is the far-end talker for the other. In each case, the near-end talker is the "closest" talker, and the far-end talker is the talker generating the echo which is being canceled.

The number of coefficients for each of the transversal filters depends on the length of the impulse response of the hybrid, which is relatively short, as well as the transmission delay from the canceler to the hybrid. The latter parameter depends on the physical location of the canceler, with locations nearer the hybrid being more desirable. However, both transversal filters cannot be near the hybrids at the two ends of a long connection unless they are "split" apart, as

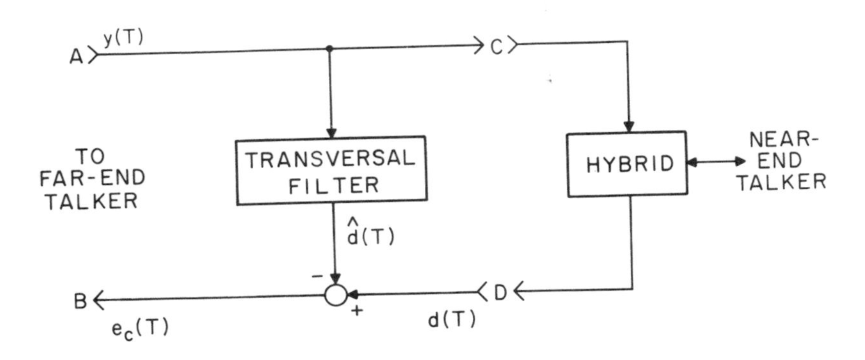

Figure 2-10. Echo canceler for one direction of transmission.

shown in figure 2-11. This split echo canceler configuration is especially important for channels with particularly long delays, such as a satellite channel. In the split configuration, the number of transversal filter coefficients need only compensate for the end delay between hybrid and canceler, which is over terrestrial facilities, and not the much longer delay through the satellite.

Echo cancellation is also used in a related application in data transmission. This is illustrated in figure 2-12, where it is desired to transmit data in both directions simultaneously (so-called full duplex) on a single two-wire facility. Applications include digital transmission on the subscriber loop and voiceband data transmission with interface to the network on a two-wire basis. There are a couple of methods besides echo cancellation that can be used to separate the two directions of transmission, including using two non-overlapping frequency bands and alternating the transmission in the two directions. The latter is not viable on long-delay channels.

The echo cancellation method shown in figure 2-12 uses an echo canceler on the two ends of the connection. It has the advantage that the bandwidth is not doubled relative to the bandwidth required for transmission in a single direction as in the alternative methods. It differs from the voice echo cancellation application in that the input signal is a data signal, and it is therefore possible if the

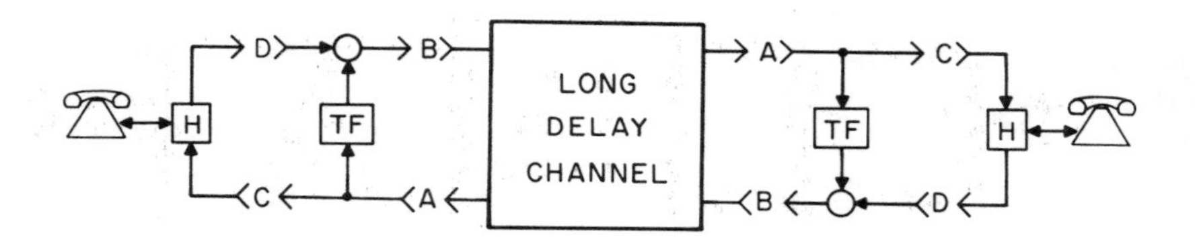

Figure 2-11. Split echo canceler configuration for two directions.

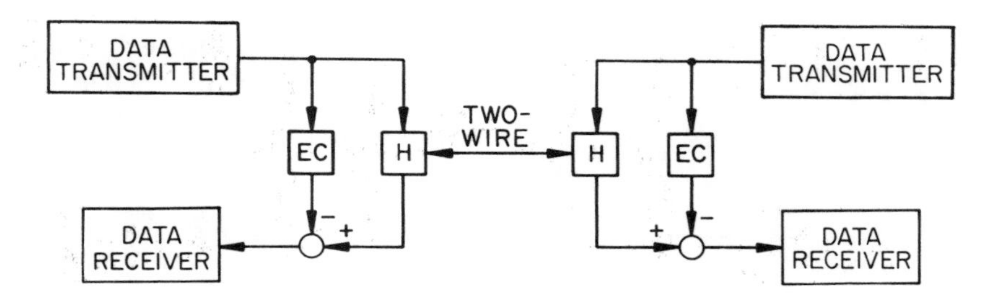

Figure 2-12. Echo cancellation for full-duplex data transmission.

sampling rate is made synchronous with the data rate to apply a binary-valued signal to the canceler. The delays in the canceler transversal filter can thereby be implemented by a simple shift register, and multiplies by the filter coefficients can be avoided. These facts simplify the implementation, but the need for much larger echo attenuations for this application makes the implementation problem actually somewhat more challenging [24].

Echo cancellation for speech transmission gives us an opportunity to illustrate a use for the generalized continuous-time filter structure of figure 1-3 [25]. This structure, which is illustrated in figure 2-13, is applicable where the cancellation is done at the hybrid itself, and is called an adaptive hybrid. This also gives an opportunity to illustrate how the hybrid itself works. The idea is to take advantage of the limited variability of the input impedance of a two-wire transmission line to which the hybrid is connected to reduce the complexity of the canceler. Since the primary input to the canceler $d(t)$ is generated by applying the reference input $y(t)$ to a voltage divider, two simple filters $H_0(s)$ and $H_1(s)$ are built from a similar voltage divider using the two impedances which are approximations to the two-wire line impedance at its extremes. The outputs of these two filters are weighted and summed and used to cancel the far-end talker component in $d(t)$. (Note that t is a continuous-time variable.)

In some applications this simple structure with only two coefficients can achieve adequate cancellation, whereas a transversal filter would require many more coefficients. This illustrates that the transversal filter may sometimes have many more degrees of freedom than is really necessary for a particular application.

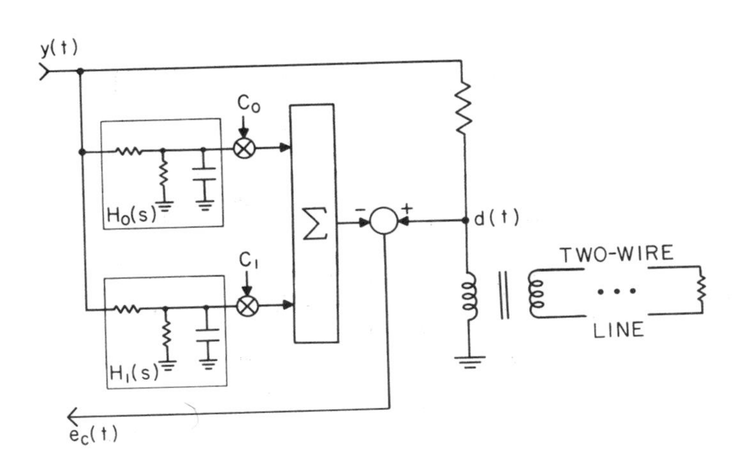

Figure 2-13. Adaptive hybrid utilizing continuous-time adaptive filtering.

2.6. CONCLUSIONS

This chapter has illustrated some practical applications of adaptive filtering. These applications illustrate several configurations for transversal filter structures in adaptive filtering: as a parameter estimator, a predictor, and a joint process estimator. None of the algorithms for adaptive filtering have been specified in these applications; that will be covered in chapters 3, 4, and 6, with performance discussed in chapter 7. In addition, sensitivity and stability issues will be covered in chapter 5.

2.7. FURTHER READING

The applications discussed in this chapter represent only a portion of the situations where adaptive filtering has been used. Other applications which can be explored by the reader are adaptive antenna arrays [26, 27], adaptive inverse modeling [28], and adaptive control [29, 30]. In addition, adaptive filtering touches on many fields, including antennas and control as already mentioned, spectrum estimation theory [31], and system identification theory [32, 33]. A more detailed treatment of speech encoding, including many frequency-domain techniques not touched on here, can be found in [34] and [2].

In this book, applications are emphasized where the signals are real-valued scalars. In many of the other applications there arise signals which are complex-valued, vector-valued, or matrix-valued. This extension requires relatively minor modification to the adaptation algorithms, but does increase the complexity of the notation and tends to obscure the underlying principles. For this reason, only the scalar case and applications which are satisfied by the scalar case are covered in this book, and the reader is referred to the references for extensions to the vector and matrix cases.

PROBLEMS

2-1. The transversal filter predictor was appropriate for modeling a speech signal which is autoregressive.

a. Specify a filter structure which would be appropriate for analyzing a signal which is modeled as a moving average (MA) random process. This structure should have the property that if the filter coefficients were adapted or chosen so that the output random process is white (uncorrelated samples), then the spectrum of the input signal can be determined from the filter coefficients.

b. Repeat part a. for an autoregressive moving average (ARMA) input random process.

2-2. Suppose that for some reason we wanted the filter in an ADPCM decoder to be an all-zero filter rather than all-pole filter. Draw the structure of both the encoder and decoder for this case.

2-3. Consider the ADPCM encoder with an added feedback loop as shown in figure P2-1.

a. Find the transfer function of this system from input to output in terms of the filters $F(z)$ and $P(z)$. For purposes of calculating this transfer function, you can assume that there is no quantization noise; i.e., that the output of the D/A is equal to the input of the A/D.

b. Find the power spectrum of the quantization error referenced to the output of the decoder. You can model the quantization error at the output of the D/A as an additive white noise which is uncorrelated with the signal.

2-4. A transversal finite impulse response filter

$$H(z) = \sum_{m=1}^{n} f_m z^{-m}$$

is known to have linear phase if n is even and

$$f_{n+1-m} = f_m \;, \quad 1 \leqslant m \leqslant \frac{n}{2} \;.$$

Draw a linear phase transversal filter realization which requires only $n/2$ multiplications by filter coefficients.

2-5. An input signal is known to be the samples of a sinusoid with known radian frequency ω and unknown amplitude A and phase θ,

$$y_T = A\cos(T\omega\tau + \theta) \;,$$

where the sampling rate is $1/\tau$. Define an adaptive filter structure which can be used to find the amplitude A and phase θ. That is, if this structure is adapted to minimize the output signal in some sense, the coefficients will give us an estimate of the amplitude and phase of the sinusoid. Be specific as to how the amplitude and phase would be determined from the filter coefficients.

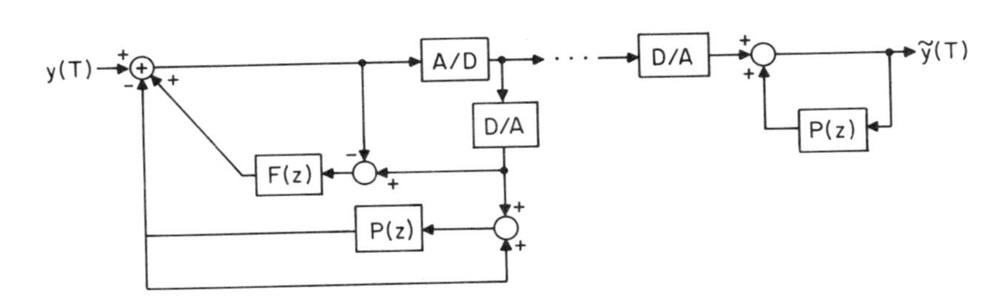

Figure P2-1. ADPCM with quantization error feedback added.

2-6. For an adaptive equalizer, suppose the channel has a response $H(z)$ to a single input data symbol. Find $H(z)$ and the coefficients of a transversal filter equalizer which eliminates the intersymbol interference at the input to the slicer for the following two cases:

a. $h_k = 0$ for $k < 0$ and $h_k = a^k$ for $k \geqslant 0$.

b. $h_k = a^{|k|}$.

2-7. Assume that the response of a channel to a single input data symbol is $H(z)$. Further assume that this transfer function can be factored in the form

$$H(z) = H^+(z)H^+(z^{-1})$$

where $H^+(z)$ corresponds to a causal time function (can be expanded in terms of negative powers of z), and $H^+(z) \neq 0$ for $|z| > 1$. Let the forward equalizer of a decision feedback equalizer have transfer function $\dfrac{1}{h_0^+ H^+(z^{-1})}$.

a. Show that this forward equalizer is a stable anti-causal filter (can be expanded in positive powers of z).

b. Show that the response at the output of the forward equalizer is causal as required if we are to cancel the remaining intersymbol interference with a decision feedback filter.

c. Find the coefficients of the feedback filter which exactly cancel all remaining intersymbol interference.

(It will be shown in problem 3-25 that the equalizers derived in this problem are the optimum infinite order forward and feedback transversal filter equalizers.)

2-8. Apply the results of problem 2-7 to find the coefficients of the forward and backward equalizer of a DFE for the channel response of problem 2-6b.

2-9. For a voice echo canceler, what would you expect the filter coefficients c_k to be in terms of the transfer function of the hybrid $H(z)$ or the impulse response of the hybrid h_k? Under what conditions can we get exact cancellation?

REFERENCES

1. J. Makhoul , "Linear Prediction: A Tutorial Review," *Proc. IEEE* **63, 4,** pp. 561-580 (April 1975).
2. L.R. Rabiner and R.W. Schafer, *Digital Processing of Speech Signals,* Prentice-Hall, Englewood Cliffs, N.J. (1978).
3. R.W. Schafer and L.R. Rabiner, "Digital Representations of Speech Signals," *Proceedings of the IEEE* **63** pp. 662-677 (April 1975).

4. M.R. Schroeder, "Vocoders: Analysis and Synthesis of Speech," *Proceedings of IEEE* **54** p. 720 (May 1966).
5. A.H. Gray and J.D. Markel, *Linear Prediction of Speech,* Springer- Verlag Berlin (1976.).
6. N.S. Jayant, "Digital Coding of Speech Waveforms: PCM, DPCM, and DM Quantizers," *Proc. IEEE* **62**(5) pp. 611-632 (May 1974).
7. B.M. Oliver, J. Pierce, and C.E. Shannon, "The Philosophy of PCM," *Proceedings IRE* **36** p. 1324 (Nov. 1948).
8. K.W. Cattermole, *Principles of Pulse Code Modulation,* Iliffe Books LTD, London (1969).
9. D.L. Cohn and J.L. Melsa, "The Residual Encoder - An Improved ADPCM System for Speech Digitization," *IEEE Trans. Comm.* **COM-23** pp. 935-941 (Sept. 1975).
10. B.S. Atal and M.R. Schroeder, "Adaptive Predictive Coding of Speech Signals," *Bell Syst. Tech. J.* **49**(8) pp. 1973-1986 (Oct. 1970).
11. B.S. Atal and M.R. Schroeder, "Predictive Coding of Speech Signals and Subjective Error Criteria," *IEEE Trans. ASSP* **ASSP-27** pp. 247-254 (June 1979).
12. R.W. Lucky, J. Salz, and E.J. Weldon, Jr., *Principles of Data Communication,* McGraw-Hill, New York (1968).
13. J.G. Proakis, *Digital Communications,* McGraw-Hill, New York (1983).
14. Viterbi, A.J. and Omura, J.K., *Principles of Digital Communication and Coding,* McGraw-Hill, New York (1979).
15. Proakis, J.G., "Advances in Equalization for Intersymbol Interference," *Advances in Communication Systems* **4**(1975).
16. Austin, M.E., *Decision-Feedback Equalization for Digital Communication Over Dispersive Channels,* M.I.T. Lincoln Laboratory, Lexington, Mass (August 1967).
17. Salz, J., "Optimum Mean-Square Decision Feedback Equalization," *Bell System Tech. J.* **52** pp. 1341-1373 (October 1973).
18. Duttweiler, D.L., Mazo, J.E., and Messerschmitt, D.G., "Error Propagation in Decision-Feedback Equalizer," *IEEE Trans. Information Theory* **IT-20** pp. 490-497 (July 1974).
19. Abend, K and Fritchman, B.D., "Statistical Detection for Communication Channels with Intersymbol Interference," *Proc. IEEE* **58** pp. 779-785 (May 1970).
20. Forney, G.D., Jr., "Maximum-Likelihood Sequence Estimation of Digital Sequences in the Presence of Intersymbol Interference," *IEEE Trans. Information Theory* **IT-18** pp. 363-378 (May 1972).
21. B. Widrow et. al., "Adaptive Noise Canceling: Principles and Applications," *IEEE Proceedings* **63**(12) pp. 1692-1716 (Dec. 1975).

22. D.G. Messerschmitt, "Echo Cancellation in Speech and Data Transmission," *Transactions on Special Topics in Communications* **COM-32**(Feb. 1984).

23. M. Sondhi and D.A. Berkley, "Silencing Echos on the Telephone Network," *IEEE Proceedings* **8**(August 1980).

24. O. Agazzi, D.A. Hodges, and D.G. Messerschmitt, "Large-Scale Integration of Hybrid-Method Digital Subscriber Loops," *IEEE Trans. Communications* **COM-30** p. 2095 (Sep. 1982).

25. D.G. Messerschmitt, "An Electronic Hybrid with Adaptive Balancing for Telephony," *IEEE Trans. on Communications* **COM-28** p. 1399 (Aug. 1980).

26. B. Widrow, P.E. Mantey, and L.J. Griffiths, "Adaptive Antenna Systems," *Proc. IEEE* **55**(12) pp. 2143-2159 (Dec. 1967).

27. R.A. Monzingo and T.W. Miller, *Introduction to Adaptive Arrays,* Wiley, New York (1980).

28. B. Widrow, J. McCool, and B. Medoff, "Adaptive Control by Inverse Modeling," in *Thirteenth Annual Assilomar Conference on Circuits, Systems, and Computers*, (Oct. 1979).

29. R.B. Asher, D. Andrisani, and P. Dorato, "Bibliography on Adaptive Control Systems," *IEEE Proceedings* **64**(8) p. 1226 (Aug. 1976).

30. G.C. Goodwin and K.S. Sin, *Adaptive Filtering Prediction and Control,* Prentice-Hall, Englewood Cliffs N.J. (1984).

31. E. Robinson, "A Historical Perspective of Spectrum Estimation," *IEEE Proceedings* **70**(9) p. 885 (Sept. 1982).

32. P. Eykhoff, *System Parameter and State Estimation,* Wiley, London (1974).

33. L. Ljung and T. Soderstrom, *Theory and Practice of Recursive Identification,* MIT Press, Cambridge (1983).

34. N.S. Jayant and P. Noll, *Digital Coding of Waveforms,* Prentice-Hall, Englewood Cliffs N.J. (1984).

3

TRANSVERSAL FILTERS

Chapter 2 described a few of the numerous applications of adaptive linear prediction and joint process estimation. This chapter will review some of the older and most widely applied techniques for adaptive filtering, including both block and recursive adaptation algorithms, in the context of the transversal filter structure. The approach of this chapter will be to fix the filter structure to be the transversal, and examine a variety of adaptation algorithms.

The starting point will be the stationary case with known statistics in section 3.1. While the assumption of stationarity and particularly the assumption of known statistics is unrealistic in most adaptive filter applications, this represents a starting point for the algorithms described later in the chapter. The solution for this case is known as the minimum mean-square error (MMSE) solution. The orthogonality principle, with particular attention to the autoregressive case, and the LMS gradient algorithm are also discussed in this chapter.

The most common approach to adaptive filtering, the stochastic gradient (SG) algorithm, is discussed in section 3.2. This is sometimes called the least mean-square (LMS) adaptive filtering algorithm. Several modifications to the SG algorithm which simplify its implementation and satisfy practical constraints are discussed. A variation on the SG approach, called stochastic approximation and discussed in section 3.2, is capable of achieving exact convergence for a truly stationary environment.

The remainder of the chapter discusses the class of least-squares (LS) algorithms. These come in two flavors: block LS and recursive LS algorithms. The block LS approaches, including the autocorrelation, covariance, and prewindowed methods, are considered in section 3.3. These methods are commonly used in speech analysis-synthesis, as previously discussed in section 2.1. The recursive LS algorithms are discussed in section 3.4. The relatively simple approaches discussed in this section are forerunners to the more conceptually difficult but computationally more efficient recursive algorithms considered in chapter 6.

Since the predictor and joint process estimator cases are so similar, the approach here will be to consider the more general joint process estimator case first, and then briefly reformulate the results for the predictor case.

As in chapter 2 define the input random processes as $y(T)$ and $d(T)$. For purposes of the remainder of the book it will be convenient to define a vector notation for the vector of n filter coefficients

$$\mathbf{f}' = [f_1 \, f_2 \cdots f_n] \, , \tag{3.0.1a}$$

$$\mathbf{c}' = [c_1 \, c_1 \cdots c_n] \, , \tag{3.0.1b}$$

for the predictor and joint process estimator transversal filters respectively. Throughout the remainder of the book, $\mathbf{c}'$ will denote the transpose of a vector $\mathbf{c}$, bold lower case letters will denote vectors, $\mathbf{A}'$ will denote the transpose of a matrix $\mathbf{A}$, and a bold upper case letter will denote a matrix.

Also define a vector of the current and $n-1$ past input samples

$$\mathbf{y}'(T) = [y(T) \, y(T-1) \cdots y(T-n+1)] \, . \tag{3.0.2}$$

With this notation in hand, the error signal for a transversal joint process estimator in (2.0.2) can be written as

$$e_c(T) = d(T) - \mathbf{c}'\mathbf{y}(T) \, . \tag{3.0.3}$$

Various methods of minimizing this error by the choice of the filter coefficients will be considered in this chapter.

A similar relation holds for the predictor case, where the prediction error can be written from (2.0.1) as

$$e_f(T) = y(T) - \mathbf{f}'\mathbf{y}(T-1) \, . \tag{3.0.4}$$

In later sections the coefficient vector will be a function of time, and the notation of (3.0.1) will have to be modified to reflect that fact. In particular, the coefficient vector at time T in analogy to (3.0.3) will be denoted $\mathbf{c}(T)$ rather than $\mathbf{c}$, and the j-th component of this vector will be called $c_j(T)$.

3.1. MINIMUM MEAN-SQUARE ERROR SOLUTION

In this section it is assumed that the input signals $y(T)$ and $d(T)$ are wide-sense stationary discrete-time random processes with known autocorrelation and power spectrum. Although this is an unrealistic assumption for the class of applications discussed in this book, it is useful to examine the solution for this case as a motivation for the subsequent techniques which are oriented toward the unknown statistics or nonstationary case. The minimum mean-square error (MMSE) n-th order joint process estimator for this input random process will be derived.

Explicitly evaluating the mean-square error from (3.0.3),

$$\begin{aligned} E[e_c^2(T)] &= E[d^2(T)] - 2\mathbf{c}'E[d(T)\mathbf{y}(T)] \\ &\quad + \mathbf{c}'E[\mathbf{y}(T)\mathbf{y}'(T)]\mathbf{c} \\ &= E[d^2(T)] - 2\mathbf{c}'\mathbf{p} + \mathbf{c}'\mathbf{\Phi}\mathbf{c} \, , \end{aligned} \tag{3.1.1}$$

where $\mathbf{p}$ and $\mathbf{\Phi}$ are defined below. This is a quadratic form in the coefficient vector $\mathbf{c}$, and hence there is a unique minimum which can be obtained by completing the square or setting the gradient equal to zero.

In (3.1.1) the undefined matrices are

$$\mathbf{p} = E[d(T)\mathbf{y}(T)] \tag{3.1.2}$$

$$\mathbf{\Phi} = E[\mathbf{y}(T)\mathbf{y}'(T)] = \begin{bmatrix} \phi_0 & \phi_1 & \cdots & \phi_{n-1} \\ \phi_1 & \cdot & & \cdot \\ \vdots & & \ddots & \vdots \\ \phi_{n-1} & \cdots & & \phi_0 \end{bmatrix}, \tag{3.1.3}$$

where neither of these quantities is a function of time due to the wide-sense stationarity assumption. In particular, $\mathbf{\Phi}$ is an autocorrelation matrix for the reference input process, since its elements are the autocorrelation coefficients,

$$\phi_j = E[y(T)y(T+j)] \,. \tag{3.1.4}$$

The $\mathbf{\Phi}$ matrix has several important properties. First, it is symmetric and the i,j element is a function of $(i-j)$. This property will play an important role throughout the book; a matrix with this property is known as a *Toeplitz* matrix [1, 2]. Secondly, since the matrix is an autocorrelation matrix, it is positive semidefinite. In most but not all applications it can be assumed that this autocorrelation matrix is positive definite, and hence nonsingular. Instances where this is not true will be discussed in section 3.2.2.4, but as long as it is true then the autocorrelation matrix has positive real eigenvalues.

Completing the square in (3.1.1),

$$E[e_c^2(T)] = E[d^2(T)] - \mathbf{p}'\mathbf{\Phi}^{-1}\mathbf{p} + (\mathbf{\Phi}^{-1}\mathbf{p}-\mathbf{c})'\mathbf{\Phi}(\mathbf{\Phi}^{-1}\mathbf{p}-\mathbf{c}) \,. \tag{3.1.5}$$

Since $\mathbf{\Phi}$ is positive definite, the third term is non-negative and can be minimized by the choice

$$\mathbf{c}_{\text{opt}} = \mathbf{\Phi}^{-1}\mathbf{p} \,. \tag{3.1.6}$$

This choice also minimizes the mean-square error, which has a resultant minimum value

$$\begin{aligned} E[e_c^2(T)] &= E[d^2(T)] - \mathbf{p}'\mathbf{\Phi}^{-1}\mathbf{p} \\ &= E[d^2(T)] - \mathbf{p}'\mathbf{c}_{\text{opt}} \,. \end{aligned} \tag{3.1.7}$$

Finding $\mathbf{c}_{\text{opt}}$ from (3.1.6) requires the solution of a system of linear equations. A particularly efficient algorithm for solving these equations, which is a consequence of the Toeplitz property of the autocorrelation matrix, is the *Levinson-Durbin algorithm* [3]. This important algorithm will be derived in chapter 4 in the context of the LMS lattice solution.

For the predictor case the previous results must be modified slightly (problem 3-2). In particular, from (3.0.4) the $\mathbf{p}$ vector assumes a particular form

$$\begin{aligned} \mathbf{p} &= y(T)\mathbf{y}(T-1) \\ &= [\phi_1 \; \phi_2 \cdots \phi_n]' \,. \end{aligned} \tag{3.1.8}$$

In addition, the matrix $\mathbf{\Phi}$ is an n by n matrix for this case.

3.1.1. Orthogonality Principle

The preceding derivation of the optimum coefficient vector for the predictor and joint process estimator was based on simple formal matrix manipulations. An interpretation of this result known as the *orthogonality principle* will now be demonstrated. In addition to being important in its own right, this principle will give a preview of the mathematical techniques extensively used in chapters 4 and 6.

Rather than developing (3.1.5) and then finding the optimum coefficient vector, the mean-square error from (3.0.3) can be differentiated directly and set to zero,

$$\begin{aligned} 0 &= \frac{\partial}{\partial c_m} E[e_c^2(T)] \\ &= 2\,E[e_c(T)\frac{\partial}{\partial c_m} e_c(T)] \\ &= -2\,E[e_c(T)y(T-m+1)], \quad 1 \leqslant m \leqslant n\ , \end{aligned} \tag{3.1.9}$$

where the interchange in the order of expectation and differentiation is justified by the linearity of both operators. Since the mean-square error is a quadratic function of the coefficients, the coefficient vector corresponding to the optimum is unique and (3.1.9) is a necessary and sufficient condition for optimality of the coefficients. Substituting for $e_c(T)$ in (3.1.9) results in a set of linear equations equivalent to (3.1.6), thereby giving the same results (problem 3-3). However, (3.1.9) demonstrates that for the optimum coefficient vector, the estimation error is uncorrelated with all the data samples which are used in the estimate. This is the orthogonality principle, a terminology which suggests that this result can somehow be given a geometric interpretation. In fact this is the case, as will be demonstrated in chapter 4.

It is easily established (problem 3-4) that a similar orthogonality principle holds for the predictor case, where $e_f(T)$ is uncorrelated with $y(T-1), \ldots, y(T-n)$.

It was stated without justification in section 2.1 that the prediction error samples at the output of an optimum linear predictor are uncorrelated (the prediction error process is white). This is true as long as the number of predictor coefficients (the length of the transversal filter) is adequate. This result will now be shown for a predictor of an arbitrary wide-sense stationary input process as $n \to \infty$ as a consequence of the orthogonality principle. In the next section it will be shown further that for an autoregressive input process this white prediction error property holds for finite predictor order n.

When $n \to \infty$, $e_f(T)$ is uncorrelated with all $y(j)$, $j < T$, by the orthogonality principle. Since all $e_f(j)$, $j < T$ are linear combinations of these samples, it follows that

$$E[e_f(T)e_f(j)] = 0, \quad j < T\ , \tag{3.1.10}$$

and since the order of the arguments in the expectation is arbitrary the condition $j < T$ can obviously be replaced by $j \neq T$.

3.1.2. MMSE Solution for An Autoregressive Input Process

Recall from section 2.1 that speech is often modeled as an autoregressive (AR) random process. Specifically, if $\eta(T)$ is a white noise (its samples are uncorrelated), then $y(T)$ is autoregressive if it can be generated by the recursion (2.1.2), which is repeated here for convenience as

$$y(T) = \eta(T) + \sum_{j=1}^{m} a_j y(T-j) , \tag{3.1.11}$$

where m is called the order of the process. As in figure 2-3a, an equivalent formulation for the generation of $y(T)$ is white noise driving an m-pole filter with transfer function

$$H(z) = \frac{1}{1 - \sum_{j=1}^{m} a_j z^{-j}} . \tag{3.1.12}$$

Some statements were made in section 2.1 which will now be justified. In particular, it was stated that the optimum linear predictor for an AR process has coefficients

$$f_j = \begin{cases} a_j, & 1 \leqslant j \leqslant m \\ 0, & m < j \leqslant n \end{cases} , \tag{3.1.13}$$

as long as the order of the predictor, n, is greater than or equal to the order of the process, m. The implication is that for an AR process there is no point to using a predictor with order greater than the order of the process. Said another way, a finite order predictor (with sufficiently large order) performs as well as an infinite order predictor. It was shown as well in section 2.1 that if this optimum predictor is used the prediction error process is white. This was shown in section 3.1.1 to be true for an infinite order predictor; the special fact about an AR process is that the prediction error process is white for an appropriate optimum finite order predictor.

From (3.1.11) it is evident that the all-pole filter of (3.1.12) must be causal. If the additional important restriction is imposed that the filter be stable (has all its poles interior to the unit circle), then the transfer function can be written in the form

$$H(z) = \sum_{j=0}^{\infty} h_j z^{-j} , \tag{3.1.14}$$

where h_j is the unit pulse response of the filter. This alternative characterization of the filter will be the key to justifying the earlier statements.

Writing the AR process as a convolution of input white noise with the unit pulse response,

$$y(T) = \sum_{j=0}^{\infty} h_j \eta(T-j) \ , \qquad \textbf{(3.1.15)}$$

an AR process can be considered as a moving average (MA) process with a particular infinite set of coefficients. Further, from the predictor of (3.1.13) as applied to the process generated by the recursion (3.1.11), it is easily shown that for $m \leqslant n$

$$e_f(T) = \eta(T) \ . \qquad (3.1.16)$$

The prediction error for the predictor of (3.1.13) is just the AR white noise input. It should be reiterated that this result applies only where the order of the predictor is high enough! The optimality of the predictor can then be checked by seeing if the orthogonality principle is satisfied. Calculating the crosscorrelation between the prediction error and a delayed version of the input process,

$$\begin{aligned} E[e_f(T)y(T-l)] &= E[\eta(T)y(T-l)] \\ &= \sum_{j=0}^{\infty} h_j E[\eta(T)\eta(T-l-j)] \\ &= 0, \ \ 1 \leqslant l \leqslant n \ , \end{aligned} \qquad (3.1.17)$$

because the noise is assumed to be white. Since the orthogonality principle is satisfied, the optimality of (3.1.13) is verified.

While this demonstrates that a finite order predictor is optimum for an AR process, obviously a process which is not AR can also be predicted using a finite order predictor. In this case, as the order n of the predictor is increased, the prediction error variance diminishes indefinitely. (For some random processes the prediction error can become zero for a finite order predictor; this is known as a *deterministic* random process since the current sample is perfectly predictable.) Using a finite order predictor for a non-AR process is in some sense equivalent to modeling the process as closely as possible by an AR or all-pole model. This approximation is very closely related to a very popular technique known as maximum-entropy spectral estimation. See [4] for an explanation of this connection.

3.1.3. The MMSE Gradient Algorithm

The previous results indicate that a system of linear equations must be solved in order to find the optimum MMSE coefficient vector. One efficient method for solving these linear equations, the Levinson-Durbin algorithm, has been mentioned. Much of the remainder of this book is devoted to finding alternate ways of solving this and similar sets of linear equations, whether they be explicit like the Levinson-Durbin algorithm or implicit. A common objective will be to find methods which reduce the computational load.

In this section, an alternative to Levinson-Durbin, a *gradient algorithm*, will be derived. This algorithm is of little interest in and of itself, but will lead directly to the widely used *stochastic gradient (SG) algorithm* in section 3.2. Further, understanding of the convergence properties of the gradient algorithm is very helpful to the understanding of the SG algorithm.

The approach in the gradient algorithm is to define a sequence of coefficient vectors which is guaranteed to converge to the optimum coefficient vector. As a starting point to the derivation, $y(T)$ is again assumed to be wide-sense stationary with known autocorrelation matrix (3.1.3). The output mean square error (MSE) given by (3.1.1) is a quadratic form in the coefficient vector which has a unique global minimum. The approach is to adjust the weights iteratively to minimize the MSE by descending along this surface with the objective of getting to the minimum.

Since the algorithm is iterative in nature, a notation for the coefficient vector which reflects this is needed. Thus, call the j-th iteration of the coefficient vector $\mathbf{c}_j$. Given the present coefficient vector $\mathbf{c}_j$, by subtracting off a term proportional to the error gradient, $\nabla_{\mathbf{c}}\{E[e_c^2(T)]\}$, the resultant tap vector should be closer to $\mathbf{c}_{opt}$. This is because the gradient of the error is a vector in the direction of maximum increase of the error. Moving a short distance in the opposite (negative) direction of the gradient should therefore reduce the error. On the other hand, moving too far in that direction might actually overshoot the minimum, and result in instability.

The gradient algorithm is illustrated in figure 3-1 for the order two case. As will be shown later, the contours in the plane of filter coefficients of constant mean-square error have an elliptical shape. The negative of the gradient points in the direction of maximum decrease of the mean-square error. When the step-size is small, the mean-square error is reduced at each step of the algorithm, and approaches the minimum of (3.1.7) asymptotically. When the step-size is too large, as shown in figure 3-1, the mean-square error can actually increase, and the algorithm becomes unstable.

The gradient algorithm is explicitly

$$\mathbf{c}_{j+1} = \mathbf{c}_j - \frac{\beta}{2}\nabla_{\mathbf{c}_j}\{E[e_c^2(T)]\}\ , \tag{3.1.18}$$

where β is a small adaptation constant or step size which controls the size of the change in $\mathbf{c}_j$ at each update. The division by two is included to avoid a factor of two in the subsequent adaptation algorithm. Referring back to (3.1.5), this algorithm becomes

$$\begin{aligned}\mathbf{c}_{j+1} &= \mathbf{c}_j + \beta(\mathbf{p} - \mathbf{\Phi}\mathbf{c}_j) \\ &= (\mathbf{I} - \beta\mathbf{\Phi})\mathbf{c}_j + \beta\mathbf{p}\end{aligned} \tag{3.1.19}$$

where $\mathbf{I}$ is the identity matrix. Hopefully, if this algorithm is simply iterated from some arbitrary initial guess $\mathbf{c}_0$ it will converge to $\mathbf{c}_{opt}$ of (3.1.6). Because understanding of the convergence of the gradient algorithm is crucial to the understanding of the SG adaptation algorithm covered in section 3.2, this convergence will now be considered in some detail.

If $\mathbf{c}_{opt}$ from (3.1.6) is subtracted from both sides of (3.1.19) and

$$\mathbf{q}_j = \mathbf{c}_j - \mathbf{\Phi}^{-1}\mathbf{p} \tag{3.1.20}$$

is defined as the error between the actual and optimal coefficient vector, then

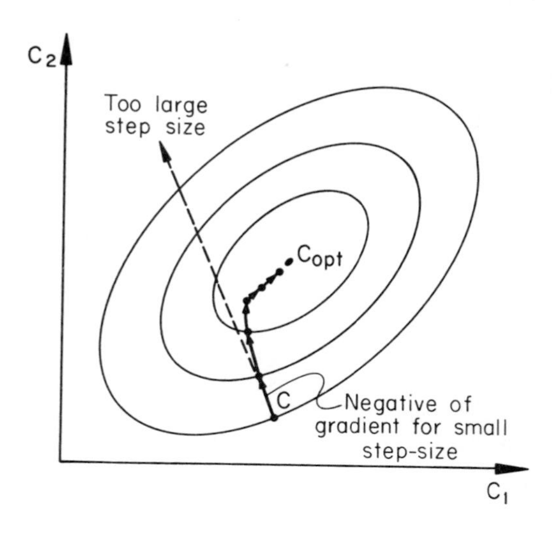

Figure 3-1. Illustration of gradient algorithm.

an iterative equation for the error results,

$$\mathbf{q}_{j+1} = (\mathbf{I} - \beta\mathbf{\Phi})\mathbf{q}_j \ , \tag{3.1.21}$$

and simply iterating this equation gives

$$\mathbf{q}_j = (\mathbf{I} - \beta\mathbf{\Phi})^j \mathbf{q}_0 \ . \tag{3.1.22}$$

The question becomes whether this error converges to zero.

The behavior of (3.1.22) depends critically on the eigenvalues of the matrix $\mathbf{\Phi}$, which we denote by $\lambda_1, \ldots, \lambda_n$, and which are explored in problem 3-5. As mentioned earlier, when $\mathbf{\Phi}$ is positive definite (semi-definite), these eigenvalues are positive (non-negative) real-valued. Further, make the simplifying assumption that these eigenvalues are distinct, and denote the corresponding eigenvectors by $\mathbf{v}_1, \ldots, \mathbf{v}_n$. Assume these eigenvectors to be normalized to unity length, i.e. $\mathbf{v}_j{}'\mathbf{v}_j = 1$, $1 \leqslant j \leqslant n$. These eigenvectors are orthogonal,

$$\mathbf{v}_i{}'\mathbf{v}_j = \begin{cases} 0, & i \neq j \\ 1, & i = j \end{cases} , \tag{3.1.23}$$

since they correspond to distinct eigenvalues (problem 3-5) Then it is easily established (problem 3-6) that $\mathbf{\Phi}$ can be written in the form

$$\mathbf{\Phi} = \sum_{i=1}^{n} \lambda_i \mathbf{v}_i \mathbf{v}_i{}' \ . \tag{3.1.24}$$

Further, it is easily shown (problem 3-7) that

$$(\mathbf{I} - \beta\mathbf{\Phi})^j = \sum_{i=1}^{n} (1 - \beta\lambda_i)^j \mathbf{v}_i \mathbf{v}_i{}' \ . \tag{3.1.25}$$

This demonstrates that the error vector $\mathbf{q}_j$ obeys a trajectory which is the sum of n modes, the ith of which is proportional to $(1 - \beta\lambda_i)^j$. The speed of convergence of each of these modes is governed by β. If β is made too large, then one or more of the $(1 - \beta\lambda_i)$ terms will be larger than unity in magnitude, and the error vector in (3.1.22) will actually increase in size with time. This is quite consistent with the intuitive behavior exhibited in figure 3-1 since the large β causes an overshoot of the minimum and actually increases the error.

This acceptable range of β can be investigated further if we order the eigenvalues from smallest to largest, denoting the smallest as λ_{min} and the largest as λ_{max}. Then the $(1 - \beta\lambda_i)$ term which governs how large β can get is the one corresponding to the largest eigenvalue, and hence the condition for $\mathbf{q}_j$ decaying exponentially to zero is

$$0 < \beta < \frac{2}{\lambda_{max}} \ . \tag{3.1.26}$$

This determines the largest value of β, but of more interest is the β corresponding to the fastest convergence of the gradient algorithm. For a fixed β, the speed of convergence of the algorithm can be considered to be dominated by the slowest converging mode in (3.1.25). This slowest mode corresponds to the largest value of $|1 - \beta\lambda_i|$. The two extreme cases are plotted in figure 3-2, where this term is calculated for λ_{min} and λ_{max}. The corresponding curves for the other eigenvalues lie in between these two curves. The value of β which results in the fastest convergence is the point labeled β_{opt} in the figure. Choice of any other value of β results in a slower convergence of the mode corresponding to either the maximum or minimum eigenvalue. This optimum value of β is easily shown to be

$$\beta_{opt} = \frac{2}{\lambda_{min} + \lambda_{max}} \ , \tag{3.1.27}$$

and for this choice of β the modes corresponding to both minimum and maximum eigenvalues converge at the same rate, namely proportional to

$$\left(\frac{\dfrac{\lambda_{max}}{\lambda_{min}} - 1}{\dfrac{\lambda_{max}}{\lambda_{min}} + 1} \right)^j \ . \tag{3.1.28}$$

The quantity in the parenthesis is plotted in figure 3-3 as a function of the parameter $\dfrac{\lambda_{max}}{\lambda_{min}}$. This parameter, the ratio of largest to smallest eigenvalue, is seen to be of fundamental importance; it is called the *eigenvalue spread*. The eigenvalue spread has a minimum value of unity, and can be arbitrarily large. The larger the eigenvalue spread of the autocorrelation matrix, the slower the convergence of the gradient algorithm. As seen in figure 3-3, the convergence

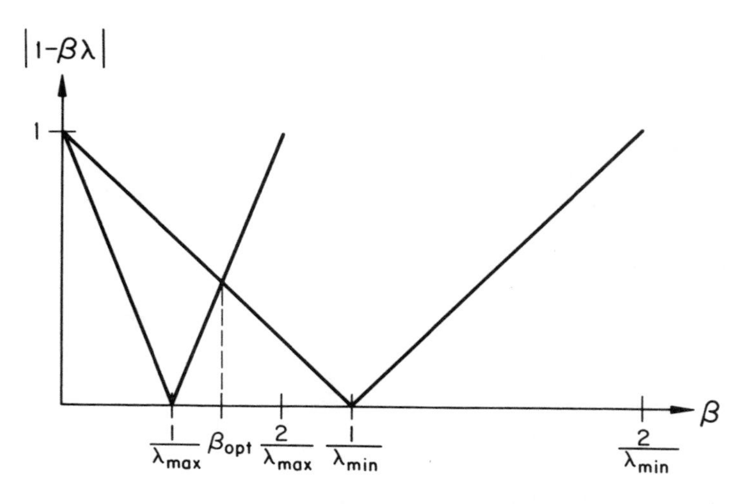

Figure 3-2. Choice of step-size for fastest convergence.

becomes arbitrarily slow as the eigenvalue spread approaches infinity since the quantity in parentheses in (3.1.28) approaches unity.

The reason that the convergence slows as the eigenvalue spread increases can be understood intuitively as follows. From (3.1.5), the difference for a given set of filter coefficients between the mean-square error and the minimum value is (problem 3-8)

$$E[e_c^2(T)] - E[e_c^2(T)]_{\min} = (\mathbf{c} - \mathbf{c}_{\text{opt}})'\mathbf{\Phi}(\mathbf{c} - \mathbf{c}_{\text{opt}})$$
$$= \sum_{i=1}^{n} \lambda_i \left[(\mathbf{c} - \mathbf{c}_{\text{opt}})'\mathbf{v}_i\right]^2, \tag{3.1.29}$$

using expansion (3.1.23). To determine the mean-square error for any coefficient vector **c**, this relation says that we should find the component of this vector in the direction of each of the eigenvectors, square that component, and multiply by the corresponding eigenvalue. This implies that the mean-square error increases most rapidly in the direction of the eigenvector corresponding to $\lambda_{\max}$ and most slowly in the direction corresponding to $\lambda_{\min}$. The largest β is determined by the maximum eigenvalue, since the gradient will be the largest in the direction of the eigenvector corresponding to the largest eigenvalue, and the correction of the gradient algorithm will thus be largest in that direction. As β gets larger, that is the direction in which the increment in the algorithm will first be so large as to actually increase the mean-square error.

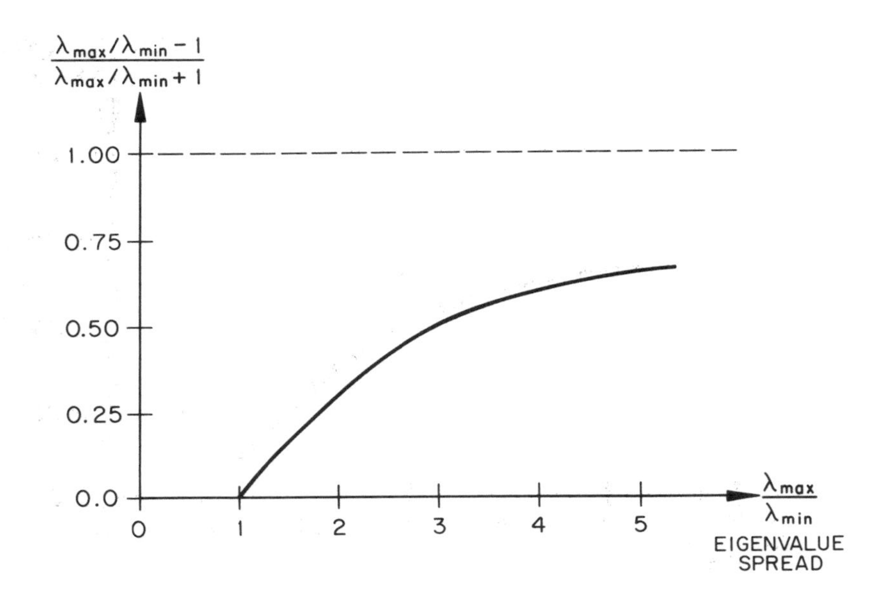

Figure 3-3. Relation of fastest convergence rate to eigenvalue spread.

The convergence of the gradient algorithm can be interpreted graphically by plotting the contours of equal mean-square error as in figure 3-4. Equation (3.1.29) illustrates that the contours of equal mean-square error are elliptical in shape, with the axes in the direction of the eigenvectors. The eccentricity of the ellipses are directly related to the relative sizes of the eigenvalues. This is illustrated in figure 3-4 for the $n = 2$ case. It is assumed that $\lambda_2 > \lambda_1$, in which case the mean-square error increases more rapidly in the direction of $\mathbf{v}_2$. The direction of the two orthogonal eigenvectors is shown. The major axis of the ellipse is in the direction of $\mathbf{v}_1$, and the minor axis in the direction of $\mathbf{v}_2$.

The case where the eigenvalue spread is small (eigenvalues approximately equal) is shown in figure 3-4a, and the ellipse is close to being a circle. The opposite case, where the eigenvalue spread is large, is shown in figure 3-4b, and for this case the ellipse is highly eccentric. In the case of a small eigenvalue spread the gradient correction is always nearly in the direction of the minimum mean-square error, and the length of the gradient vector is always approximately the same. For the large eigenvalue spread, the direction of the negative gradient can be quite different from the direction of the minimum, although for small steps the mean-square error still gets smaller. Since each step does not go directly toward the minimum, the number of required steps will be increased for some starting conditions.

More importantly, the length of the gradient vector will be much smaller in the direction of the major axis of the ellipse, since the MSE is not varying as rapidly in that direction. The step size is therefore governed by the largest eigenvalue, so that the steps do not overshoot in the direction of the corresponding eigenvector, which is the minor axis of the ellipse. A step-size which maintains stability along the minor axis results in very small increments in the direction of the major axis.

Figure 3-4 also leads to an intuitive interpretation of figure 3-2. Consider the case where the starting coefficient vector is on the minor axis of the ellipse, so that convergence is in the direction of the eigenvector corresponding to the largest eigenvalue. If β is chosen to be smaller than $1/\lambda_{max}$, then each step of the algorithm in this direction does not overshoot the minimum, and the MSE gets smaller. When $\beta = 1/\lambda_{max}$, the algorithm converges to the minimum in one iteration. When β is greater than $1/\lambda_{max}$, the algorithm overshoots on each iteration, but as long as β is smaller than $2/\lambda_{max}$ the MSE still decreases and the algorithm converges. It is advantageous from the point of view of maximizing the worst-case convergence rate to choose $\beta = \beta_{opt}$, a choice which results in the algorithm overshooting the minimum in the direction of the eigenvector corresponding to λ_{max} in order that the algorithm converge faster in the direction of the eigenvector corresponding to λ_{min}.

Since the eigenvalue spread plays such an important role in the adaptation speed, it is instructive to relate it to the power spectral density of the wide-sense stationary reference random process. It is a classical result of Toeplitz form theory [2] that the eigenvalues of (3.1.3) are bounded by

$$\min_{\omega} S(e^{j\omega}) < \lambda_i < \max_{\omega} S(e^{j\omega}) , \tag{3.1.30}$$

where $S(e^{j\omega})$ is the power spectral density of the reference random process defined as the Fourier transform of the autocorrelation function (the elements of the matrix),

$$S(e^{j\omega}) = \sum_{T=-\infty}^{\infty} \phi_T e^{-j\omega T} . \tag{3.1.31}$$

While the eigenvalues depend on the order of the matrix, n, as $n \rightarrow \infty$ the maximum eigenvalue

$$\lambda_{max} \rightarrow \max_{\omega} S(e^{j\omega}) . \tag{3.1.32}$$

and the minimum eigenvalue

$$\lambda_{min} \rightarrow \min_{\omega} S(e^{j\omega}) . \tag{3.1.33}$$

See [2] for more precise statements of these results. This interesting relationship between the eigenvalues and the power spectrum is explored in problem 3-9.

It follows that the spectra which result in slow convergence of the gradient algorithm are those for which the ratio of the maximum to minimum spectrum is large, and spectra which are almost flat (have an eigenvalue spread near

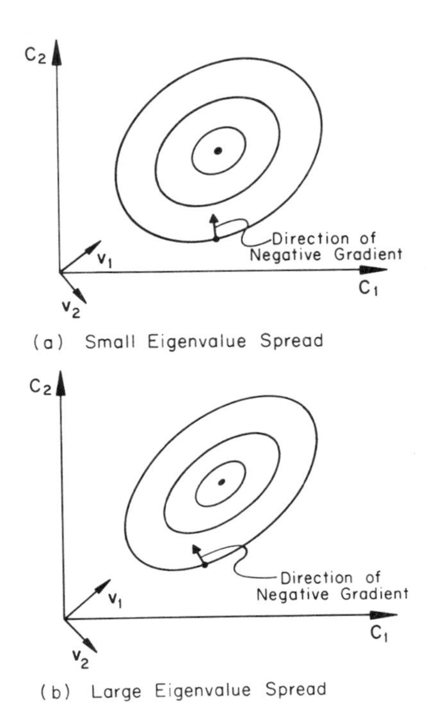

Figure 3-4. Effect of eigenvalue spread on convergence.

unity) result in fast convergence.

Since the modes of convergence of the gradient algorithm are all of the form of γ^j where γ is a positive real number less than unity and j is the iteration number, the error in decibels can be determined by taking the logarithm of the square (problem 3-13),

$$10 \log_{10}(\gamma^{2j}) = [10 \log_{10}(\gamma^2)]\, j \ , \tag{3.1.34}$$

and thus the error expressed in decibels decreases linearly with iteration number (the constant factors multiplying these exponentially decaying terms give a constant factor in decibels). The convergence of a gradient algorithm is thus often expressed in units of *dB per iteration*, which is the number of decibels of decrease in the error power per iteration.

3.1.4. Applications -- Echo Cancellation and Noise Cancellation

This section has determined the MMSE solution to the joint process estimation (and hence prediction) problem. That solution is determined in terms of the solution of a set of linear equations involving the autocorrelation matrix of the reference process, assumed to be known.

As an illustration of how to use these results, consider the echo canceler for voice described in section 2.5. Assume the discrete-time unit pulse response of the echo channel is h_T and write the primary input to the canceler in the form

$$d(T) = \sum_{m=0}^{\infty} h_m y(T-m) + x(T) , \tag{3.1.35}$$

where $x(T)$ is a near-end talker signal plus possibly noise. Then this primary input can be written as

$$d(T) = \mathbf{h}'\mathbf{y}(T) + v(T) , \tag{3.1.36}$$

where

$$\mathbf{h} = [h_0 \cdots h_{n-1}]' \tag{3.1.37}$$

is a portion of the echo impulse response within the range where it can be canceled by an n-coefficient joint process estimation transversal filter, and

$$v(T) = \sum_{m=n}^{\infty} h_m y(T-m) + x(T) \tag{3.1.38}$$

can be considered the uncancelable portion of the primary input; namely, the echo response with delays too large for the n-coefficient echo canceler plus the near-end talker signal. Substituting into (3.1.2), $\mathbf{p}$ can be determined to be

$$\mathbf{p} = E[\mathbf{y}(T)\mathbf{y}'(T)]\mathbf{h} + E[v(T)\mathbf{y}(T)] , \tag{3.1.39}$$

and the optimum echo canceler coefficients are, from (3.1.6),

$$\mathbf{c}_{\text{opt}} = \mathbf{h} + \mathbf{\Phi}^{-1}E[v(T)\mathbf{y}(T)] . \tag{3.1.40}$$

For the case where $v(T)$ and $\mathbf{y}(T)$ are uncorrelated, the second term would be zero. For this case

$$\mathbf{c}_{\text{opt}} = \mathbf{h} \tag{3.1.41}$$

and the echo canceler replicates the first n coefficients of the echo channel as we would expect. This result would apply, for instance, when the near-end and far-end talkers were uncorrelated (as expected) and the number of echo canceler coefficients was large enough to exactly replicate the echo channel impulse response.

Next, consider noise cancellation as described in section 2.4. For this case a typical element of the autocorrelation matrix $\mathbf{\Phi}$ is

$$\phi_i = E[\eta_2(T)\eta_2(T+i)] , \tag{3.1.42}$$

which is of course the discrete-time autocorrelation function of $\eta_2(T)$. Similarly, a typical element of the vector $\mathbf{p}$ is

$$p_l = E[d(T)y(T-l+1)]$$
$$= E[(s(T) + \eta_1(T))\eta_2(T-l+1)] \qquad (3.1.43)$$
$$= E[\eta_1(T)n_2(T-l+1)] \ ,$$

because of the assumption that the signal and the reference noise are uncorrelated. This element is recognized as the discrete-time crosscorrelation between the primary noise and the reference noise.

The conclusion for this application is that the MMSE solution requires knowledge of the autocorrelation function of $\eta_2(T)$ and the crosscorrelation function of the two noises (which must be jointly wide-sense stationary). The statistics of the signal are not necessary (other than knowledge that the signal is uncorrelated with both noises).

In most applications, of course, these quantities are not known. This is where the adaptive filters to be discussed in the remainder of this chapter become applicable.

3.2. LMS STOCHASTIC GRADIENT ALGORITHM

The most widely used method for adaptation of a transversal filter (both prediction and joint process estimation) is the *stochastic gradient (SG)* algorithm. This adaptation algorithm is also sometimes called the LMS adaptive transversal filter. The term LMS stands for least-mean square, and the algorithm does not provide an exact solution to the problem of minimizing the mean-square error but rather only approximates the solution. This approximation is the price to be paid for not requiring a stationary input or knowledge of the ensemble statistics.

The SG algorithm overcomes a problem of the MMSE solution described in section 3.1; namely, that $\nabla_{\mathbf{c}}\{E[e_c^2(T)]\}$ is usually unavailable since taking the expectation requires knowledge of the ensemble statistics. Thus, the MMSE solution is not applicable to the given data case. The approach taken in this section is to circumvent this problem by in effect substituting a time average for the ensemble average.

It was shown in section 3.1.1 that an alternate expression for the gradient is

$$\nabla_{\mathbf{c}}\{E[e_c^2(T)]\} = -2\, E[e_c(T)\mathbf{y}(T)] \ . \qquad (3.2.1)$$

The troublesome part of this expression is the expectation operator. The principle behind the SG algorithm is to ignore the expectation operator. The quantity which is left, while random, has an expected value equal to the desired gradient. Thus, it is an unbiased estimate of the gradient. This "noisy" or "stochastic" gradient is substituted for the actual gradient in the algorithm of (3.1.18) resulting in the SG algorithm

$$\mathbf{c}(T+1) = \mathbf{c}(T) - \frac{\beta}{2}\nabla_{\mathbf{c}}[e_c^2(T)] \ , \qquad (3.2.2)$$

or

$$\begin{aligned}\mathbf{c}(T+1) &= \mathbf{c}(T) + \beta e_c(T)\mathbf{y}(T) \\ &= [\mathbf{I} - \beta\, \mathbf{y}(T)\mathbf{y}'(T)]\, \mathbf{c}(T) + \beta\, d(T)\, \mathbf{y}(T) .\end{aligned} \tag{3.2.3}$$

A corresponding SG algorithm can be derived for a predictor (problem 3-27).

There is another rather subtle but important difference between this SG algorithm and the algorithm of (3.1.19). In the former algorithm, the index j corresponded to the iteration number for the iterative algorithm for solving a system of linear equations. In (3.2.3), on the other hand, the iteration number T corresponds to the sample number (or time index) of the given data at the input to the algorithm. Thus each iteration corresponds to a new given data sample. The algorithm is in effect performing a time average in order to determine the gradient.

It is not surprising to note the similarity between the SG algorithm of (3.2.3) and the gradient algorithm of (3.1.19). The former substitutes the stochastic matrix $\mathbf{y}(T)\mathbf{y}'(T)$ for $\mathbf{\Phi}$ and the stochastic vector $d(T)\mathbf{y}(T)$ for the vector $\mathbf{p}$. In each case the deterministic matrix or vector corresponds to the ensemble average of the stochastic matrix or vector for the stationary case.

The nature of adaptation algorithm (3.2.3) is illustrated in figure 3-5. What is shown is a single filter coefficient and how it contributes both to the transversal filter structure as well as how it is adapted. The first form of (3.2.3) is assumed, as this is a simpler form for implementation. Specifically, $y(T-j+1)$ is taken from the output of the $(j-1)$-th unit delay and multiplied by the j-th coefficient at time T, which is denoted by $c_j(T)$. The resultant value contributes to the summation which is subtracted from the second input $d(T)$ to obtain the filter output $e_c(T)$. Using this notation for the j-th coefficient, the adaptation algorithm of (3.2.3) can be rewritten for the j-th component as

$$c_j(T+1) = c_j(T) + \beta e_c(T) y(T-j+1) . \tag{3.2.4}$$

In accordance with this equation, $\mathbf{c}_j(T)$ is obtained by crosscorrelating (using a time average) the estimation error $e_c(T)$ with the delayed input $y(T-j+1)$. This crosscorrelation consists of taking the product of $e_c(T)$ with $y(T-j+1)$ and step-size β and accumulating the result. This accumulation is the discrete-time analog of integration in continuous-time.

This algorithm is not surprising in light of the orthogonality principle of (3.1.9). In particular, when all the filter coefficients are optimum, the orthogonality principle says that the input to the accumulator in figure 3-5 will average to zero. Thus, under these conditions, the output of the accumulator will maintain the same average value (namely the optimum filter coefficient). What the orthogonality principle does not tell us directly is that when the coefficients are not optimum, the non-zero average of the accumulator input is of the correct sign so as to move each coefficient toward the optimum.

An intuitive explanation of why this is true goes as follows. The term in the product $e_c(T)y(T-j+1)$ which depends on $c_j(T)$ is $-c_j(T)y^2(T-j+1)$. Consider for example the case where the optimum $c_j(T)$ is positive. Since the $y^2(T-j+1)$ is positive, if $c_j(T)$ is too large this term is more negative than it should be, which makes the average correction to $c_j(T)$ negative and on

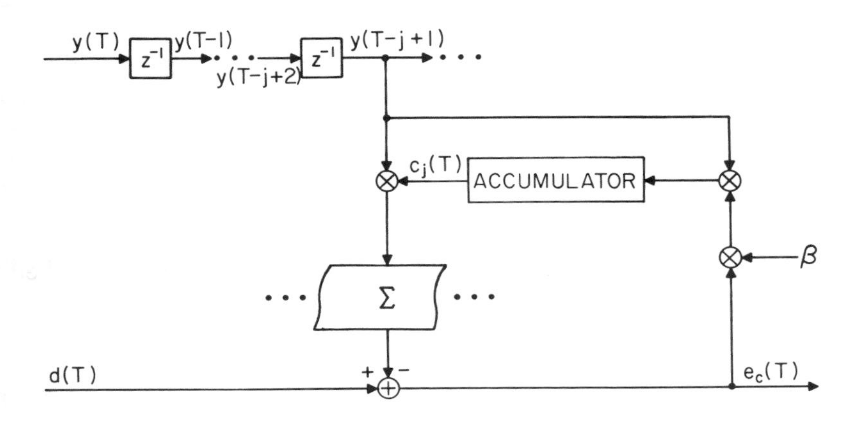

Figure 3-5. Stochastic gradient algorithm for one coefficient.

average $c_j(T+1)$ is smaller than $c_j(T)$.

This explanation does not answer the important question as to the effect of the interaction among the adaptation of the different coefficients. This interaction occurs because all the coefficients affect the error $e_c(T)$, which in turn affects the coefficient adaptation. This question will be considered in detail in chapter 7, but some simple preliminary answers in terms of the trajectory of the average filter coefficient vector will be considered here.

One difference between this algorithm and (3.1.19) is that in (3.1.19) the coefficient vector follows a deterministic and predictable trajectory, while the trajectory in (3.2.3) is random or stochastic. The cause of this random fluctuation is the use of the time average in place of ensemble average, or alternatively using the given data samples in place of the ensemble averages.

A common and useful method for analyzing the convergence behavior of a given data adaptive filtering algorithm is to assume (often unrealistically) that the given data can be modeled as a wide-sense stationary random process with known statistics (correlation function or power spectrum). The coefficient vector can then be expected to converge in some sense to be determined to the appropriate value for the known statistics. The speed of this convergence, while not directly applicable to a case where the input statistics are actually changing with time, is a good indication of the convergence performance of the algorithm.

If the assumption is made that the given data can be modeled as a wide-sense stationary random process and the expectation of the coefficient vector in (3.2.3) is taken,

$$\begin{aligned} E[\mathbf{c}(T+1)] &= E[(\mathbf{I} - \beta\, \mathbf{y}(T)\mathbf{y}'(T))\, \mathbf{c}(T)] + \beta\ E[d(T)\mathbf{y}(T)] \\ &\approx (\mathbf{I} - \beta\, \boldsymbol{\Phi})\ E[\mathbf{c}(T)] + \beta\, \mathbf{p}\ , \end{aligned} \tag{3.2.5}$$

where the approximation has been made that the coefficient vector is independent of the input data samples. The nature of the approximation in (3.2.5) is discussed further in chapter 7. It is valid because of the slow trajectory of $\mathbf{c}(T)$ which results when β is small, making the coefficient vector and the given data samples approximately uncorrelated. Comparing (3.2.5) with (3.1.19), it is seen that this approximate average trajectory precisely obeys the earlier deterministic algorithm (3.1.19). It can be asserted without further analysis that within the accuracy of this approximation the average trajectory of the stochastic gradient converges to the optimum coefficient vector under the same condition that guarantees convergence of (3.1.19), and the nature of the convergence is identical to that discussed in section 3.1.3.

This does not mean, however, that any particular coefficient vector trajectory itself converges to the optimum, but only that the average of all trajectories converges to the optimum. In fact, the coefficient vector does not converge to the optimum. To see this, observe that even after convergence of the coefficient vector in the mean-value sense, difference equation (3.2.3) still has a stochastic driving term, and therefore the coefficient vector continues to fluctuate about the optimum coefficient vector randomly. The larger the value of the step-size β, the larger this fluctuation. The size of this fluctuation is considered in detail in chapter 7. Keeping it reasonably small generally requires a much smaller step size than the value given by (3.1.27). Therefore, in practice the range of step sizes encompasses a region where the convergence rate is increasing as β is increasing, while the asymptotic fluctuation about the optimum is also increasing.

As β is increased, the average trajectory has a convergence rate which increases, then decreases, and eventually goes unstable, as seen in section 3.1.3. Choosing a β which results in faster convergence also enables the algorithm to track more rapid variations in the statistics in a nonstationary environment. The price that is paid is the larger fluctuation of the coefficient vector about its optimum value after average convergence of the coefficient vector, as considered in chapter 7. Consideration of this fluctuation will generally dictate that the step-size be chosen in a region where the rate of convergence of the average trajectory is increasing as β is increased.

The SG algorithm has been used and studied extensively in the context of all the applications mentioned in chapter 2 [5, 6, 7, 8].

3.2.1. Stochastic Approximation

It is not surprising that when the given data input is stationary, a condition which is realistic in a few applications, a form of the SG algorithm which actually converges exactly to the optimum coefficient vector can be found. One such algorithm is called the Robbins-Monro *stochastic approximation* algorithm [9].

The aspect of (3.2.3) which causes the asymptotic fluctuation is the fixed step size in the gradient correction term. The modification which results in convergence of every trajectory is the replacement of this fixed step-size by a step size which decreases with time. When the step-size approaches zero, then asymptotically the random driving term in (3.2.3) approaches zero and there is asymptotically no fluctuation of the coefficient vector. However, this decreasing step size must be chosen carefully to insure that the mean-value of the coefficient vector converges to the optimum. Replacing β by $\beta(T) \geqslant 0$ in (3.2.3) does result in an algorithm which converges as long as this step-size obeys the relations

$$\sum_{i=0}^{\infty} \beta(T) = \infty \tag{3.2.6}$$

$$\sum_{i=0}^{\infty} \beta^2(T) < \infty \tag{3.2.7}$$

where the usual example of such a sequence is $\beta(T) = \beta\, T^{-1}$. It can be shown that for such a step-size, when the input is stationary in the wide sense, then the coefficient vector converges to the optimum (3.1.6) in the mean-square sense (i.e. the mean-square error converges to zero) [9].

It should be emphasized that this result holds only for the stationary case, and this algorithm has no merit in situations where the input statistics are varying with time, since the algorithm will not be able to track these changes. Stochastic approximation does however suggest a modification to the SG algorithm which is applicable to a common practical situation: the adaptive filter environment is initially unknown but varies very slowly with time (as is often the case in adaptive equalization for example). What is sometimes done is to put a *gear shift* in the adaptation algorithm of (3.2.3). That is, the step size is initially made relatively large for rapid initial convergence, and after some fixed period of time the step size is reduced to yield small asymptotic fluctuations of the coefficient vector.

3.2.2. Illustrative Applications

The SG algorithm is the most widely used of the adaptation algorithms. Several of the applications discussed in chapter 2 will be reviewed to illustrate the use of the algorithm and to show how modifications to the algorithm can be made to fit a particular circumstance.

3.2.2.1. Adaptive Prediction in ADPCM

One application of the SG algorithm is ADPCM, which was discussed in section 2.2. In this case the objective is to define an adaptation algorithm for the predictor which will be performed in both the encoder and decoder. In order to ensure that the two adaptation algorithms track each other precisely, they must be driven by the same signals. This requires that a couple of modifications be made to the adaptation algorithm because of the constraint of the limited information that is available in the decoder for the adaptation.

First, the decoded quantized sample $\tilde{y}(T)$ must replace the input sample $y(T)$ since the latter is not available in the decoder. Second, for the same reason the quantized error $\tilde{e}_f(T)$ must replace the actual error $e_f(T)$. The adaptation algorithm of (3.2.3) therefore becomes

$$\mathbf{f}(T+1) = \mathbf{f}(T) + \beta\, \tilde{e}_f(T)\tilde{\mathbf{y}}(T-1) \tag{3.2.8}$$

where

$$\tilde{\mathbf{y}}(T) = \left[\, \tilde{y}(T) \cdots \tilde{y}(T-n+1) \,\right]' \tag{3.2.9}$$

is the vector made up of the past quantized versions of the input signal. What is the effect of substituting these quantized values in the adaptation algorithm? Experience and intuition suggest that the adaptation algorithm still works fine as long as the quantization is relatively accurate.

There are some special problems introduced due to the fact that there are actually two versions of (3.2.8) running at the same time, one in the encoder and the other in the decoder. Unless special precautions are taken, the algorithm must be prepared for different initial conditions in the encoder and decoder. Even if the initial conditions are the same, the two versions will occasionally get out of step due to transmission errors which cause the decoder to momentarily have incorrect input data. Thus the two versions must converge to the same state from different initial conditions so that any mistracking does not persist indefinitely.

To investigate the effect of the initial conditions on the behavior of (3.2.8), it can simply be iterated to yield

$$\mathbf{f}(T) = \mathbf{f}(0) + \beta \sum_{l=0}^{T-1} \tilde{e}_f(l)\tilde{\mathbf{y}}(l-1)\ . \tag{3.2.10}$$

If two versions of the algorithm are running, both with the same inputs but with different initial conditions, then the second term will be the same and mistracking will persist forever because of the difference in the $\mathbf{f}(0)$ term. It would be desirable if the algorithm could be modified so that the effect of the initial condition died away with time.

There is another problem with the SG algorithm for the ADPCM algorithm which can also be addressed. Specifically, when the speech goes away (as is inevitable during intervals of silence), the update to the algorithm represents only noise and the predictor coefficients follow an unpredictable trajectory. Since the optimum predictor coefficients when the speech resumes is likely to have little to do with the previous optimum coefficients, it would be desirable if the coefficients approached some compromise value during intervals of silence.

Both the aforementioned advantages can be obtained if the gradient of $\tilde{e}_f^2(T)$ could be replaced by the gradient of

$$\tilde{e}_f^2(T) + \mu(\mathbf{f} - \mathbf{f}_b)'(\mathbf{f} - \mathbf{f}_b)\ , \tag{3.2.11}$$

where $\mathbf{f}_b$ is the compromise coefficient vector toward which we would like the algorithm to converge during silent intervals. This cost function penalizes the algorithm for coefficient vectors with are different from the compromise, although for small μ the penalty is not large. Taking the gradient, and substituting in the stochastic gradient of (3.2.1) gives,

$$\mathbf{f}(T+1) = (1 - \mu\beta)\,\mathbf{f}(T) + \mu\beta\,\mathbf{f}_b + \beta\,\tilde{e}_f(T)\tilde{\mathbf{y}}(T-1)\,. \tag{3.2.12}$$

When $\mu = 0$ this algorithm reduces to the earlier SG algorithm. Iterating (3.2.12), it can be shown (problem 3-28) that

$$\mathbf{f}(T) = (1 - \mu\beta)^T\,\mathbf{f}(0) + \left[1 - (1 - \mu\beta)^T\right]\mathbf{f}_b + \beta \sum_{i=1}^{T-1} (1 - \mu\beta)^{T-i}\,\tilde{e}_f(i)\tilde{\mathbf{y}}(i-1)\,. \tag{3.2.13}$$

Note two properties of this algorithm: as $T \rightarrow \infty$ and for $\mu > 0$ the effect of the initial condition $\mathbf{f}_0$ dies away as desired, and in the absence of an input the coefficient vector approaches the compromise vector $\mathbf{f}_b$.

A modification to the predictor similar to that illustrated in (3.2.12) has been proposed [10].

3.2.2.2. Nonlinear Correlation Multipliers

One of the first applications of adaptive filtering in telecommunications was the use of SG algorithms in channel equalization as discussed in section 2.3. A more recent application of this same algorithm is to echo cancellation as discussed in section 2.5. An interesting modification which is often made to the algorithm in both applications is to replace the error signal $e_c(T)$ by the sign of this signal $\text{sgn}[e_c(T)]$. The motivation for doing this is simplification of the hardware: the adaptation equation of (3.2.4) becomes

$$\mathbf{c}(T+1) = \mathbf{c}(T) + \beta\ \text{sgn}[e_c(T)]\,\mathbf{y}(T)\,, \tag{3.2.14}$$

where the only multiplication is by β or $-\beta$. This limited multiplication is particularly simple to implement if β is chosen to be a power of 2^{-1} in a digital implementation.

One would expect intuitively that the value of the crosscorrelation would not be affected in a major way by using the sign of the error. It has been shown [11] that the convergence of the algorithm is not compromised by this simplification, but that the speed of adaptation is reduced somewhat. In fact, it is shown more generally that any nonlinearity in the multiplication adversely affects convergence to some extent.

In spite of the simplification in using the sign of the error, in a digital implementation multiplications are still inevitable in implementing the transversal filter itself. A rare exception to this is in the echo canceler for data transmission shown in figure 2-12. As discussed in section 2.5, the binary input to the canceler allows multiplications to be avoided.

3.2.2.3. Normalization of Step Size

The SG algorithm displays an undesirable dependence of speed of convergence on input signal power. This can be seen from (3.2.4), where if the input signal is increased in size by a factor γ, then this is equivalent to increasing the step size β by the factor γ^2 to $\beta\gamma^2$. (This same effect can be observed by looking at the eigenvalues of $\mathbf{\Phi}$.) Thus, the speed of convergence of the SG algorithm is strongly affected, by the size of the input signal. A serious consequence is that if the input signal grows too large the adaptation algorithm becomes unstable. This is a particular problem in applications such as speech where the input signal power varies considerably.

A frequent solution to this difficulty is to normalize the step-size of the algorithm. The size of the updates can be kept approximately the same size on average if the update is normalized by an estimate of the input signal power, which is equivalent to choosing a step-size in (3.2.4) equal to

$$\beta(T) = \frac{a}{\sigma^2(T) + b} , \tag{3.2.15}$$

where $\beta(T)$ is the step-size at time T, a and b are some appropriately chosen constants, and $\sigma^2(T)$ is an estimate of the input signal power at time T. The purpose of the b in the denominator is to prevent $\beta(T)$ from getting too large (causing instability) when the input signal power gets very small.

As an example of how the estimate of input signal power can be developed, it is advantageous to use an exponentially weighted time average of the input signal power, as in

$$\sigma^2(T) = (1 - \alpha) \sum_{j=0}^{\infty} \alpha^j y^2(T-j) \tag{3.2.16}$$

where α is an appropriately chosen constant and the $(1 - \alpha)$ factor normalizes the estimate to be an unbiased estimate of the input signal power. The reason for choosing this estimate is that it can be written recursively as

$$\sigma^2(T) = \alpha\sigma^2(T-1) + (1 - \alpha)y^2(T) \tag{3.2.17}$$

(this is a preview of some recursive LS algorithms to be considered in section 3.4).

The normalization of the step-size is frequently applied to adaptive filtering of speech, as in ADPCM and echo cancellation. The adaptation properties of the SG algorithm will be considered in detail in chapter 7, where the effects of this normalization will be considered further. This normalization is usually not necessary in applications like channel equalization where the signal level is relatively fixed and known in advance.

3.2.2.4. Conditions for Uniqueness of Filter Coefficients

In this section we will determine the conditions under which the filter coefficients of a MMSE predictor or joint process estimator are unique, and discuss how the SG algorithm can be modified if the coefficients are not unique.

The existence of a unique solution to the MMSE problem, as well as the arguments for the convergence of the gradient and SG algorithms, depended on the nonsingularity of the autocorrelation matrix of the reference signal $y(T)$, $\mathbf{\Phi}$. Since $\mathbf{\Phi}$ is always positive semidefinite, the singular case corresponds to one or more zero eigenvalues. From the argument in section 3.1.3, the case of concern is where the spectrum of the reference input vanishes at some frequency(s), since (3.1.33) would then predict that one or more eigenvalues would approach zero as $n \rightarrow \infty$. That the vanishing of the input spectrum would cause problems is not surprising, since the adaptive filter transfer function in the regions of zero spectrum could be anything without affecting the quality of the joint process estimate, and the filter coefficients might then not be unique.

$\mathbf{\Phi}$ is singular when

$$\mathbf{q}'\mathbf{\Phi}\mathbf{q} = 0 \tag{3.2.18}$$

for some nonzero vector $\mathbf{q}$, since this condition implies that $\mathbf{\Phi}$ is not positive definite. Writing out this relation and taking into account that the elements of $\mathbf{\Phi}$ are given by (3.1.4), (3.2.18) becomes

$$\sum_{i=1}^{n}\sum_{j=1}^{n} q_i q_j \phi_{i-j} = 0 \,. \tag{3.2.19}$$

Defining the power spectrum of the reference input as in (3.1.31), the inverse is given by

$$\phi_T = \int_{-\pi}^{\pi} S(e^{j\omega}) e^{j\omega T} \frac{d\omega}{2\pi} \,. \tag{3.2.20}$$

Substituting (3.2.20) into (3.2.19),

$$\int_{-\pi}^{\pi} S(e^{j\omega}) |Q(e^{j\omega})|^2 \frac{d\omega}{2\pi} = 0 \,, \tag{3.2.21}$$

where

$$Q(z) = \sum_{i=0}^{n-1} q_{i+1} z^{-i} \tag{3.2.22}$$

is the Z-transform of the vector $\mathbf{q}$. Since the integrand of (3.2.21) is non-negative, the integral can vanish only if the integrand is identically zero,

$$S(e^{j\omega}) |Q(e^{j\omega})|^2 = 0, \;\; 0 \leqslant \omega \leqslant \pi \,. \tag{3.2.23}$$

Necessary and sufficient conditions for $\mathbf{\Phi}$ to be nonsingular are now at hand. Since $Q(z)$ is an $(n-1)$-th order polynomial in z, unless it is identically zero (which is ruled out) is has at most $n-1$ zeros. Thus (3.2.23) cannot be satisfied as long as the spectrum $S(e^{j\omega})$ is nonzero at n or more frequencies. Further, since the spectrum is symmetric about d.c. ($\omega = 0$), as long as the spectrum is nonzero at $n/2$ or more positive frequencies, $\mathbf{\Phi}$ will be nonsingular. For example, if the input signal consists of $n/2$ or more sinusoids, then $\mathbf{\Phi}$ will be nonsingular and the MMSE solution will be unique, whereas if the input signal consists of fewer sinusoids the MMSE solution will not be unique. This makes sense intuitively, since $n/2$ sinusoids have n unknown parameters, the amplitude and phase of each sinusoid, and the adaptive filter has n degrees of freedom (filter coefficients). Using these degrees of freedom, the adaptive filter transfer function (amplitude and phase) can be independently adjusted at the frequencies of $n/2$ sinusoids, and if there are fewer input sinusoids there are degrees of freedom left over after the filter transfer function is adjusted at all the relevant frequencies.

An input signal consisting of a small number of sinusoids is a legitimate concern in voice echo cancellation, where it is possible that the voice channel can contain a tone (for example for test purposes) or a small number of tones (for example during the startup sequence of a voiceband data set). For such a case, the canceler can still operate, canceling the signal at that frequency, but the coefficients are not unique and the transfer function of the adaptive filter is indeterminate at any frequency other than the frequency of the input sinusoid. This illustrates that the echo cancellation error can be small even though the filter coefficients are a poor match to the unit pulse response of the echo channel!

In practice there will always be some noise at the input to the adaptive filter at all frequencies. As a result, mathematically the filter coefficients corresponding to the minimum mean-square error will be unique. However, there are still two important practical concerns. First, when the input signal spectrum is very small at some frequencies, then (3.1.33) predicts that one or more eigenvalues will be very small, particularly as the filter order increases. This implies a numerical instability for the filter, which must be accounted for in the choice of the number of bits of precision for a digital implementation. Second, in any implementation, analog or digital, there will be a maximum value that a filter coefficient can assume. As one or more eigenvalues get very small, it becomes more likely that this maximum value will be inadequate, and the proper operation of the filter comes into question.

This latter point is illustrated in figure 3-6. The region of allowed filter coefficients for a two coefficient filter is usually a square centered at the origin as constrained by implementation considerations. As an eigenvalue approaches zero, the sensitivity of the mean-square error in the direction of the corresponding eigenvector to the filter coefficients becomes very small. In practice, as explained in chapter 7, even after convergence of the filter coefficients to the optimum in a mean-value sense, there will continue to be a fluctuation of the filter coefficients about that optimum, with the adaptation algorithm continually bringing the coefficients back toward the optimum. The fluctuation of the coefficients in the direction of least sensitivity (the direction of the eigenvector corresponding to the minimum eigenvalue) will tend to be larger. As the eigenvalue gets smaller, the probability of that fluctuation taking the coefficients out of the allowed region gets large.

The conclusion is that we usually cannot depend on noise alone to stabilize a filter where the input signal does not cover the full bandwidth. There are a couple of potential solutions to this problem. One is the use of saturation arithmetic which maintains the filter coefficients on the boundary of the allowed region as the adaptation algorithm attempts to take the coefficients out of that region. A second solution is to put a leakage in the adaptation that tends to force the coefficients toward the origin, thus keeping them smaller. This latter algorithm has been proposed for a so-called fractionally spaced channel equalizer [12]. This leakage can be obtained by changing the criterion which the adaptation algorithm is minimizing to

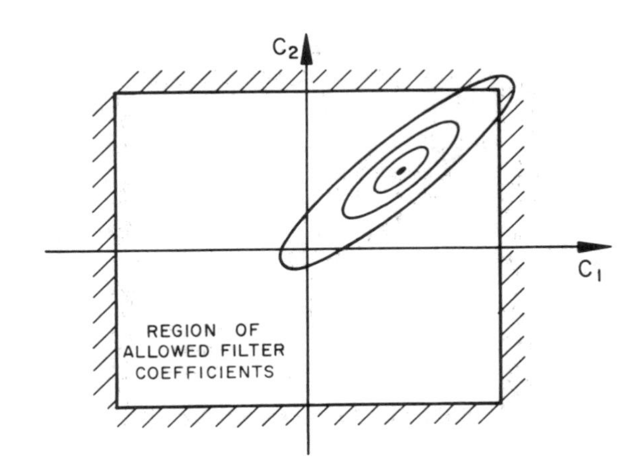

Figure 3-6. Contours of equal MSE for a large eigenvalue spread.

$$E[e_c^2(T)] + \mu\, \mathbf{c}'\mathbf{c}\ , \tag{3.2.24}$$

where μ is another small constant. Instead of simply minimizing the MSE, the criterion tries in addition to minimize the length of the coefficient vector. This criterion results in some compromise in the asymptotic mean-square error and coefficient vector, which are no longer optimum in the sense of (3.1.6). The optimum coefficient vector corresponding to criterion (3.2.24) is investigated in problem 3-26. The second term in (3.2.24) biases the solution in the direction of keeping the coefficients small. The SG algorithm corresponding to minimizing (3.2.24) is

$$\mathbf{c}(T+1) = (1 - \beta\, \mu)\mathbf{c}(T) + \beta\, e_c(T)\mathbf{y}(T)\ , \tag{3.2.25}$$

where $\mu = 0$ corresponds to the previous algorithm without leakage (3.2.3). A similar predictor algorithm can also be derived (problem 3-30). The operation of this algorithm is evident, since the coefficient vector is multiplied by a constant slightly less than unity at each step before adding in the correction term. When the corrections are small, as when the coefficient vector is wandering in the direction of a eigenvector corresponding to a small eigenvalue, this leakage decreases the size of the vector over time.

This leakage is quite similar to the leakage that was introduced in ADPCM to counter the effects of transmission errors. In fact, criterion (3.2.24) is identical to (3.2.11) for a compromise vector $\mathbf{f}_b = \mathbf{0}$.

3.2.2.5. Decision Feedback Equalization

Not all applications fit precisely within the model of the joint process estimator governed by (3.0.3). However, the other applications can usually be treated with minor modifications to the development thus far. One such important application is the decision feedback equalizer (DFE), discussed in section 2.3.

The DFE is illustrated in figure 3-7. If the current input is $y(T)$ as before, and there are L forward filter coefficients and M feedback filter coefficients (and these numbers are often quite different), then the current output of the forward filter is summed with the current output of the feedback filter to form the estimate of the data symbol L symbols in the past. Thus,

$$\hat{d}(T-L) = \sum_{i=1}^{L} c_i y(T-i+1) + \sum_{i=L+1}^{L+M} c_i \tilde{d}(T-i) , \tag{3.2.26}$$

where the difference from the adaptive equalizer in (2.3.3) is twofold. First, the number of coefficients before and after the center coefficient corresponding to the joint process estimate is no longer the same, and second the decisions are substituted for the given data samples for delays large enough so that the decisions are available. Note that (3.2.26) can be written in a vector form similar to (3.0.3) if the vectors

$$\tilde{\mathbf{d}}(T-L) = [\tilde{d}(T-L-1) \cdots \tilde{d}(T-L-M)]' \tag{3.2.27}$$

and

$$\mathbf{v}(T) = \begin{bmatrix} \mathbf{y}(T) \\ \tilde{\mathbf{d}}(T-L) \end{bmatrix} \tag{3.2.28}$$

are defined, since the joint process estimation error is then

$$e_c(T-L) = d(T-L) - \mathbf{c}'\mathbf{v}(T) \tag{3.2.29}$$

in analogy to (3.0.3). The gradient algorithm then becomes simply

$$\mathbf{c}(T+1) = \mathbf{c}(T) + \beta\, e_c(T-L)\mathbf{v}(T) , \tag{3.2.30}$$

in analogy to (3.2.3).

Of course, there is the same problem with the decision feedback equalizer that was faced with the adaptive equalizer: where do we get the data symbols $d(T)$ required to generate the estimation error $e_c(T)$? The solution is the same: either use a training sequence known to both transmitter or receiver, or alternatively use the receiver decisions in a decision directed fashion.

The analysis of the convergence of this algorithm is troublesome, since the past decisions depend on the given data samples in a complicated nonlinear fashion. However, it is a reasonable approximation to substitute the actual data symbol $d(T)$ for the corresponding decision $\tilde{d}(T)$ in (3.2.26) in the analysis (but not implementation) of the algorithm. This approximation is quite valid as long as the error rate is low. The calculation of the corresponding correlation matrices then becomes straightforward (problem 3-20).

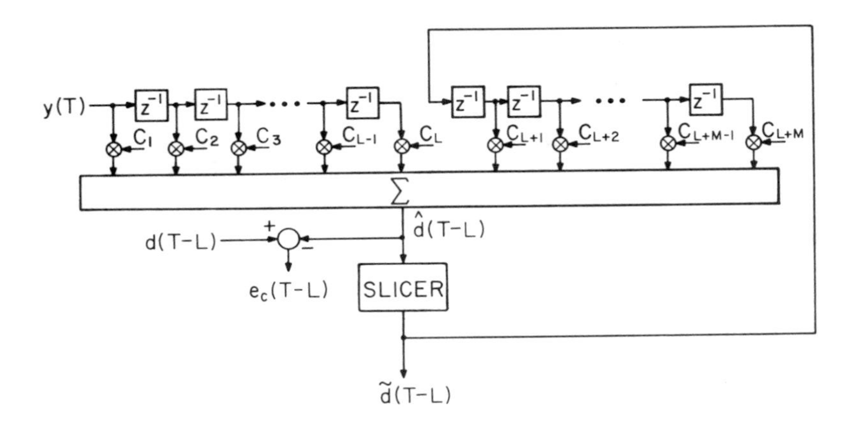

Figure 3-7. Structure of a decision feedback equalizer.

3.2.2.6. Adaptive Hybrid as an Example of Analog Adaptive Filtering

The adaptive hybrid of figure 2-12 gives an opportunity to show how a SG algorithm could be derived in a continuous time context. For this structure, the error signal at the output is (note that in this section t is continuous time)

$$e(t) = d(t) - [c_1 x_1(t) + c_2 x_2(t)] \ , \tag{3.2.31}$$

where due to the special application only two coefficients are required. In fact, the number of coefficients can be further reduced by limiting the range of adaptation. In particular, take $c_1 = \theta$ and $c_2 = (1-\theta)$ where $0 \leqslant \theta \leqslant 1$. As θ is varied, in some sense the filter is adapting to a point intermediate to the two filters $H_1(s)$ and $H_2(s)$. Taking the derivative of $e(t)$ with respect to the single coefficient θ,

$$\frac{\partial}{\partial \theta} \, e_c^2(t) = 2e(t)[x_2(t) - x_1(t)] \ . \tag{3.2.32}$$

The continuous-time analog of incrementing the coefficient by the gradient would be to make an adjustment $d\theta$ to θ proportional to the partial derivative of (3.2.32) times a small increment of time dt. Doing this,

$$d\theta(t) = \beta \, e(t)[x_2(t) - x_1(t)] \, dt \tag{3.2.33}$$

and integrating both sides of (3.2.33),

$$\theta(t) = \theta(0) + \beta \int_0^t e(\tau)(x_2(\tau) - x_1(\tau)) \, d\tau \ . \tag{3.2.34}$$

The complete adaptive hybrid including adaptation algorithm is illustrated in figure 3-8. It is analogous to the algorithm of figure 3-5 except that the accumulator is replaced by the continuous-time integrator. As in the discrete-time case, the step-size β cannot be chosen too large or instability results, and the choice of β affects the tradeoff between speed of adaptation and asymptotic fluctuation of the coefficient $\theta(t)$.

3.3. BLOCK LEAST-SQUARES METHODS

An alternative to the SG algorithm which is attractive particularly for signals with rapidly changing statistics like speech, is to use one of the *block least-squares* (LS) methods. In these methods ensemble averages are replaced by time averages, thus overcoming the hurdle of not knowing the ensemble averages, and the nonstationarity of the input samples is overcome by performing the time averages over short time frames during which the data can be assumed to be approximately stationary. These are called least-squares algorithms since they minimize the sum of squares of an error signal, as opposed to the mean of the square of an error signal as in the MMSE algorithm.

The two most popular block LS methods, the *autocorrelation* and *covariance* methods, form a least-squares estimate of the prediction coefficients based upon a block of M data samples, where M represents the block length over which the data is assumed to be approximately stationary. In both methods a so-called *window* function is applied in some fashion to the given data samples. This window is non-zero only over the M samples, thereby limiting the processing to the block of interest. The two methods differ in the details of how the window is applied in the least-squares criterion. In particular, they utilize different methods for coping with the end effects at the endpoints of the block of

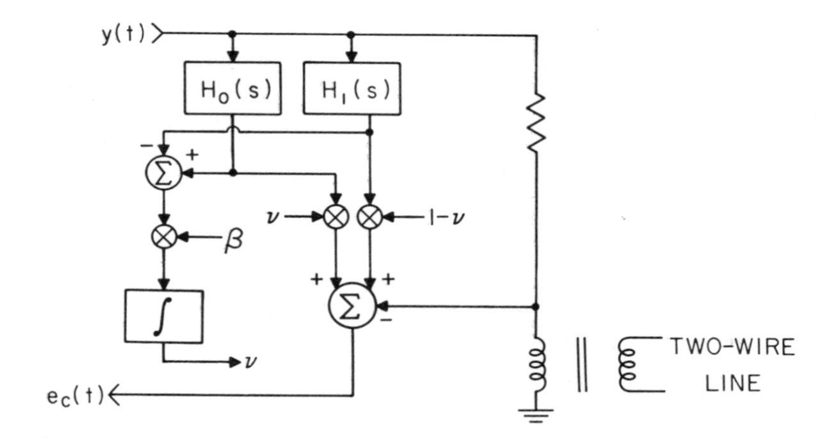

Figure 3-8. Complete adaptive hybrid.

samples. In anticipation of the recursive least-squares algorithms to be discussed in section 3.4 and chapter 6, another method called the *prewindowed* method will also be mentioned.

In all the methods, a block of nominally M samples is used to estimate the filter coefficients. For convenience, reference this block to the samples 0 through $M-1$. The difference in the methods is how the end-points are handled: at both ends, the error inherently involves samples outside the block of M samples. In the covariance method, the coefficient vector $\mathbf{c}$ is selected to minimize the sum

$$\epsilon \equiv \sum_{T=0}^{M-1} w_T \left(d(T) - \mathbf{c}' \mathbf{y}(T) \right)^2 , \tag{3.3.1}$$

where w_T is a *window function* usually chosen to be rectangular (i.e. $w_T = 1$). The coefficient vector $\mathbf{c}$ which results from this minimization is associated with the block of given data, and a new vector is calculated for the next block. In this covariance method, the actual given data samples outside the block are used, and in particular the samples y_{-n} through y_{-1} are actually included in the estimate for this block, as can be seen from (3.3.1). One result is that when the covariance method is applied to successive blocks of M samples, some of the samples near the ends of the blocks will be included in the estimates associated with two blocks.

The use of window functions is common in signal processing as a vehicle for picking out a small portion of a much longer waveform for purposes of further processing. In block LS methods the choice of the window can have a considerable impact on the quality of the resulting spectral estimation. The issue of choice of a window function is explored further in [3].

In the autocorrelation method, the window is applied to the data samples prior to the calculation of the error. The window is assumed to be zero outside the range of M given data samples, thus forcing the estimate to be based on the samples only within the block. The error for this method is written as

$$\epsilon \equiv \sum_{T=0}^{M-2+n} \left(w_T d(T) - \sum_{m=1}^{n} c_m w_{T-m+1} y(T-m+1) \right)^2 , \tag{3.3.2}$$

where the summation could be written from minus to plus infinity but has been restricted to only the non-zero terms. Unlike the covariance method the summation is also allowed to run over the end of the block to include errors at times M to $M+n-2$, because these errors are in terms of given data samples within the block.

In effect, in the autocorrelation method the samples outside the block are replaced by samples which are all identically zero, and then the time averaged square error is calculated over all time. This artificial choice for the samples outside the block results in some artifacts at the end-points of the block. For example, at the beginning of the block the given samples $d(T)$ are estimated on the basis of artificially chosen zero samples outside the block, and at the end of the block artificially chosen $w_T d(T) = 0$ zero samples are being estimated on the basis of the given samples within the block. To prevent these artifacts from affecting the estimate appreciably, the window is usually chosen to approach zero smoothly near the ends of the block so that the end-point samples are given less weight in the estimate. So that all the given data samples are used in the estimation, it is also common to overlap the blocks.

Finally, in the prewindowed method the coefficient vector is again chosen to minimize sum (3.3.1). Thus, by terminating the sum at $M-1$ there are no end effects at the end of the block. Unlike the covariance method, however, the given data samples outside the window, namely $y(T)$ for $T < 0$, are artificially chosen to be zero. This method can therefore be viewed as in some sense half-way between the covariance and autocorrelation methods; the autocorrelation method is used at the beginning of the block, and the covariance method is used at the end of the block!

The prewindowed method is not often used in block least-squares, but is the most common approach in the recursive least-squares approach discussed in section 3.4 and chapter 6. The prewindowed recursive least-squares approach assumes that the given data starts at sample zero, and actually uses a block which is growing with time (i.e. utilizes the data from 0 to T at time T). In effect the upper limit in the summation of (3.3.1) is replaced by T at time T rather than being fixed at $M-1$. A growing computation is avoided by finding recursive relationships for the new estimate based on the most recent estimate and the new given data sample. The assumption of zero data samples prior to sample zero is merely a convenience for initializing the recursive algorithm. Because of the growing memory of the algorithm, the effect of whatever initialization method is chosen quickly dies away.

Another recursive least squares approach which will be described in chapter 6 is the *sliding window* criterion. This is analogous to the block least squares, where the criterion considers a block of M samples, but in this case the optimum coefficient vector is determined at each sample time and the block is shifted by only one sample rather than M samples. This technique has also been referred to as *recursive LS with finite memory*, since data samples outside the window of interest are completely forgotten.

The covariance and autocorrelation methods both require the solution of a set of linear equations to find the optimum coefficient vector, just as in the MMSE case of section 3.1. The autocorrelation and crosscorrelation ensemble averages in the MMSE case are replaced by time averages. The method of solution is to differentiate (3.3.1) or (3.3.2) with respect to the components of $\mathbf{c}$ and set the result equal to zero. The result in both cases is an optimum coefficient vector

$$\mathbf{c}_{\text{opt}} = \hat{\mathbf{\Phi}}^{-1}\hat{\mathbf{p}} , \tag{3.3.3}$$

where the manner in which the matrix and vector are calculated depends on the method. Note the similarity to the relation for the optimum coefficient vector in the MMSE case of (3.1.6). For the covariance and prewindowed methods,

$$\hat{\mathbf{\Phi}} = \sum_{T=0}^{M-1} w_T \mathbf{y}(T)\mathbf{y}'(T) \tag{3.3.4}$$

and

$$\hat{\mathbf{p}} = \sum_{T=0}^{M} w_T d(T)\mathbf{y}(T) . \tag{3.3.5}$$

For the autocorrelation method, the matrix and vector are

$$\hat{\mathbf{\Phi}} = \left[\sum_j w_{j-m} w_{j-l} y(j-m) y(j-l)\right]_{0 \leqslant m,l \leqslant n} \tag{3.3.6}$$

$$\hat{\mathbf{p}} = \left[\sum_j w_j w_{j-l} d(j) y(j-l)\right]_{0 \leqslant l \leqslant n} . \tag{3.3.7}$$

If $d(T)$ and $y(T)$ were to be modeled as jointly stationary random processes, then $\hat{\mathbf{\Phi}}$ and $\hat{\mathbf{p}}$ are respectively time-average estimates of $M\mathbf{\Phi}$ and $M\mathbf{p}$ in (3.1.2-3). The extra factors of M cancel in (3.3.3).

The covariance method seems to be the most natural, and the question arises as to what motivates interest in the autocorrelation method. The answer lies in the structure of the two autocorrelation matrices in (3.3.4) and (3.3.6). If the autocorrelation method is used, the matrix $\hat{\mathbf{\Phi}}$ is Toeplitz and advantageously the Levinson-Durbin algorithm can be used to evaluate $\mathbf{c}$ just as in the MMSE solution of section 3.1. This similarity between the autocorrelation matrices for the MMSE and autocorrelation block LS solutions suggests that in some sense the MMSE and autocorrelation block LS algorithms are actually the solution to the same problem. This is verified in chapter 4, where it is shown that the MMSE and autocorrelation block LS problems can be put into the same geometrical framework and solved simultaneously.

Another important feature of the autocorrelation block LS solution (as well as the MMSE solution) for the prediction case is that the resulting coefficient vector $\mathbf{f}$ corresponds to a predictor filter which has all its zeros within the unit circle. This will be shown in chapter 5. Thus, the all-pole filters of the LPC and ADPCM decoders described in sections 2.1 and 2.2 will be guaranteed to be stable if infinite precision arithmetic is used (of course finite precision arithmetic can result in instability).

The covariance block LS method, in addition to being somewhat more natural, results often in "better" estimates due to its improved handling of block end effects. Unfortunately, its time-average autocorrelation matrix is not Toeplitz (problem 3-31) and hence requires somewhat more computation to invert since the Levinson-Durbin algorithm is not applicable. There is an alternate but somewhat more computationally intensive algorithm available for this case called Cholesky decomposition [3]. In addition, the all-zero prediction error filter is not guaranteed to have all its zeros interior to the unit circle, resulting in the potential for stability problems.

For a detailed comparison of the autocorrelation and covariance methods in the context of speech processing see [13, 3]. Block LS methods have been used extensively in speech processing applications. Their main disadvantage is the large amount of computation required to evaluate the autocorrelation or covariance matrix for every frame analyzed. In addition, the matrix elements must be calculated with a great deal of precision to obtain a reasonable estimate of the optimal coefficient vector after matrix inversion.

3.4. RECURSIVE LEAST-SQUARES METHODS

One approach to deriving a recursive LS algorithm for prediction and joint process estimation is to substitute for a window which covers a block of samples a window which decays into the infinite past, or at least back to the algorithm initialization, and which can be generated recursively (for example decays exponentially). Thus, the entire past of the given data input is considered (the algorithm has infinite memory) for each new estimate, but input samples far into the past receive a lot less weight than recent input samples. (Another approach is the sliding window recursive least squares algorithms considered in chapter 6.)

Naively one would think that this approach would not be viable since the amount of computation would be growing with time. However, this growing computational load can be avoided by finding a *recursive* solution. That is, the current coefficient estimate can be generated from the past coefficient estimate, the current given data sample, and perhaps a finite number of internal state variables to the algorithm. As a result, a new estimate is generated for each new sample (as opposed to each new block of samples).

In this section two simply derived algorithms of this type will be displayed. While these algorithms will not have many of the nice features of more sophisticated algorithms derived in chapter 6 (and in particular still require an explicit solution of a system of linear equations), they amply demonstrate some of the principles behind the recursive least-squares algorithms.

The two algorithms which will be derived are based on autocorrelation and covariance type of criteria. These two algorithms could easily be derived in similar fashion. However, in order to illustrate the different interpretations which can be applied to these algorithms, they will deliberately be derived in different terms. The two algorithms follow in the next two subsections.

3.4.1. Recursive LS Using an Autocorrelation Criterion

This section will derive a recursive LS prediction algorithm using an autocorrelation criterion (see problem 3-32 for a corresponding joint process estimation algorithm). This algorithm has the structure of figure 3-9, which consists of n crosscorrelation estimates which are obtained by multiplying the given data samples by delayed samples and passing the result through a low-pass filter (LPF). The LPF for each autocorrelation lag is in general different. The output of the LPFs are the autocorrelation estimates, which must be processed by the Levinson-Durbin algorithm to yield the predictor or joint process estimator filter coefficients. An algorithm based solely on intuition would probably not be much different (the most likely difference would be identical LPFs).

The algorithm of figure 3-9 is based on the autocorrelation method, in which the data sequence $y(T)$ is first windowed and then the squared prediction error is minimized. Define the windowed data sequence as

$$x(i) = w_{T-i}y(i) \ , \tag{3.4.1}$$

where T is the current sample number, and where w_l is a causal window function of infinite extent, i.e.,

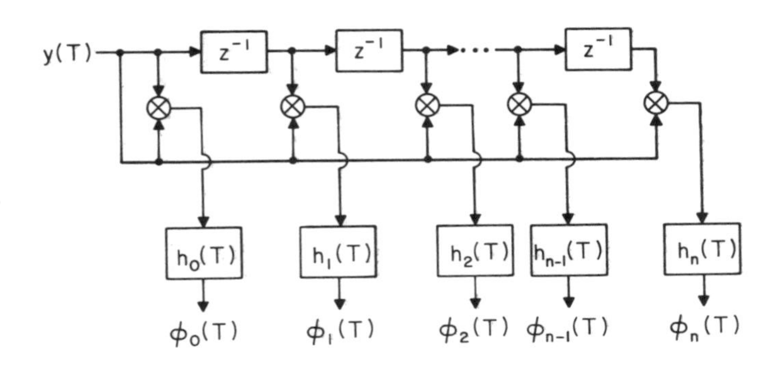

Figure 3-9. Autocorrelation estimate for recursive LS autocorrelation algorithm.

$$w_l = 0, \quad l < 0 . \tag{3.4.2}$$

The window function would generally be decaying to zero; the result of the window in (3.4.1) is therefore to force the $x(i)$ data sequence to be zero for $i > T$ and to give more weight to the recent given data (near sample T).

In order to achieve a simple recursive algorithm, the window itself must be representable by a simple recursion. Since the solution of a linear difference equation with constant coefficients is a sum of exponentials (possibly multiplied by a power of the time index), the window must be of that form. The simplest example of a suitable window function would thus be the single exponential

$$w_l = \alpha^l, \quad l \geqslant 0 , \tag{3.4.3}$$

for a constant α less than unity.

In (3.4.1) notice that the window can and will be moved one sample at a time, as determined by the constant T, rather than in blocks. Further, the current time T is implicit in $x(i)$ through the positioning of the window. Considering the predictor case, the least-squares criterion minimizes the quantity

$$\epsilon(T) = \sum_{l=-\infty}^{\infty} [x(l) - \sum_{j=1}^{n} f_j(T)x(l-j)]^2 . \tag{3.4.4}$$

The argument T denotes the time at which the estimate is formed, and note again that T does not explicitly appear on the right hand side in the windowed signal $x(T)$, but is implicit through the positioning of the window. The upper limit of summation has been made infinite for convenience, although it is actually finite due to the effect of the causal window.

As before, the optimum coefficient vector can be found by differentiating (3.4.4). Defining an estimate of the m-th autocorrelation coefficient at sample T as

$$\phi_m(T) = \sum_{l=-\infty}^{\infty} x(l)x(l-m) , \tag{3.4.5}$$

it is readily verified that the minimum is found by solving the set of linear equations

$$\phi_m(T) = \sum_{j=1}^{n} f_j(T)\phi_{m-j}(T), \quad 1 \leqslant m \leqslant n \tag{3.4.6}$$

which is identical to (3.1.6) and (3.3.3) (the MMSE and block LS cases respectively) except for the particulars of how the autocorrelation estimate is obtained. Thus, the final solution can be found as before by the Levinson-Durbin algorithm since matrix (3.4.6) is Toeplitz.

Obviously it is impractical to calculate autocorrelation estimate (3.4.5) as shown because of the infinite summation. However, it can be converted to a recursive relationship for windows which are themselves defined recursively. In particular, from (3.4.1) and (3.4.5)

$$\phi_m(T) = \sum_{l=-\infty}^{\infty} w_{T-l} w_{T-l+m} y(l)y(l-m), \; m \geqslant 0 . \tag{3.4.7}$$

Defining an impulse response corresponding to the mth autocorrelation lag as

$$h_m(T) = w_T w_{T+m} \tag{3.4.8}$$

where this filter is also causal for $m \geqslant 0$, the autocorrelation can be expressed as

$$\phi_m(T) = \sum_{l=-\infty}^{\infty} h_m(T-l)y(l)y(l-m) . \tag{3.4.9}$$

$$= h_m(T) * y(T)y(T-m) ,$$

where "*" denotes convolution. Thus, the structure of figure 3-9 is confirmed, wherein the LPF for the m-th filter coefficient has impulse response given by (3.4.8). Surprisingly, in general the LPF's for different autocorrelation lags are not identical.

For the simple example of (3.4.3),

$$h_m(T) = \alpha^m \alpha^{2T} , \; T \geqslant 0 , \tag{3.4.10}$$

which corresponds to a LPF with transfer function

$$H_m(z) = \frac{\alpha^m}{1-\alpha^2 z^{-1}} . \tag{3.4.11}$$

For this simple example, the difference in the LPF's is just a gain factor which depends on autocorrelation lag m. In more complicated cases, the LPF's will depend in a less trivial way on m.

This algorithm includes a recursive solution for the autocorrelation estimates, followed by the solution of a system of linear equations to determine the coefficient vector. A similar class of recursive LS algorithms is derived in chapter 6. In these more sophisticated algorithms, this recursive structure is extended to include the determination of the filter coefficients directly (or equivalent parameters), thus avoiding the explicit solution of the linear equations. Those algorithms use a finite but expanding block, and thus properly initialize the recursions, whereas the algorithm derived here is assumed to have always been running and is never initialized.

The essential features of the algorithm derived here and those of chapter 6 are that a new coefficient-vector is generated at each input sample rather than on a block basis. While this increases the computational load, it has the advantage that significant changes in input statistics can be detected at any time, rather than at arbitrary block boundaries. On the other hand, the computational load can be reduced to be comparable to block LS algorithms by performing the Levinson-Durbin solution of (3.4.6) only every M samples rather than every sample.

3.4.2. Prewindowed Recursive LS

This section will derive a recursive LS algorithm which uses a criterion similar to the covariance criterion. As in the last section, to take account of the infinite window extent it is necessary to assume that the given data samples are zero prior to time zero in order to get the recursive algorithm going. This modification to the covariance method is called the *prewindowed* method.

This algorithm was first derived as an application of Kalman filtering theory to adaptive linear prediction [14]. The resulting recursive LS algorithm had received little attention in the adaptive signal processing community due to its computational complexity. By demonstrating the improvement in performance of the Kalman algorithm over the conventional SG transversal algorithm, however, this algorithm stimulated substantial interest in recursive LS techniques. This work was followed by the development of the fast Kalman and LS lattice algorithms discussed in chapter 6.

The Godard LS transversal algorithm will be derived in this section. It differs from the recursive LS algorithm of section 3.4.1 only in the windowing technique. The method of derivation will be somewhat different, although the resulting algorithm will turn out to have a very similar structure.

The time-average square error to be minimized for a joint process estimator will be (see problem 3-34 for a predictor)

$$\epsilon(T) = \sum_{j=0}^{T} w_{T-j} \left(d(j) - \mathbf{c}'(T)\mathbf{y}(j) \right)^2 , \tag{3.4.12}$$

and this quantity will be minimized by the choice of the coefficient vector $\mathbf{c}(T)$. It is important to note this minimization problem will in effect be resolved at every time T, resulting in a new coefficient vector $\mathbf{c}(T)$. Also note that it is assumed that in the minimization the latest coefficient vector estimate is used for the entire past. The coefficient vectors determined in previous iterations are discarded entirely. The window function w_T is convolved with the square-error sequence, so that the window slides along with time. The criterion involves given data samples prior to $T = 0$, and the prewindowed criterion assumes that those samples are zero.

In analogy to (3.3.4), the time-average autocorrelation matrix at sample T is defined as

$$\hat{\mathbf{\Phi}}(T) = \sum_{j=0}^{T} w_{T-j} \mathbf{y}(j)\mathbf{y}'(j) . \tag{3.4.13}$$

Define a crosscorrelation vector estimate similar to (3.4.5) as

$$\hat{\mathbf{p}}(T) = \sum_{j=0}^{T} w_{T-j} d(j)\mathbf{y}(j) . \tag{3.4.14}$$

In terms of these estimates, it is readily verified by differentiation that (3.4.12) is minimized by the choice

$$\mathbf{c}(T) = \hat{\mathbf{\Phi}}^{-1}(T)\hat{\mathbf{p}}(T) , \tag{3.4.15}$$

which in principle completes the LS algorithm.

As they are expressed in (3.4.13) and (3.4.14), the calculation of the autocorrelation and crosscorrelation estimates requires a growing computation, which is impractical. Fortunately, as long as the window function can be written recursively, both the autocorrelation and crosscorrelation estimates can also be written recursively, and the computation does not grow with time. For example, taking the simple exponential window of (3.4.3), simple recursive relationships result directly from (3.4.13) and (3.4.14) as

$$\hat{\mathbf{\Phi}}(T) = \alpha\hat{\mathbf{\Phi}}(T-1) + \mathbf{y}(T)\mathbf{y}'(T) \tag{3.4.16}$$

and

$$\hat{\mathbf{p}}(T) = \alpha\hat{\mathbf{p}}(T-1) + d(T)\mathbf{y}(T) . \tag{3.4.17}$$

The final algorithm therefore requires the calculation of these two recursive relations followed by a solution of set of equations (3.4.15) to yield the coefficient vector. This structure is similar to that of the autocorrelation algorithm of section 3.4.1. Recursions (3.4.16-17) correspond to the low-pass filters of figure 3-9, except that in this case each of the low-pass filters is identical (problem 3-33).

The algorithm can be put into a different form which illustrates its close relationship to the SG algorithm by combining (3.4.15) and (3.4.17) into a single recursion for the coefficient vector. The result is an update equation for the coefficient vector,

$$\mathbf{c}(T) = \mathbf{c}(T-1) + \hat{\mathbf{\Phi}}^{-1}(T)e_c(T)\mathbf{y}(T) \tag{3.4.18}$$

where the error at time T is defined in terms of the coefficient vector at time T as

$$e_c(T) = d(T) - \mathbf{c}'(T-1)\mathbf{y}(T) . \tag{3.4.19}$$

This relation will be derived shortly, but first of all it is instructive to interpret it in light of the SG algorithm (3.2.3). It is virtually identical, except that the constant scalar step-size β of the gradient algorithm has been replaced by the inverse of the correlation matrix estimate. This modification is the source of the improved convergence performance of the LS algorithm as discussed in chapter 7. One interpretation is that the inverse of the correlation matrix "orthogonalizes" the updates to the coefficient vector, eliminating the interaction among the adaptation of the filter coefficients (see problem 3-35).

The modified form of the algorithm is rather simple to derive. Substituting (3.4.15) into (3.4.17),

$$\hat{\mathbf{\Phi}}(T)\mathbf{c}(T) = \alpha\hat{\mathbf{\Phi}}(T-1)\mathbf{c}(T-1) + d(T)\mathbf{y}(T) \tag{3.4.20}$$

and substituting in from (3.4.16)

$$\hat{\mathbf{\Phi}}(T)\mathbf{c}(T) = [\hat{\mathbf{\Phi}}(T) - \mathbf{y}(T)\mathbf{y}'(T)]\mathbf{c}(T-1) + d(T)\mathbf{y}(T) \tag{3.4.21}$$

which is easily manipulated into the form of (3.4.18).

This algorithm is known in the literature as the *Kalman/Godard algorithm*, in reference to the fact that it was first derived by applying Kalman filtering theory to the channel equalization algorithm by Godard. The primary disadvantage of the Kalman/Godard algorithm given by (3.4.16), (3.4.18), and (3.4.19) is that the n by n covariance matrix, $\hat{\mathbf{\Phi}}(T)$, must be calculated, stored, and inverted every iteration where n may typically be on the order of fifty (i.e. in channel equalization). Thus on the order of n^2 operations must be performed per iteration in contrast to the approximately n operations required by gradient algorithms. An efficient method for computing the Kalman gain vector, $\mathbf{g}(T) = \hat{\mathbf{\Phi}}^{-1}(T)\mathbf{y}(T)$, that only requires on the order of n operations per update [15, 16] is derived in chapter 6, and is called the *fast Kalman algorithm*. While computationally efficient, the fast Kalman uses the same criterion as the Kalman/Godard algorithm just derived and hence will have identical performance except for finite precision effects.

One derivation of the fast Kalman algorithm relies upon the following shifting property of the data,

$$\mathbf{y}(T-1) = \begin{bmatrix} y(T-1) \\ \cdot \\ \cdot \\ \cdot \\ y(T-n) \end{bmatrix} \rightarrow \mathbf{y}(T) = \begin{bmatrix} y(T) \\ \cdot \\ \cdot \\ \cdot \\ y(T-n+1) \end{bmatrix}.$$

At each new time instant only one new data value is received. This translates to the following shifting property of the n-th order covariance matrix $\hat{\mathbf{\Phi}}(T)$,

$$\hat{\mathbf{\Phi}}(T) = \left[\begin{array}{c|c} q & \mathbf{q}' \\ \hline \mathbf{q} & \hat{\mathbf{\Phi}}(T-1)-1 \end{array}\right] = \left[\begin{array}{c|c} \hat{\mathbf{\Phi}}(T-1) & \hat{\mathbf{p}} \\ \hline \hat{\mathbf{p}}' & p \end{array}\right],$$

where $\mathbf{q}$ and $\hat{\mathbf{p}}$ are related $n-1$ dimensional vectors and p and q are scalars. Since $\hat{\mathbf{\Phi}}(T)$ and $\mathbf{y}(T)$ do not change a great deal from one time interval to the next, this shifting property can be exploited to obtain an efficient recursive method for computing the Kalman gain, $\mathbf{g}(T)$.

The fast Kalman algorithm was first derived using matrix manipulations of this type. In chapter 6 the fast Kalman algorithm will be derived using an alternative geometric approach based on the derivation of a similar lattice algorithm.

3.4.3. Applications of Recursive Least Squares

The recursive LS algorithms described here have generally been superseded by the more computationally efficient algorithms considered in chapter 6. Either type of algorithm could be considered in any application requiring adaptive prediction or joint process estimation. The major advantage of these LS algorithms is their faster convergence as compared to the transversal filter adaptive gradient algorithm, as discussed in chapter 7.

Only adaptation algorithms based on the transversal filter structure have been considered in this chapter. In chapter 4 algorithms based on the lattice filter structure will be described. Adaptation algorithms for this structure can also be broken down into the SG and LS types. In this instance, there is a much smaller difference in performance between the two types of algorithms, since the lattice SG algorithm is in some sense already orthogonalized.

In practical applications it is often necessary or advantageous to modify the algorithms to fit the particular situation. To illustrate this, the design of an adaptive quantizer for the ADPCM encoder and decoder will now be considered. This algorithm does not use the LS predictor or joint process estimator, but rather uses a squared signal estimate of the power in the quantizer input signal.

3.4.3.1. Adaptive Quantization

A widely used algorithm for adapting the step-size of a quantizer is derived in this section. This algorithm addresses the problem of adjusting the signal level at the input to a quantizer properly. If the signal level is too low, only a small range of the quantizer is utilized, and the step-size of the quantizer is unnecessarily large in relation to the signal size. On the other hand, if the signal is too large, the signal samples are often too large to be accommodated by the finite range of the quantizer steps. The problem is particularly acute for signals with a wide dynamic range, such as speech or music.

A couple of solutions are available: one is to use nonuniform quantization, or instantaneous companding, and the other considered here is to use an adaptive quantizer. The algorithm derived here was proposed in [17] and modified to be unsusceptible to transmission errors in [18]. The derivation follows reference [19].

The approach that will be followed is illustrated in figure 3-10. What is shown is just the adaptive quantizer portion of the ADPCM coder and decoder in figure 2-4 (actually this adaptive quantizer could also be used by itself without the predictor). The transmission of step-size information explicitly is avoided by using the output of the A/D converter to drive the step-size adaptation circuit. This same signal is available at the decoder for driving an identical step-size adaptation circuit there. A fixed uniform A/D converter is used, and the input signal level is controlled by dividing by an estimate of signal level in front of the A/D converter in the coder and multiplying by the same estimate after the D/A converter in the decoder. The same effect could be achieved by controlling the step-size of each converter directly, if they were implemented with an adjustable step-size.

In figure 3-10 the signal level estimate at sample T is called $\sigma(T)$, and the code word passing between the A/D converter and D/A converter is called $q(T)$. For example, if the A/D converter is a four-bit converter, $q(T)$ assumes one of 16 values. The output of the D/A converter is called $Q[q(T)]$.

As in the design of the ADPCM adaptive predictor, the trick to avoid mistracking between encoder and decoder is to derive the step-size adaptation in the decoder (because that is where the least information is available) and then simply replicate the algorithm in the coder. In (3.2.17) one way to estimate the power of a signal using a recursively generated time average was shown in the context of normalization of the step-size of the SG algorithm. Repeating that equation in the present context, using the output signal $\tilde{e}_f(T)$ for generation of the power estimate,

$$\sigma^2(T) = (1-\alpha) \sum_{j=0}^{\infty} \alpha^j \tilde{e}_f^2(T-1-j) , \tag{3.4.22}$$

where the new wrinkle is that the power estimate at sample T is based on only past outputs (the current output can only be determined once the current power estimate is known!). The normalization of the signal is then performed using the square root of this power, $\sigma(T)$, which is an estimate of the rms power (or standard deviation) of the signal. Of course, (3.4.22) can be written recursively as

$$\sigma^2(T) = (1-\alpha)\, \tilde{e}_f^2(T-1) + \alpha\sigma^2(T-1) . \tag{3.2.23}$$

Since it is desired to relate this step-size estimate back to $q(T)$, the output can be written as

$$\tilde{e}_f(T) = \sigma(T) Q[q(T)] , \tag{3.4.24}$$

and substituting (3.4.24) into (3.4.23) the update in terms of $q(T)$ becomes

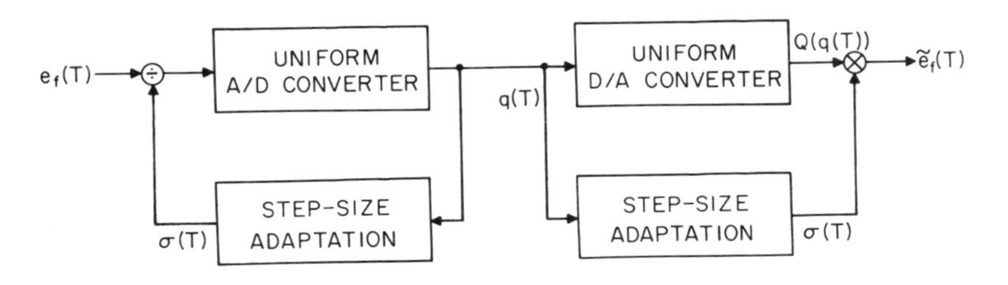

Figure 3-10. Adaptive quantization in ADPCM.

$$\sigma(T) = M[q(T-1)]\,\sigma(T-1)\ . \tag{3.4.25}$$

where the "step-size multiplier" is specifically

$$M^2[q(T-1)] = \alpha + (1-\alpha)\,Q^2[q(T-1)]\ . \tag{3.4.26}$$

The intuitive understanding of (3.4.25) is simple. Whenever a sample is near the origin of the quantizer, this is an indication that the step-size is too large, at least if this happens often. For this $q(T)$ the multiplier $M[q(T)]$ will be less than one and the step-size will be decreased slightly. Conversely, when the sample is near the overload of the quantizer, the step-size will be increased slightly because the indication is that the step size is too small. Since the increments are small (the multiplier is near unity) the cumulative effect will be to adjust the step-size in the appropriate direction based on the signal level in relation to the current step-size.

As in the design of an adaptive predictor, there is concern that the encoder and decoder quantizer adaptations do not mistrack in the presence of different initial conditions or transmission errors. Unfortunately, (3.4.25) suffers from this problem, as can be seen if the effect of an initial condition is displayed explicitly by iteration,

$$\sigma(T) = \prod_{j=0}^{T-1} M[q(j)]\,\sigma(0)\ . \tag{3.4.27}$$

The two versions of the recursion will permanently mistrack if the initial conditions are different but the inputs are the same. This problem can easily be repaired in much the same manner as in (3.2.10) if the logarithm of both sides of (3.4.25) is taken,

$$\log\sigma(T) = \beta\log\sigma(T-1) + \log M[q(T-1)]\ , \tag{3.4.28}$$

where the value $\beta = 1$ corresponds to (3.4.25). If some leakage is introduced in the algorithm, by choosing $0 < \beta < 1$, the initial condition will decay away, and the permanent mistracking will be eliminated. Writing (3.4.27) in the form of (3.4.23), the algorithm with leakage becomes

$$\sigma(T) = M[q(T-1)]\,\sigma^{\beta}(T-1)\ . \tag{3.4.29}$$

3.5. CONCLUSIONS

This chapter has presented a variety of algorithms for adapting a predictor or joint process estimator constructed from a transversal filter structure. Each of these algorithms has evolved naturally from the ones before it.

Fundamentally, three types of algorithms have been proposed: the MMSE, the LS and the SG. The MMSE algorithm is useful theoretically and as a starting point for the SG algorithms, but is of little practical use because of the unrealistic assumptions underlying it. The LS algorithms include the block LS algorithms of section 3.3, and the recursive LS algorithms of section 3.4. These algorithms have a solid theoretical foundation, which is the replacement of time average statistics for ensemble average statistics in the MMSE algorithm, as well as a limitation of the memory of the algorithm to handle the possible nonstationarity of the environment. Finally, the SG and approximation algorithms of sections 3.2 are based on gradient search techniques using a stochastic approximation to the MMSE gradient. The LS criteria with a properly chosen weighting function are able to handle the nonstationary statistics case, whereas the MMSE criterion applies inherently to the stationary case and the SG approximation is necessary to extend it to the nonstationary case.

Of these adaptation algorithms, the block LS and stochastic gradient algorithms have been by far the most widely applied in practice. However, there is increasing momentum toward the algorithms yet to be covered in chapters 4 and 6. Those algorithms have two significant advantages over those presented in this chapter. First, SG algorithms based on the lattice filter structure to be derived in chapter 4 generally converge significantly faster for similar asymptotic fluctuation in the coefficient vector. The lattice structure also has other advantages which will be elaborated in chapter 5 as well as chapter 4. Second, the recursive LS algorithms to be derived in chapter 6 require significantly reduced computation relative to those of this chapter. This advantage becomes particularly significant for large filter orders n. There is no difference in the convergence performance of LS algorithms based on the transversal and lattice structures (since they minimize an identical criterion), but LS algorithms based on the lattice structure still offer significant advantages.

PROBLEMS

3-1 Establish (3.1.6) by setting the gradient of $E[e_c^2(T)]$ equal to zero.

3-2 Recalculate (3.1.1) for the case of a predictor rather than joint process estimator, and verify (3.1.8).

3-3 Rederive (3.1.6) starting from orthogonality principle (3.1.9).

3-4 Derive the orthogonality principle for a linear predictor.

3-5 Let $\mathbf{A}$ be a real symmetric matrix. Then establish the following facts:

a. The eigenvalues of $\mathbf{A}$ are real-valued.

b. Let λ_i and λ_j be two distinct eigenvalues of $\mathbf{A}$. Show that the associated eigenvectors $\mathbf{v}_i$ and $\mathbf{v}_j$ are orthogonal; that is, $\mathbf{v}'_i\mathbf{v}_j = 0$.

c. Assume that the eigenvalues of $\mathbf{A}$ are all distinct, and that the eigenvectors are normalized to unit length ($\mathbf{v}_i'\mathbf{v}_i = 1$). Define the matrix

$$\mathbf{V} = [\mathbf{v}_1, \mathbf{v}_2, \cdots, \mathbf{v}_n]$$

and show that

$$\mathbf{A} = \mathbf{V}\boldsymbol{\Lambda}\mathbf{V}'$$

where $\boldsymbol{\Lambda}$ is a diagonal matrix of eigenvectors,

$$\mathbf{\Lambda} = \mathrm{diag}(\lambda_1, \lambda_2, \cdots, \lambda_n) \ .$$

d. Show that if $\mathbf{A}$ is positive definite (semi-definite) then its eigenvalues are all positive (non-negative).

3-6 Based on the results of problem 3-5, establish (3.1.24).

3-7 Show that the eigenvectors of the matrix $(\mathbf{I} - \beta \mathbf{\Phi})$ are the same as the eigenvectors of $\mathbf{\Phi}$, and that the eigenvalues are $(1 - \beta\lambda_i)$, $1 \leqslant i \leqslant n$, and thus establish the validity of (3.1.25).

3-8 Verify (3.1.29).

3-9 This problem will attempt to make plausible the relationship between the eigenvalues of an autocorrelation matrix and the power spectrum displayed in (3.1.32-33). Let $\mathbf{\Phi}$ be a $(2m+1)\times(2m+1)$ autocorrelation matrix, and let the components of an eigenvector of this matrix be

$$\mathbf{v}' = [v_{-m}, v_{-m+1}, \cdots, v_m] \ .$$

a. Show that the eigenvector and associated eigenvalue λ satisfies the relationship

$$\sum_{i=-m}^{m} \phi_{j-i} v_i = \lambda v_j \ , \ \ -m \leqslant j \leqslant m \ .$$

b. Let $m \rightarrow \infty$ and take the Fourier Transform to show that

$$S(e^{j\omega}) V(e^{j\omega}) = \lambda V(e^{j\omega})$$

where $S(e^{j\omega})$ is the power spectrum defined by (3.1.31) and

$$V(e^{j\omega}) = \sum_{i=-\infty}^{\infty} v_i e^{-j\omega i} \ .$$

c. For the case where $S(e^{j\omega})$ is a single valued function (that is, it doesn't assume the same value at two different frequencies), argue that the infinite eigenvectors are complex exponentials $(e^{j\omega_0 i})$ with corresponding eigenvalues equal to the power spectrum at the same frequency.

d. Use these results to argue the validity of (3.1.32-33).

3-10 Consider an input wide-sense stationary random process with autocorrelation function $\phi_T = \alpha^{|T|}$.

a. Find the power spectrum of this random process.

b. Find the asymptotic minimum and maximum eigenvalues of the autocorrelation matrix.

c. Find, as a function of α, the eigenvalues and eigenvectors of the 2×2 autocorrelation matrix.

d. Find the eigenvalue spread of the autocorrelation matrix as predicted by approximate relations (3.1.32-33) and compare to the results of c.

e. Find, as a function of α and as $n \to \infty$, the step-size β and resulting dominant mode of convergence of the MMSE gradient algorithm. Interpret this result intuitively.

3-11 Show that for the MMSE gradient algorithm, the error vector is given by

$$\mathbf{c}_j - \mathbf{c}_{opt} = \sum_{i=1}^{n} (1 - \beta\lambda_i)^j (\mathbf{v}_i'(\mathbf{c}_0 - \mathbf{c}_{opt}))\mathbf{v}_i ,$$

and interpret this equation.

3-12 Using the results of problem 3-11 and (3.1.29), show that the excess MSE is given, for the MMSE gradient algorithm, by the relation

$$E[e_c^2(T) - E[e_c^2(T)]_{min} = \sum_{i=1}^{n} \lambda_i (1 - \beta\lambda_i)^{2j} (\mathbf{v}_i'(\mathbf{c}_0 - \mathbf{c}_{opt}))^2 ,$$

and interpret this equation.

3-13 Consider the dominant mode of the MMSE gradient algorithm.

a. The excess MSE as a function of time expressed in decibels is approximately given by $\gamma_1 - \gamma_2 j$ and find the constants γ_1 and γ_2.

b. Evaluate these constants for the particular case where β is very small, and discuss the tradeoff between speed of convergence and step-size for this case. Thus, the MSE expressed in decibels decreases linearly with time. This result applies particularly to the stochastic gradient algorithm, where in practice β is usually chosen to be much smaller than the optimum β predicted by (3.1.27) (see chapter 7).

3-14 For the echo canceler as described in section 3.1.4, assume that the near-end talker $x(T)$ is a white noise process uncorrelated with the far-end talker $y(T)$, and that the latter has the autocorrelation function of problem 3-10. Determine the MMSE solution of (3.1.40) for this case, and show that the optimum coefficient vector approaches the echo impulse response as the number of coefficients increases.

3-15 Consider an ADPCM encoding with a first order predictor, and define

$$\rho = \frac{\phi_1}{\phi_0} .$$

a. What is the optimum predictor coefficient?

b. What is the prediction gain; that is, the improvement in signal-to-quantization noise ratio relative to no predictor at all?

3-16 In problem 3-15, let the predictor coefficient be fixed at unity. Over what range of ρ is the signal-to-quantization ratio better than with no predictor at all?

3-17 For an input process with mean value $\mu \neq 0$, show that the MMSE first order predictor is

$$\hat{y}(T) = \rho y(T-1) + (1-\rho)\mu \ .$$

where ρ is defined in problem 3-15.

3-18 An input WSS process has power spectral density

$$\Phi(z) = \frac{A}{(1-\alpha z)(1-\alpha z^{-1})}, \ 0 < \alpha < 1$$

and the region of convergence includes the unit circle.

a. Find the autocorrelation function ϕ_T.

b. Find the predictor coefficients for a MMSE n-th order predictor.

3-19 Consider the modified ADPCM algorithm shown in figure P3-1. Only the decoder is shown, and there is an all-zero filter followed by an all-pole filter. Both the FIR filters $C(z)$ and $F(z)$ are strictly causal. Assume that considerations of guaranteed stability cause us to fix predictor $F(z)$, and adapt only $C(z)$.

a. Draw the ADPCM encoder corresponding to this decoder.

b. Show that the overall error is the same as the quantization error, as in ordinary ADPCM.

c. Specify the stochastic gradient algorithm which is appropriate to adapt $C(z)$.

d. We can give the all-pole filter adaptive characteristics as shown in figure P3-2 by implementing two fixed FIR filters $F_1(z)$ and $F_2(z)$, and combining the the output of the two filters with constants $\theta(T)$ and $(1 - \theta(T))$, $0 \leqslant \theta(T) \leqslant 1$. The two filters are chosen to guarantee stability for all allowed values of $\theta(T)$. Specify a stochastic gradient algorithm which adapts $\theta(T)$ as well as the all-zero filter.

e. How would you insure that $\theta(T)$ never left the allowed range?

3-20 For the decision feedback equalizer governed by (3.2.26) find the equations which when solved will give the MMSE solution for the

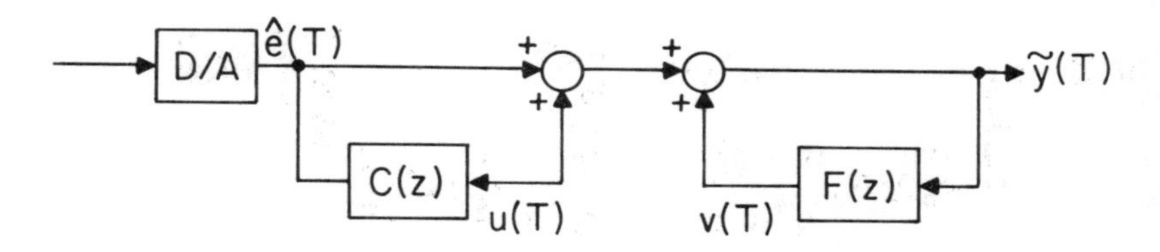

Figure P3-1. ADPCM decoder with both poles and zeros.

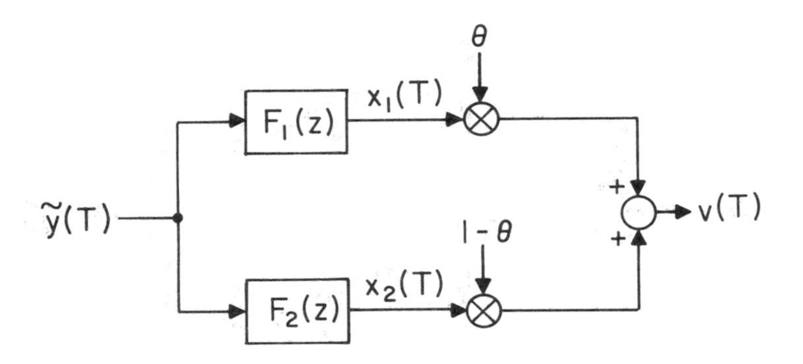

Figure P3-2. An adaptive predictor for the ADPCM decoder of figure P3-1.

coefficients. Assume that the data digits are independent and identically distributed, and that no decision errors are made.

3-21 For the signal power estimation algorithm of (3.2.17), assume that $y(T)$ is a white Gaussian process.

a. Show that $\sigma^2(T)$ is an unbiased estimator of the signal power.

b. Find the variance of this estimate. Interpret how this variance depends on step-size α.

3-22 A MMSE prediction problem is solved, and the solution turns out to be

$$e_c(T) = y(T) + y(T-1) - 0.25y(T-2)$$

a. Can we infer uniquely the power spectrum of the input process which is used to arrive at this solution? Depending on whether your answer is positive or negative, answer part b. or c.

b. Give the unique power spectrum.

c. Give an example of a power spectrum which would result in this solution.

3-23 Consider a first order AR process given by

$$y(T) = \eta(T) + ay(T-1), \ |a| < 1$$

where $E[\eta(i)\eta(j)]$ is σ^2 for $i = j$ and zero otherwise.

a. Calculate the impulse response of the all-pole filter.

b. Plot the frequency response of the all-pole filter.

c. Calculate the auto-correlation function of the process, $R_y(T)$.

d. Calculate the power spectrum $S_y(e^{j\omega})$.

3-24 Repeat problem 3-23 for a MA process

$$y(T) = \sum_{j=0}^{m} b_j \eta(T-j) .$$

3-25 In this problem we will show that the DFE forward and feedback filters of problem 2-7 are optimum. Consider the channel model shown in figure P3-3, where $G(\omega)$ is the transfer function of the continuous-time channel, there is additive noise $\eta(t)$ which is white with power spectrum N , and the channel output is passed through a matched filter with transfer function $G^*(\omega)$ and sampled at the rate of transmission of data symbols $1/\tau$. (For reasons beyond the scope of this book this is the optimum processing.)

a. Calculate the overall channel discrete-time transfer function from data symbol input to sampler output, and show that it is

$$H(e^{j\omega}) = \frac{1}{\tau} \sum_k |G(\frac{\omega - 2\pi k}{\tau})|^2 .$$

b. Show that the noise power spectrum at the sampler output is $N_0 H(e^{j\omega})$.

c. Show that the zero-forcing equalizer ($C(z)$ in figure 2-7) is

$$C(e^{j\omega}) = \frac{1}{H(e^{j\omega})} .$$

c. Show that the noise power spectrum at the input to the slicer will be white if the predictor is chosen to be

$$1 - F(z) = \frac{H^+(z)}{h_0^+} ,$$

where $H^+(z)$ is defined in problem 2-6.

d. Show that the forward equalizer transfer function is then the one given in problem 2-6; namely, $\dfrac{1}{h_0^+ H^+(z^{-1})}$.

3-26 Suppose the MMSE criterion is modified to minimize the quantity

$$E[e_c^2(T)] + \mu \mathbf{c}'\mathbf{c}$$

in similar fashion to (3.2.24).

a. Find the coefficient vector $\mathbf{c}_\mu$ which minimizes this quantity, and show that the error between this solution and the coefficient vector $\mathbf{c}_{opt}$ of (3.1.6) is given by

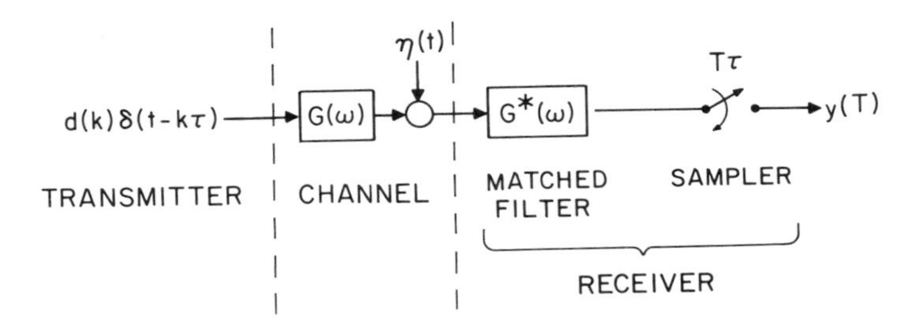

Figure P3-3. Channel model for a decision-feedback equalizer.

$$\mathbf{c}_\mu - \mathbf{c}_{\text{opt}} = -\mu \sum_{i=1}^{n} \frac{\mathbf{v}_i{}'\mathbf{p}}{\lambda_i(\lambda_i + \mu)} \mathbf{v}_i ,$$

and that the resulting excess MSE of (3.1.29) is given by

$$E[e_c^2(T)] - E[e_c^2(T)]_{\min} = \mu^2 \sum_{i=1}^{n} \frac{(\mathbf{v}_i{}'\mathbf{p})^2}{\lambda_i(\lambda_i + \mu)^2} .$$

How does this MSE increase as we vary μ?

b. Find the MMSE gradient algorithm which iteratively minimizes this error.

c. Find the criterion on the step-size of this algorithm which guarantees stability.

d. Find the step-size which maximizes the rate of convergence.

e. Investigate how the maximum rate of convergence can be altered by the choice of μ, particularly where the eigenvalue spread is large. Discuss the tradeoff between "excess MSE" and rate of convergence.

f. How do these results apply to the stochastic gradient algorithm of (3.2.25)?

3-27 a. Show that the stochastic gradient algorithm for adapting a linear predictor analogous to (3.2.3) is

$$\mathbf{f}(T+1) = \mathbf{f}(T) + \beta e_f(T)\mathbf{y}(T-1) .$$

b. Determine the update equation for a single coefficient analogous to (3.2.4) and draw a realization analogous to figure 3-5.

c. Make an approximation similar to (3.2.5) to find the average trajectory, and show that the optimum step size for fastest convergence of the average trajectory is the same as for the joint process estimator and the speed of convergence is the same.

3-28 Verify equation (3.2.13).

3-29 The step size normalization of (3.2.15) is not appropriate if the nonlinear correlation multiplier of (3.2.14) is used. What would be an appropriate normalization?

3-30 Apply a similar technique to (3.2.24) to a predictor, and find the resulting algorithm analogous to (3.2.25).

3-31 Show that the time average correlation matrix of (3.3.4) is not Toeplitz, while the matrix of (3.3.6) is Toeplitz.

3-32 Derive a joint process estimator algorithm analogous to the recursive autocorrelation LS predictor of section 3.4.1.

3-33 Interpret (3.4.16) in terms of a bank of filters analogous to figure 3-9.

3-34 Derive a predictor recursive LS algorithm analogous to the joint process estimation algorithm of section 3.4.2.

3-35 This problem will illustrate how the LS algorithm of (3.4.18-19) can achieve a faster convergence than the SG algorithm.

a. Assuming the input $y(T)$ is wide-sense stationarity, what is the average value of the time average autocorrelation matrix calculated via (3.4.16)?

b. Substitute this average value into (3.4.18), and making the usual approximations calculate the average trajectory of the coefficient vector.

c. How does this trajectory depend upon the eigenvalue spread of the correlation matrix?

3-36 An adaptive quantizer develops the estimate

$$\sigma(T) = (1-\alpha)\,|y(T-1)| + \alpha\sigma(T-1)$$

of the input signal standard deviation. Derive an algorithm similar to the one derived in section 3.4.3.1 for this standard deviation estimate.

3-37 Derive the prewindowed recursive LS algorithm using the window function

$$w_T = Ta^T,\ T > 0\,.$$

Hint: The window function can be generated with the recursion

$$w_T = 2aw_{T-1} - a^2 w_{T-2}\,.$$

REFERENCES

1. Robert M. Gray, "On the Asymptotic Eigenvalue Distribution of Toeplitz Matrices," *IEEE Trans. on Inform. Theory* **IT-18** pp. 725-730 (Nov. 1972).

2. U. Grenander and G. Szego, *Toeplitz Forms and Their Applications,* University of California Press (1958).

3. L.R. Rabiner and R.W. Schafer, *Digital Processing of Speech Signals,* Prentice-Hall, Englewood Cliffs, N.J. (1978).

4. A. Papoulis, "Maximum Entropy and Spectral Estimation: A Review," *IEEE Trans. on Acoustics, Speech, and Signal Processing* **ASSP-29**(6) p. 1176 (Dec. 1981).

5. G. Ungerboeck, "Adaptive Equalization Techniques in Voice-Band Data Transmission," *Proc. 1980 IEEE International Communications Conference*, pp. 8.4.1-8.4.6. (1980).

6. B. Widrow et. al., "Adaptive Noise Canceling: Principles and Applications," *IEEE Proceedings* **63**(12) pp. 1692-1716 (Dec. 1975).

7. D.G. Messerschmitt, "Echo Cancellation in Speech and Data Transmission," *Transactions on Special Topics in Communications* **COM-32**(Feb. 1984).

8. B. Widrow, "Adaptive Filters I: Fundamentals," *(Tech. Rept. 6764-6)*, (Dec. 1966).

9. D.J.Sakrison, "Stochastic Approximation: A Recursive Method for Solving Regression Problems," in *Advances in Communication Systems*, ed. A.V.Balakrishnan,Academic Press, New York (1966).

10. D.L. Cohn and J.L. Melsa, "The Residual Encoder - An Improved ADPCM System for Speech Digitization," *IEEE Trans. Comm.* **COM-23** pp. 935-941 (Sept. 1975).

11. D.L. Duttweiler, "Adaptive Filter Performance with Nonlinearities in the Correlation Multiplier," *IEEE Trans. Acoustics, Speech and Signal Processing* **ASSP-30** p. 578 (Aug. 1982).

12. R.D. Gitlin, H.C. Meadors, and S.B. Weinstein, "The Tap Leakage Algorithm: An Algorithm for the Stable Operation of a Digitally Implemented, Fractionally Spaced Adaptive Equalizer," *B.S.T.J.* **61**(8) p. 1817 (Oct. 1982).

13. J. Makhoul , "Linear Prediction: A Tutorial Review," *Proc. IEEE* **63, 4,** pp. 561-580 (April 1975).

14. D. Godard, "Channel Equalization Using Kalman Filter for Fast Data Transmission," *IBM J. of Res. and Dev.*, pp. 267-273 (May 1974).

15. D.D. Falconer and L. Ljung, "Application of Fast Kalman Estimation to Adaptive Equalization," *IEEE Trans. Comm.* **COM-26** pp. 1439-1446 (Oct. 1978).

16. L. Ljung, M. Morf, and D. Falconer, "Fast calculation of gain matrices for recursive estimation schemes," *Int. J. Control* **27**(1) pp. 1-19 (1978).

17. N.S. Jayant, "Adaptive Quantization With a One-Word Memory," *Bell Syst. Tech. J.* **52** pp. 1119-1144 (Sept. 1973).

18. D.J. Goodman and R.M. Wilkinson, "A Robust Adaptive Quantizer," *IEEE Trans. on Communications* **COM-23**(11) p. 1362 (Nov. 1975).

19. D.L. Cohn and J.L. Melsa, "The relationship between an adaptive quantizer and a variance estimator," *IEEE Trans. Inform. Theory* **IT-21** pp. 669-671 (Nov. 1975).

4

LATTICE FILTERS

In recent years the transversal filter has been largely displaced in applications of adaptive linear prediction in the context of speech processing by the all-zero lattice filter shown in fig. 4-1. The all-zero (fixed coefficient) lattice filter has the same transfer function as the (fixed coefficient) transversal filter; however, the lattice structure offers significant advantages which should become evident in this chapter.

In the remainder of this book, a geometric approach is taken to the derivation of the algorithms. There are several reasons for this. First, the lattice filters and algorithms built around lattice filters are naturally expressed in geometric terms. In particular, the lattice filter itself is simply the result of doing a Gram-Schmidt orthogonalization procedure on the data samples entering into the prediction. Second, the geometric approach lends an intuitive interpretation to many of the results which would otherwise seem to be merely mathematical manipulations. Finally, the geometric approach allows us to treat the different cases, such as the LMS and LS criteria, in a common framework and notation. This avoids the rederivation of results for the different criteria.

The purpose of section 4.1 is therefore to review some mathematical preliminaries necessary to the understanding of the later geometric derivations. Lest some readers be deterred at this point, we emphasize that the necessary background is limited to the simple and easy to understand concepts of inner product, projections, and subspaces. The only difficulty the reader might experience is in casting such concrete items as signals and random variables into the slightly more abstract objects of vectors in a linear space. This section attempts to make this connection clear.

After these preliminaries, in section 4.2 the LMS lattice filter is derived from first principles. In section 4.3 the Levinson-Durbin algorithm (widely used for solving the linear equations which arise in the LMS and block LS algorithms of sections 3.1 and 3.2) is derived in a straightforward way from the lattice filter. In section 4.4, an LMS joint process estimator based on the lattice filter structure is derived. The analogous algorithm to the SG algorithm for the transversal filter based on the lattice filter structure is derived in section 4.5. section 4.6 discusses the lattice filter in the context of the applications introduced in chapter 2.

4.1. MATHEMATICAL PRELIMINARIES

In this chapter and chapter 6, there are two alternative approaches that could be used in the derivation of filter structures such as the lattice filter and adaptive filtering algorithms such as the LS algorithms. One would be to utilize matrix manipulations, and the second would be to use geometrical concepts such as orthogonality, projection, subspaces, and the Gram-Schmidt orthogonalization procedure. Many or most of these latter geometrical concepts should be quite familiar to many readers, and they are used here because of the additional understanding and intuition which they evoke. The application of these concepts, most commonly thought of in the context of n-dimensional Euclidean space, in the more abstract vein of discrete-time signals and random processes may be new to some readers. The purpose of this section is to make that connection, and to define some notation which will be used throughout the remainder of this book.

4.1.1. Notation Which Includes Filter Order

In the sequel it will be important to explicitly display the filter order n in the notation because we will frequently be developing recursive equations in filter order as well as in time. Therefore, where it is important, the filter order will be displayed after the time index with an intermediate vertical slash. For example, the prediction error for an n-th order predictor becomes

$$e_f(T) \to e_f(T|n) . \tag{4.1.1}$$

Occasionally it will be important to display the filter order in the vector of filter coefficients, as in

$$\mathbf{f}(T) \to \mathbf{f}(T|n) . \tag{4.1.2}$$

In this case the n indicates not only the order of the predictor filter but also the number of components in the vector. An analogous notation for the j-th coefficient of the filter becomes

$$f_j(T) \to f_j(T|n) . \tag{4.1.3}$$

As an example of the use of this notation, an explicit relation for the prediction error for an n-th order filter is

$$e_f(T|n) = y(T) - \mathbf{f}(T|n)\mathbf{y}(T) , \tag{4.1.4}$$

or equivalently

$$e_f(T|n) = y(T) - \sum_{i=1}^{n} f_i(T|n)y(T-i) . \tag{4.1.5}$$

These equations presume that the filter coefficients are functions of time. Where they are fixed with time, the corresponding notation would be $\mathbf{f}(n)$ and $f_j(n)$ for an n-th order filter.

4.1.2. Review of Linear Spaces and Inner Products

Ordinary Euclidean space is the most familiar example of a *linear space* or *vector space.* In Euclidean space, a vector is a point in the space, and is specified by its coordinates, n coordinates in an n-dimensional space. We then introduce the notation

$$\mathbf{X} \longleftrightarrow (x_1, x_2, \cdots, x_n) \ , \tag{4.1.6}$$

where $\mathbf{X}$ is the vector and $x_1, \ldots x_n$ are the n components of that vector. The notation " $\longleftrightarrow$ " means that $\mathbf{X}$ is the vector which corresponds to components $x_1, \ldots, x_n$. There are then rules for adding two vectors (sum the individual components) and multiplying a vector by a scalar (multiply each of the components by that scalar). The term *scalar* is a special term for a real or complex number as distinct from a vector.

The linear space concept can be generalized in the following fashion. Formally, a linear space is a set H of *elements* or *vectors* of the space, together with a rule for adding two vectors in the space to generate another vector and a rule for multiplying a vector by a scalar (real number) to generate another vector. A vector in the space will be denoted in this section by a bold-faced letter, as for example $\mathbf{X}$.

The addition rule associates with the sum of two vectors $\mathbf{X} + \mathbf{Y}$ another vector, and must obey the ordinary rules of arithmetic, including the commutative and associative laws,

$$\mathbf{X} + \mathbf{Y} = \mathbf{Y} + \mathbf{X} \ , \tag{4.1.7}$$

$$\mathbf{X} + (\mathbf{Y} + \mathbf{Z}) = (\mathbf{X} + \mathbf{Y}) + \mathbf{Z} \ . \tag{4.1.8}$$

Anticipating the geometric interpretation of vectors which will be developed shortly, the sum of two vectors has the interpretation illustrated in figure 4-1a (in that case in two dimensional Euclidean space). The linear space must include a zero vector $\mathbf{0}$, which has the property that

$$\mathbf{0} + \mathbf{X} = \mathbf{X} \tag{4.1.9}$$

and there must for every vector $\mathbf{X}$ be another vector $(-\mathbf{X})$ with the property that

$$\mathbf{X} + (-\mathbf{X}) = \mathbf{0} \ . \tag{4.1.10}$$

The rule for multiplication by a scalar associates with each scalar (real number) α and vector $\mathbf{X}$ another vector $\alpha \cdot \mathbf{X}$ which must obey the associative law,

$$\alpha \cdot (\beta \cdot \mathbf{X}) = (\alpha\beta) \cdot \mathbf{X} \tag{4.1.11}$$

and also follow the rules

$$1 \cdot \mathbf{X} = \mathbf{X} \tag{4.1.12}$$

and

$$0 \cdot \mathbf{X} = \mathbf{0} \ . \tag{4.1.13}$$

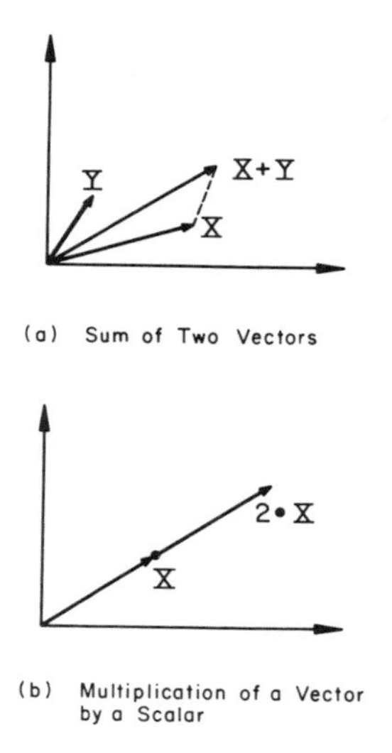

Figure 4-1. Elementary operations in a two-dimensional linear space.

The geometric interpretation of multiplying a vector by a scalar is shown in figure 4-1b, and is simply a shortening or lengthening of the vector.

Finally, addition and multiplication must obey the distributive laws,

$$\alpha\cdot(\mathbf{X} + \mathbf{Y}) = \alpha\cdot\mathbf{X} + \alpha\cdot\mathbf{Y} \tag{4.1.14}$$

$$(\alpha + \beta)\cdot\mathbf{X} = \alpha\cdot\mathbf{X} + \beta\cdot\mathbf{X} \tag{4.1.15}$$

Linear spaces can be real spaces (scalars have real values) or complex spaces (scalars have complex values). Real spaces will suffice for this book, so it will be presumed that all scalars are real numbers.

The definition of Euclidean space given earlier meets all these requirements, and is therefore a linear space. There are two other examples of linear spaces of particular importance in this book. These are the space of infinite dimensional vectors (which is a generalization of Euclidean space to infinite dimensions), and the space of random variables. In the linear estimation problem considered in chapter 3, the former space is applicable to the given data case (where the criterion is the LS), and the latter space applies to the known stationary statistics case (where the criterion is the MMSE).

For the given data case of estimation, as in the block LS criterion of section 3.3, define a linear space consisting of infinite dimensional vectors. In particular, let

$$\mathbf{Y} \longleftrightarrow (\ldots y(-1), y(0), y(1), \ldots) \tag{4.1.16}$$

be a vector of given data samples, where the components of the vector are the given data samples. This vector is very similar to Euclidean space as defined in (4.1.6), the difference being that the number of components is infinite rather than finite. The additional assumption must be made that

$$\sum_T y^2(T) < \infty , \tag{4.1.17}$$

or in words that the total energy in the given data samples is finite. This assumption is necessary for mathematical reasons and should not concern us too much since we typically window the given data sequences prior to processing, forcing this condition to be satisfied. Usually this window is finite in extent, so that only a finite number of terms contribute to the summation in (4.1.17), or the window is exponentially decreasing as in section 3.4 which also assists in forcing (4.1.17) to be satisfied.

To complete the definition of this linear space, define $\alpha \cdot \mathbf{Y}$ for a real-valued scalar α as the vector obtained by multiplying $\mathbf{Y}$ component-wise by α,

$$\alpha \cdot \mathbf{Y} \longleftrightarrow (\ldots \alpha y(-1), \alpha y(0), \alpha y(1), \ldots) . \tag{4.1.18}$$

Similarly, the sum of two vectors is defined as the vector obtained by adding the respective components, as in

$$\mathbf{X} + \mathbf{Y} \longleftrightarrow (\ldots x(-1) + y(-1), x(0) + y(0), x(1) + y(1), \ldots) . \tag{4.1.19}$$

The vector $\mathbf{0}$ is of course the vector with all zero components

$$\mathbf{0} \longleftrightarrow (\ldots, 0, 0, 0, \ldots) . \tag{4.1.20}$$

Note that all these definitions are natural extensions of the corresponding definitions in finite-dimensional Euclidean space.

A second linear space of concern here is the space of random variables with finite second moments. This space will correspond to the known stationary statistics case, as in the MMSE solution in section 3.1. Let X be a random variable with zero mean and finite second moment,

$$E(X) = 0 \tag{4.1.21}$$

$$E(X^2) < \infty , \tag{4.1.22}$$

where (4.1.22) is again a finite signal energy assumption analogous to (4.1.17). The collection of all such random variables can be considered as a linear space, where the vectors correspond to random variables,

$$\mathbf{X} \longleftrightarrow X . \tag{4.1.23}$$

To complete the definition of this space, $\mathbf{0}$ is defined as the random variable which is always zero,

$$\mathbf{0} \longleftrightarrow 0 , \tag{4.1.24}$$

and the vector $\alpha \cdot \mathbf{X}$ corresponds to the random variable αX,

$$\alpha \cdot \mathbf{X} \longleftrightarrow \alpha X \ . \tag{4.1.25}$$

The sum of two vectors corresponds to the sum of the corresponding random variables,

$$\mathbf{X} + \mathbf{Y} \longleftrightarrow X + Y \ . \tag{4.1.26}$$

The first example is just the familiar Euclidean space, except that the number of dimensions is infinite. The second example is difficult to relate to Euclidean space, but the fact that it satisfies all the properties of a linear space can easily be verified.

The definition of linear space does not capture the most important properties of Euclidean space; namely, the *geometric* structure. This structure includes such concepts as the length of a vector in the space, and the angle between two vectors. All these properties of Euclidean space can be deduced from the definition of *inner product* of two vectors. This inner product $<\mathbf{X},\mathbf{Y}>$ is a real-valued quantity defined for Euclidean space as

$$<\mathbf{X},\mathbf{Y}> = \sum_{i=1}^{n} x_i y_i \ , \tag{4.1.27}$$

and has the interpretation illustrated in figure 4-2; namely, the inner product of two vectors is equal to the product of the length of the first vector, the length of the second vector, and the cosine of the angle between the vectors.

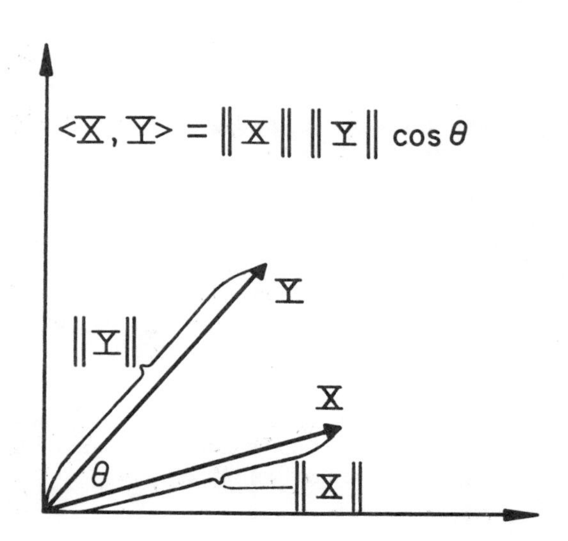

Figure 4-2. Geometrical interpretation of inner product.

In the figure, $||\mathbf{X}||$ denotes the length of a vector, which has not been defined. However, once the definition of inner product (4.1.27) has been given, we can *deduce* the length of a vector, since the inner product of a vector with itself, $<\mathbf{X},\mathbf{X}>$, is the square of the length of the vector (the angle θ is zero). A special notation

$$\begin{aligned} <\mathbf{X},\mathbf{X}> &= ||\mathbf{X}||^2 \\ &= \sum_{i=1}^{n} x_i^2 \end{aligned} \tag{4.1.28}$$

can then be introduced, where $||\mathbf{X}||$ is called the *norm* of the vector $\mathbf{X}$ and has the geometric interpretation as the length of the vector. This notation is used in figure 4-2.

The case of special interest is where the two vectors are *perpendicular* or *orthogonal*, in which case the inner product is zero. This concept of orthogonality will play a central role in the derivation of the lattice filter.

The inner product as applied to Euclidean space can be generalized to the other linear spaces of interest defined earlier. The important consequence is that the geometric concepts familiar in Euclidean space can be applied to these spaces as well. Let $\mathbf{X}$ and $\mathbf{Y}$ be vectors of a linear space, and suppose that an inner product $<\mathbf{X},\mathbf{Y}>$ of two vectors is defined on that space. This inner product is a scalar (real number), and must obey the rules

$$<\mathbf{X}+\mathbf{Y},\mathbf{Z}> = <\mathbf{X},\mathbf{Z}> + <\mathbf{Y},\mathbf{Z}> \tag{4.1.29}$$

$$<\alpha\cdot\mathbf{X},\mathbf{Y}> = \alpha<\mathbf{X},\mathbf{Y}> \tag{4.1.30}$$

$$<\mathbf{X},\mathbf{Y}> = <\mathbf{Y},\mathbf{X}> \tag{4.1.31}$$

$$<\mathbf{X},\mathbf{X}> > 0, \ \ \mathbf{X} \neq 0 \, . \tag{4.1.32}$$

These rules are all obeyed by the familiar Euclidean space inner product of (4.1.27), as can be easily verified. For the other linear spaces of interest, analogous definitions of inner product satisfying the rules can be made. In particular,

$$<\mathbf{X},\mathbf{Y}> = \sum_{T=-\infty}^{\infty} x(T)y(T) \tag{4.1.33}$$

for the infinite dimensional vectors (note the similarity to the definition of inner product usually used in Euclidean space), and

$$<\mathbf{X},\mathbf{Y}> = E(XY) \tag{4.1.34}$$

for the space of random variables with finite second moments. These inner products can be given the same interpretation as in Euclidean space; namely, as the product of the length of two vectors times the cosine of the angle between them. These definitions also induce, as in Euclidean space, the concept of the length of a vector. This length or norm $||\mathbf{X}||$ is defined formally as in (4.1.28). For the two spaces of interest, the norm becomes

$$||\mathbf{X}||^2 = \sum_{T=-\infty}^{\infty} x^2(T) \tag{4.1.35}$$

$$||\mathbf{X}||^2 = E(\mathbf{X}^2) \; . \tag{4.1.36}$$

Note that the two conditions (4.1.17) and (4.1.22) that were imposed correspond to the assumption that the vector has finite norm or length.

The geometric properties are so important that the special name *inner product space* is given to a linear space on which an inner product is defined. Thus, both of the examples defined earlier are inner product spaces. If the inner product space has the additional property of *completeness*, then it is defined to be a *Hilbert space*. Intuitively the notion of completeness means that there are no "missing" vectors that are arbitrarily close to vectors in the space but are not themselves in the space. Since the spaces used in this book are all complete and hence formally Hilbert spaces, we will not dwell on this property further. In the sequel, all linear spaces considered will be Hilbert spaces.

A concept crucial to the development of the lattice filter is *orthogonality*, which was already mentioned in the context of Euclidean spaces. In particular, two vectors are orthogonal if

$$<\mathbf{X},\mathbf{Y}> = 0. \tag{4.1.37}$$

The geometric interpretation of orthogonality is shown in figure 4-3, where the two vectors are at a 90 degree angle with respect to one another. In the case of the linear space made up of random variables with finite second moments, two vectors are orthogonal if the corresponding random variables are uncorrelated.

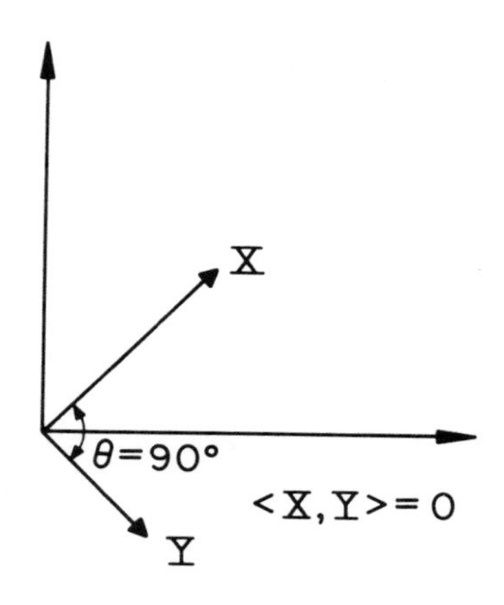

Figure 4-3. Orthogonal vectors.

Another important object is the *subspace* of a linear space. This is a subset of the linear space which is itself a linear space. Roughly speaking this means that the sum of any two vectors in the subspace must also be in the subspace, and the product of any vector in the subspace by any scalar must also be in the subspace. An example of a subspace in three dimensional Euclidean space is either a line or a plane in the space, where in either case the vector **0** must be in the subspace. Another example of a subspace is the set of vectors obtained by forming all possible weighted linear combinations of n vectors $\mathbf{X}_1, \cdots, \mathbf{X}_n$. The subspace so formed is said to be *spanned* by the set of n vectors. This is illustrated in figure 4-4 for three dimensional Euclidean space. In figure 4-4a, the subspace spanned by **X** is the dashed line which is infinite in length and colinear with vector **X**. Any vector on this line can be obtained by multiplying **X** by the appropriate scalar. In figure 4-4b, the subspace spanned by **X** and **Y** is the plane with infinite extent (depicted by the dashed line) which is determined by the two vectors. Any vector in this plane can be formed as a linear combination of the two vectors multiplied by appropriate scalars.

Two subspaces M_1 and M_2 are said to be *orthogonal subspaces* if every pair of vectors, one taken from M_1 and the other taken from M_2, are orthogonal. An

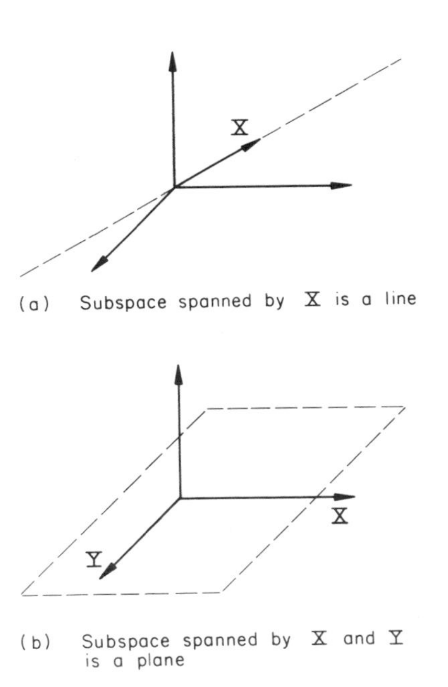

Figure 4-4. Subspaces in three dimensional Euclidean space.

example of two orthogonal subspaces would be the subspaces spanned by **X** and **Y** if these two vectors are orthogonal as in (4.1.37). The *sum of two subspaces* M_1 and M_2, denoted by $M_1 \oplus M_2$, is the subspace formed by all possible linear combinations of a vector in M_1 and a vector in M_2, and is itself a subspace. An example of a sum of two subspaces is shown in figure 4-4b, where the plane spanned by **X** and **Y** is also the sum of two subspaces, one spanned by **X** and the other spanned by **Y**.

The *projection theorem* is the basis for much of the remainder of this book, and in particular will be used to derive the lattice filter structure later in this chapter. What follows is a statement of the projection theorem, which is proven in [1]:

> **(Projection Theorem)** Given a subspace M of a Hilbert space H and a vector **X** in H there is a unique vector $\mathbf{P}_M\mathbf{X}$ in M called the *projection of* **X** *on* M which has the property that
>
> $$<\mathbf{X}-\mathbf{P}_M\mathbf{X},\mathbf{Y}> = 0 \tag{4.1.38}$$
>
> for every vector **Y** in M.

Another way of stating this result would be that M and the one dimensional subspace spanned by $(\mathbf{X} - \mathbf{P}_M\mathbf{X})$ are orthogonal.

The concept of a projection is illustrated in figure 4-5 for three dimensional Euclidean space, where the subspace M is the plane formed by the x-axis and y-axis and **X** is an arbitrary vector. The projection is the result of dropping a line from **X** down to the plane which is perpendicular to the plane (this is the dashed line in the figure). The resulting vector $(\mathbf{X} - \mathbf{P}_M\mathbf{X})$ is the vector which is shown parallel to the dashed line. It is orthogonal to the plane M, and hence

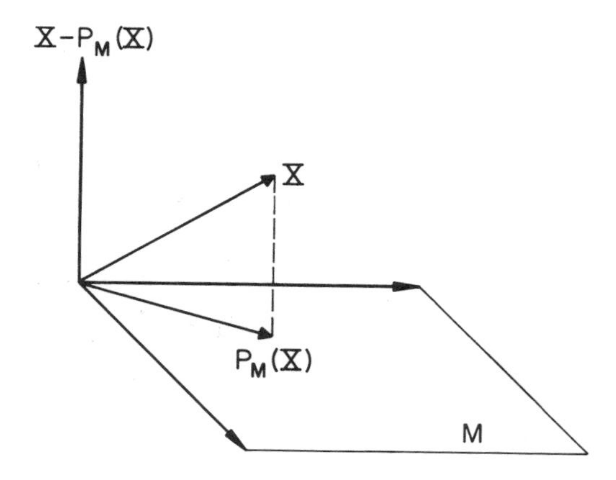

Figure 4-5. Illustration of projection for three dimensional Euclidean space.

to every vector in M.

A consequence of the projection theorem (problem 4-3) is that the projection $\mathbf{P}_M\mathbf{X}$ is the unique vector in M which is closest to $\mathbf{X}$; that is,

$$||\mathbf{X} - \mathbf{P}_M\mathbf{X}|| < ||\mathbf{X} - \mathbf{Y}|| \tag{4.1.39}$$

for every $\mathbf{Y} \neq \mathbf{P}_M\mathbf{X}$ in M. This is illustrated geometrically in figure 4-6, where the subspace M is a line. The vector on the line which is closest to another vector $\mathbf{X}$ is evidently the projection as shown, and any other vector in M is farther from $\mathbf{X}$.

Given two orthogonal subspaces M_1 and M_2 of Hilbert space H and an arbitrary vector $\mathbf{X}$ in H, the projection of $\mathbf{X}$ on $M_1 \oplus M_2$ can be expressed uniquely as (problem 4-4)

$$\mathbf{P}_{M_1 \oplus M_2}\mathbf{X} = \mathbf{P}_{M_1}\mathbf{X} + \mathbf{P}_{M_2}\mathbf{X} , \tag{4.1.40}$$

or in words the sum of the projection on M_1 and the projection on M_2. This is illustrated in figure 4-7 for three dimensional Euclidean space, where M_1 is the y-axis and M_2 is the x-axis. For this case the subspace $M_1 \oplus M_2$ is the plane determined by the x-axis and y-axis. One way to find the projection of $\mathbf{X}$ on the plane is to find the projections on the two subspaces M_1 and M_2, and then sum the two projections. It should be emphasized that this approach does not work if the two subspaces are not orthogonal.

Property (4.1.40) will be very convenient for the derivation of the lattice filter and similar structures. This completes our tutorial introduction to subspaces and projection. Those readers interested in a more detailed treatment of these topics can consult any textbook on functional analysis. A particularly good reference for engineering oriented readers is the book by Naylor and Sell [1].

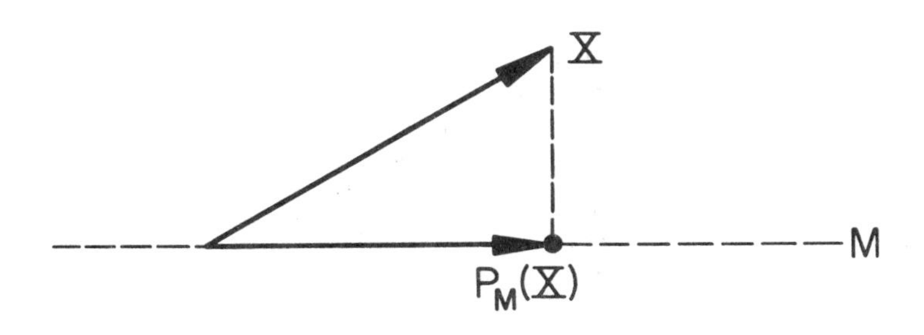

Figure 4-6. Interpretation of projection as closest vector of subspace.

4.1.3. Orthogonality Principle Revisited

In section 3.1.1 the orthogonality principle was derived for the MMSE criterion. The terminology "orthogonality principle" suggests that this principle can be given a geometric interpretation, and indeed this is the case as will be demonstrated in this section. With the mathematical machinery which has been developed so far in this chapter, many of the estimation problems of chapter 3 can be reformulated and a geometric interpretation given to their solutions.

Since the prediction and joint process estimates of chapter 3 form the weighted linear summations of delayed versions of the signal, it is appropriate at this point to define some new notation for the vector corresponding to a delayed version of the signal. For the given data case, recalling definition (4.1.16) for the signal vector, define the following notation for the new vector obtained by delaying the signal by i samples,

$$z^{-i}\mathbf{Y} \longleftrightarrow (....,y(-1-i),y(-i),y(1-i),....) . \qquad (4.1.41)$$

In this notation, $z^{-i}\mathbf{Y}$ is a new vector obtained by delaying the samples in the $\mathbf{Y}$ vector by i samples. The z^{-i} notation is simply a suggestive notation drawn from the Z-transform, but does not imply that $\mathbf{Y}$ is a Z-transform. (It is not!) As an example of the use of this notation, suppose that a new signal is formed by taking the difference between adjacent samples of the original signal $y(T)$. Then this new signal is associated with a vector $\mathbf{Y} - z^{-1}\mathbf{Y}$, which has components given by

$$\mathbf{Y} - z^{-1}\mathbf{Y} \longleftrightarrow (..,y(-1)-y(-2),y(0)-y(-1),y(1)-y(0),..) . \qquad (4.1.42)$$

This vector represents the result of this "first difference operation" for all time.

Recall from (4.1.5) the definition of an nth order prediction error as

$$e_f(T|n) = y(T) - \sum_{j=1}^{n} f_j(n)y(T-j) , \qquad (4.1.43)$$

where a fixed coefficient vector $\mathbf{f}(n)$ has been assumed. This error signal for all time can be represented as a vector as

$$\mathbf{E}_f(n) \longleftrightarrow (....,e_f(-1|n),e_f(0|n),e_f(1|n),....) . \qquad (4.1.44)$$

In terms of the notation of (4.1.41), error vector (4.1.44) can be rewritten as (problem 4-6)

$$\mathbf{E}_f(n) = \mathbf{Y} - \sum_{j=1}^{n} f_j(n)z^{-j}\mathbf{Y} . \qquad (4.1.45)$$

This notation will simplify subsequent equations, and is also very suggestive of the nature of the prediction in the time domain. Note also that for the remainder of the book, for simplicity of notation the "dot" in the multiplication of a vector by a scalar will be omitted (i.e. $f_j(n) \cdot z^{-j}\mathbf{Y}$ has been replaced by $f_j(n)z^{-j}\mathbf{Y}$).

The optimum block LS predictor using the autocorrelation criterion can now be found, and the result can be given a geometric interpretation. The block LS criterion of (3.3.2) can be reformulated for the predictor case, with the result

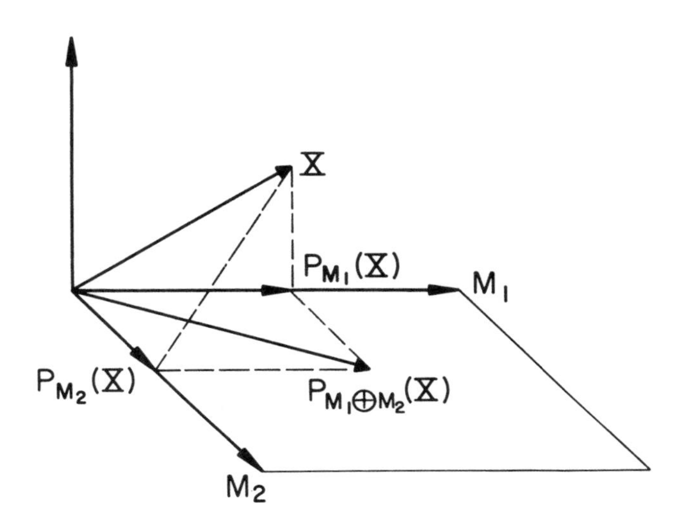

Figure 4-7. Taking the projection on the sum of two orthogonal subspaces.

that the quantity being minimized is

$$\epsilon = \sum_{T=-\infty}^{\infty} e_f^2(T|n) , \tag{4.1.46}$$

where it is presumed that the input samples have been first windowed so that the summation in (3.3.2) can be replaced by an infinite summation. In terms of the geometric interpretation, (4.1.46) is actually the square of a norm in linear space in accordance with definition (4.1.35),

$$\begin{aligned} \epsilon &= ||\mathbf{E}_f(n)||^2 \\ &= ||\mathbf{Y} - \sum_{j=1}^{n} f_j(n) z^{-j} \mathbf{Y}||^2 . \end{aligned} \tag{4.1.47}$$

Thus, the autocorrelation block LS criterion can be interpreted geometrically as choosing the estimate so as to minimize the norm of the error vector, which is the difference between **Y** and the estimate of **Y**.

A similar interpretation can be given to the MMSE criterion of section 3.1, where the input signal sequence $y(T)$ is modeled as a wide-sense stationary random process. If one of the input samples is associated with a vector **Y**, the same relation (4.1.43) defines an error sample in the time domain. The MMSE prediction error which must be minimized in this case is

$$\epsilon = E[e_f^2(T|n))] , \tag{4.1.48}$$

where the error $e_f(T|n)$ is given by (4.1.43). The MMSE error in (4.1.48) is not a function of the time index T in spite of the appearance of T on the right side because of the stationarity assumption. Since the time index is of no importance, associate the vector **Y** with one input sample (a random variable),

$$\mathbf{Y} \longleftrightarrow y(T) , \tag{4.1.49}$$

where it makes no difference what value is chosen for T. Similarly, associate the vector $z^{-m}\mathbf{Y}$ with the signal sample delayed by m samples

$$z^{-m}\mathbf{Y} \longleftrightarrow y(T-m) . \tag{4.1.50}$$

This notation then leads to a prediction error vector given by (4.1.45). It makes no difference in (4.1.49) and (4.1.50) what the time index T is, as long as the same value is chosen in both equations, since the geometric properties of the vectors are independent of T. The MSE in (4.1.48) is, in accordance with definition (4.1.37), the square of the norm of the error vector, and hence (4.1.47) also applies to the MMSE criterion.

A key advantage of the linear space notation is the fact that both the MMSE and block LS prediction problems have been expressed in a common mathematical notation and framework. In fact, both formulations have been shown to be restatements of the same problem, and the projection theorem can now be used to solve both problems simultaneously.

In the prediction problem, we are given a signal vector Y, and the object is to minimize the length of the n-th order prediction error vector given by (4.1.47). The summation term (the predictor) is a vector in the subspace spanned by $z^{-1}\mathbf{Y}, \cdots ,z^{-n}\mathbf{Y}$. This motivates us to define more generally the subspace spanned by $z^{-k}\mathbf{Y}, \cdots ,z^{-m}\mathbf{Y}$, $k \leqslant m$, as $M(k,m)$. The summation term on the right side of (4.1.45) (the predictor) is a general vector in $M(1,n)$, since it consists of a linear combination of the spanning vectors in $M(1,n)$. From the projection theorem, the vector of the subspace $M(1,n)$ which is closest to **Y** is the projection of **Y** on $M(1,n)$, and thus

$$\mathbf{E}_f(n) = \mathbf{Y} - \mathbf{P}_{M(1,n)}\mathbf{Y} \tag{4.1.51}$$

is the error corresponding to the optimum predictor. This optimum predictor has the property that

$$<\mathbf{E}_f(n),z^{-j}\mathbf{Y}> = 0 , 1 \leqslant j \leqslant n , \tag{4.1.52}$$

since the error must be orthogonal to the subspace, and hence every vector in the subspace. But (4.1.52) is just a restatement of the orthogonality condition derived for the MMSE case in (3.1.9), where the expectation is reinterpreted as an inner product.

Substituting (4.1.45) into (4.1.52), the condition for optimality of the predictor coefficients becomes

$$\phi_j = \sum_{m=1}^{n} f_m(n)\phi_{j-m} , 1 \leqslant j \leqslant n . \tag{4.1.53}$$

where ϕ_j is the inner product of the signal vector with the delayed signal vector,

$$\phi_j = <\mathbf{Y}, z^{-j}\mathbf{Y}> . \tag{4.1.54}$$

This condition is equivalent to the solution for the MMSE case given by (3.1.6) and (3.1.8). It is also equivalent for the condition derived in (3.3.3) for the autocorrelation LS criterion (again reinterpreted for the predictor case). To see this, the inner product of (4.1.54) can be evaluated explicitly from (4.1.33-34) and (4.1.41) as

$$\phi_j = \sum_{T=-\infty}^{\infty} y(T)y(T-j) \tag{4.1.55}$$

for the LS criterion, and

$$\phi_j = E[y(T)y(T-j)] \tag{4.1.56}$$

for the MMSE criterion. In both cases this inner product turns out to be a type of autocorrelation function.

In summary, in this section we have succeeded in reformulating the estimation problems of chapter 3 in the notation and framework of inner product spaces. This has accomplished two things. First, it has provided a common notational framework in which two seemingly different (but in reality identical) estimation problems can be solved simultaneously. This solution will be extended to the lattice filter in the next section. Second, it has lent an additional geometric interpretation to the solution of the estimation problems. This geometric interpretation will be particularly valuable in understanding the lattice filter.

4.2. DERIVATION OF LATTICE FILTER

The lattice filter is an alternative to the transversal filter structure for the realization of a predictor. We will see in section 4.4 that it is also applicable to the realization of a joint process estimator. This section will derive the lattice filter structure using the mathematical tools of the section 4.1, and the projection theorem in particular.

Recall that the prediction problem at hand is to find the projection of $\mathbf{Y}$ on the subspace $M(1,n)$ spanned by $z^{-1}\mathbf{Y}, \ldots, z^{-n}\mathbf{Y}$. The lattice filter is simply a consequence of finding a new set of vectors which also span the subspace $M(1,n)$, but which have the valuable property of being mutually orthogonal. A set of vectors which are mutually orthogonal and span a subspace is called an *orthogonal basis* for the subspace.

A set of n vectors which satisfy this need is

$$\mathbf{E}_b(0) = \mathbf{Y} \tag{4.2.1}$$

$$\mathbf{E}_b(m) = z^{-m}\mathbf{Y} - \mathbf{P}_{M(0,m-1)}(z^{-m}\mathbf{Y}) \ , \ \ 1 \leqslant m \leqslant n-1 \ . \tag{4.2.2}$$

In particular, it will be shown momentarily that $\mathbf{E}_b(m)$, $0 \leqslant m < n$, is an orthogonal basis for $M(0,n-1)$, or equivalently that $z^{-1}\mathbf{E}_b(m)$, $0 \leqslant m < n$ is an orthogonal basis for $M(1,n)$. However, first let us examine and interpret the time domain equivalents of these vectors.

Each of the vectors in (4.2.2) can be written in the form

$$\mathbf{E}_b(m) = z^{-m}\mathbf{Y} - \sum_{j=1}^{m} b_j(m) z^{-j+1}\mathbf{Y} \tag{4.2.3}$$

for some particular constants $b_j(m)$ corresponding to the projection in (4.2.2). The time domain equivalent of (4.2.3) is

$$e_b(T|m) = y(T-m) - \sum_{j=1}^{m} b_j(m) y(T-j+1) . \tag{4.2.4}$$

This filter is illustrated in figure 4-8. Carefully examining this figure, $e_b(T|m)$ is the error in predicting a sample m sampling intervals in the past in terms of a linear combination of m more recent samples, up to and including the current sample. This is similar to the predictor discussed in chapter 3 and section 4.1, but estimating distant past samples in terms of more recent samples is backwards from that definition of a predictor. For this reason, the predictor of (4.2.4) is called a "backward predictor", $e_b(T|m)$ is called the m-th order backward prediction error, and the $b_j(m)$ are called the backward prediction coefficients.

The vectors in (4.2.2) represent different orders of a backward predictor referenced to the current time. That is, all that is fixed is the most recent sample which is included in the prediction (namely $y(T)$). Each of the different orders of predictors is predicting a different sample in the past, and each is optimal in terms of minimizing the norm of the backward prediction error vector.

Those familiar with the "Gram-Schmidt orthogonalization procedure" may recognize the vectors in (4.2.1-2) as simply the consequence of applying this procedure to the vectors $\mathbf{Y}, z^{-1}\mathbf{Y}, \ldots, z^{-n+1}\mathbf{Y}$. These different orders of backward predictors form an orthogonal basis for $M(0,n-1)$ as will now be demonstrated. First, the successive orders of backward prediction error span $M(0,n-1)$. The reason is that $\mathbf{E}_b(m)$ contains as one component $z^{-m}\mathbf{Y}$, and hence each of $\mathbf{Y}, z^{-1}\mathbf{Y}, \ldots, z^{-n+1}\mathbf{Y}$ is included in one or more of the backward

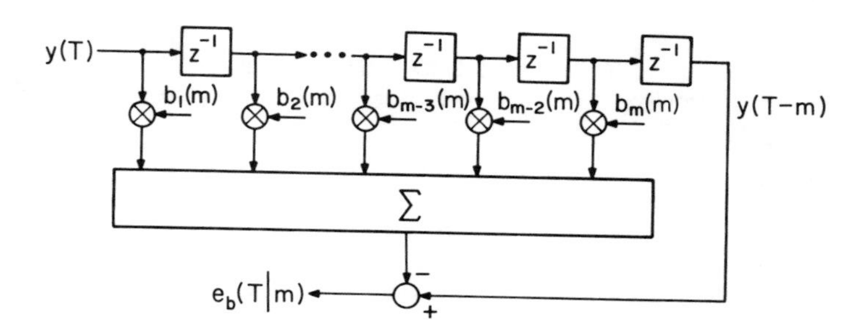

Figure 4-8. Realization of an m-th order backward predictor.

prediction errors. Second, the backward prediction errors are mutually orthogonal. By definition of the projection, $\mathbf{E}_b(m)$ is orthogonal to $M(0,m-1)$. But $\mathbf{E}_b(0)$ through $\mathbf{E}_b(m-1)$ are vectors in the subspace $M(0,m-1)$, since they can each be written as linear combinations of $\mathbf{Y}, \ldots, z^{-m+1}\mathbf{Y}$. Since each $\mathbf{E}_b(m)$ is orthogonal to $\mathbf{E}_b(0)$ through $\mathbf{E}_b(m-1)$, it follows (by induction if you are a stickler!) that the different orders of prediction error are mutually orthogonal.

The backward predictors have no important direct physical significance, except as an intermediate signal in the construction of the lattice filter. We will see that the lattice filter realizing the n-th order forward predictor will generate all lower orders of backward prediction error. Notice that we have used the terminology "forward predictor" to describe the predictor of (4.1.45). This is so that the two different predictors can be distinguished. Note also that the subscript on the prediction error and the prediction coefficients coincide with the terminology "forward predictor" and "backward predictor" to assist in keeping them straight.

Due to the symmetry of the autocorrelation function (i.e. $\phi_j = \phi_{-j}$), it is straightforward to show (problem 4-8) that the optimal backward predictor coefficients are the mirror image of the optimal forward predictor coefficients, i.e.,

$$b_j(m) = f_{m-j+1}(m), \quad 1 \leqslant j \leqslant m \ . \tag{4.2.5}$$

This is not true of all forward and backward predictors, but only the optimal ones. From this fact it follows immediately (problem 4-9) that the backward and forward prediction errors have the same norm,

$$||\mathbf{E}_f(m)||^2 = ||\mathbf{E}_b(m)||^2 \ , \quad 1 \leqslant m \leqslant n \ . \tag{4.2.6}$$

These results are intuitive, since whether prediction is done forward or backward in time, the error vector should be the same length (only the lengths will be equal, not the prediction error signals themselves).

The lattice structure is defined by the equations

$$\mathbf{E}_f(m) = \mathbf{E}_f(m-1) - k_m^b z^{-1}\mathbf{E}_b(m-1), \quad \mathbf{E}_f(0) = \mathbf{Y} \tag{4.2.7}$$

$$\mathbf{E}_b(m) = z^{-1}\mathbf{E}_b(m-1) - k_m^f \mathbf{E}_f(m-1), \quad \mathbf{E}_b(0) = \mathbf{Y} \tag{4.2.8}$$

for $1 \leqslant m \leqslant n$ where n is the order of the desired forward predictor and k_m^b and k_m^f are appropriate constants. These are called *order update equations* since they relate higher order forward and backward prediction errors to lower order prediction errors. In the time domain, the order updates become

$$e_f(T|m) = e_f(T|m-1) - k_m^b e_b(T-1|m-1) \tag{4.2.9}$$

$$e_f(T|0) = y(T)$$

$$e_b(T|m) = e_b(T-1|m-1) - k_m^f e_f(T|m-1) \tag{4.2.10}$$

$$e_b(T|0) = y(T) .$$

These update equations apply regardless of whether the block LS or MMSE criterion is used (the difference will reflect itself in how the coefficients k_m^b and k_m^f are determined).

The lattice filter structure is shown in figure 4-9. In figure 4-9a, the successive stages of the filter develop the successively higher order forward and backward prediction errors. In finally developing an n-th order forward predictor at the final output (which was after all the goal!), all lower order prediction errors are developed as intermediate signals between the stages.

One stage of the filter corresponding to (4.2.9-10) is shown in figure 4-9b. A lattice filter stage has as its input the forward and backward prediction errors using an $(m-1)$-th order filter and outputs the forward and backward prediction errors using an m-th order filter. Both inputs to the first lattice stage are **Y** (or $y(T)$ in the time domain), which also corresponds to both prediction errors of order 0.

In most formulations of the lattice filter, the coefficients k_m^b and k_m^f are equal for $1 \leqslant m \leqslant n$ (exceptions include some versions of the stochastic gradient

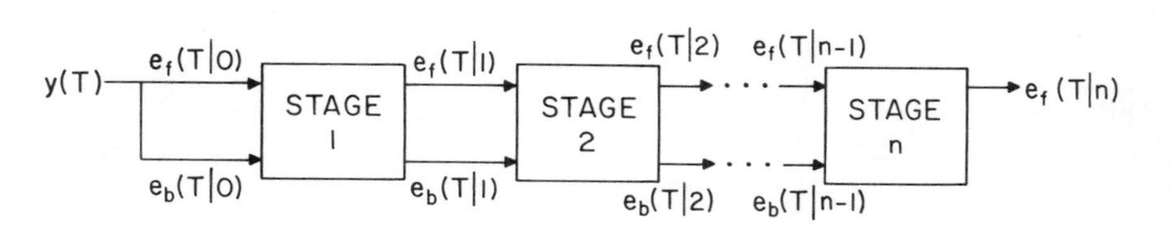

(a) Overall structure

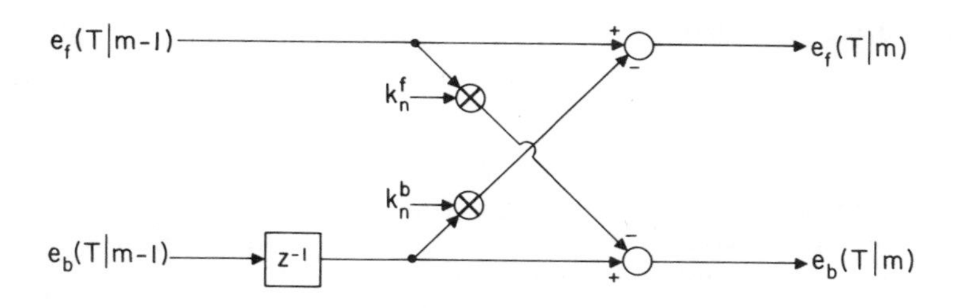

(b) One stage

Figure 4-9. A lattice filter n-th order predictor.

and LS algorithms covered later). When they are equal, the superscript will be suppressed, as in

$$k_m = k_m^b = k_m^f,\ 1 \leqslant m \leqslant n\ . \tag{4.2.11}$$

The coefficients k_m, $1 \leqslant m \leqslant n$, are called *partial correlation* or *PARCOR* coefficients and are a different representation of the corresponding transversal filter prediction coefficients in the sense that a one-to-one correspondence exists between $f_1(n), f_2(n), \ldots, f_n(n)$ and $k_1, k_2, \ldots, k_n$. Occasionally they are also called *reflection coefficients.*

The order recursions (4.2.7-8) will now be derived, and in the process, relations for the PARCOR coefficients will be developed. There are two subspace decompositions which form the basis for the lattice filter,

$$M(0,m) = M(1,m) \oplus \{\mathbf{E}_f(m)\} \tag{4.2.12}$$

$$M(1,m+1) = M(1,m) \oplus \{z^{-1}\mathbf{E}_b(m)\} \tag{4.2.13}$$

where the notation $\{\mathbf{Y}\}$ means the subspace spanned by the vector $\mathbf{Y}$, i.e., the set of vectors of the form $\alpha \cdot \mathbf{Y}$ where α is a scalar. Decomposition (4.2.12) is valid because the component which is missing in $M(1,m)$ in (4.2.12) is $\mathbf{Y}$, which is one component in $\mathbf{E}_f(m)$. Similarly (problem 4-10), the component missing in $M(1,m)$ in (4.2.13) is $z^{-m-1}\mathbf{Y}$, which is a component in $z^{-1}\mathbf{E}_b(m)$. Since both $\mathbf{E}_f(m)$ and $z^{-1}\mathbf{E}_b(m)$ are orthogonal to the subspace $M(1,m)$, the subspaces on the right of (4.2.12) and (4.2.13) are orthogonal to one another in each case. Therefore, applying (4.1.40) (a consequence of the projection theorem), to $\mathbf{Y}$ in the case of (4.2.13) and $z^{-m-1}\mathbf{Y}$ in the case of (4.2.12),

$$\mathbf{P}_{M(1,m+1)}\mathbf{Y} = \mathbf{P}_{M(1,m)}\mathbf{Y} + \mathbf{P}_{\{z^{-1}\mathbf{E}_b(m)\}}\mathbf{Y} \tag{4.2.14}$$

$$\begin{aligned} \mathbf{P}_{M(0,m)}(z^{-m-1}\mathbf{Y}) = {} & \mathbf{P}_{M(1,m)}(z^{-m-1}\mathbf{Y})\ . \\ & + \mathbf{P}_{\{\mathbf{E}_f(m)\}}(z^{-m-1}\mathbf{Y}) \end{aligned} \tag{4.2.15}$$

First the forward error order update of (4.2.7) can be derived using (4.2.14). Applying definition (4.1.51),

$$\begin{aligned} \mathbf{E}_f(m+1) &= \mathbf{Y} - \mathbf{P}_{M(1,m+1)}\mathbf{Y} \\ &= \mathbf{Y} - \mathbf{P}_{M(1,m)}\mathbf{Y} - \mathbf{P}_{\{z^{-1}\mathbf{E}_b(m)\}}\mathbf{Y} \\ &= \mathbf{E}_f(m) - \mathbf{P}_{\{z^{-1}\mathbf{E}_b(m)\}}\mathbf{Y} \\ &= \mathbf{E}_f(m) - k_{m+1}^b z^{-1}\mathbf{E}_b(m) \end{aligned} \tag{4.2.16}$$

where the last equation follows because $\mathbf{P}_{\{z^{-1}\mathbf{E}_b(m)\}}\mathbf{Y}$ is of the form $k_{m+1}^b z^{-1}\mathbf{E}_b(m)$ for some constant k_{m+1}^b. The value of k_{m+1}^b can be found using the projection theorem, (4.1.38), which says that

$$< \mathbf{Y} - k_{m+1}^b z^{-1}\mathbf{E}_b(m), z^{-1}\mathbf{E}_b(m) > \; = 0 \tag{4.2.17}$$

or

$$k_{m+1}^b = \frac{<\mathbf{Y}, z^{-1}\mathbf{E}_b(m)>}{||z^{-1}\mathbf{E}_b(m)||^2} . \tag{4.2.18}$$

A more enlightening form of this expression can be obtained when it is noted from (4.1.51) that

$$\begin{aligned} <\mathbf{Y}, z^{-1}\mathbf{E}_b(m)> &= <\mathbf{Y} - \mathbf{P}_{M(1,m)}\mathbf{Y}, z^{-1}\mathbf{E}_b(m)> \\ &= <\mathbf{E}_f(m), z^{-1}\mathbf{E}_b(m)> , \end{aligned} \tag{4.2.19}$$

since $z^{-1}\mathbf{E}_b(m)$ is orthogonal to any vector in $M(1,m)$, including $\mathbf{P}_{M(1,m)}\mathbf{Y}$. Also noting (4.2.6), k_{m+1}^b can be written in the form

$$k_{m+1}^b = \frac{<\mathbf{E}_f(m), z^{-1}\mathbf{E}_b(m)>}{||\mathbf{E}_f(m)|| \; ||z^{-1}\mathbf{E}_b(m)||} , \tag{4.2.20}$$

which demonstrates that the k_m at stage m+1 of the filter is the normalized inner product of the forward prediction error and the backward prediction error delayed by one sample.

The Schwarz inequality (problem 4-12) states that for two vectors **X** and **Y**,

$$|<\mathbf{X}, \mathbf{Y}>| \leqslant ||\mathbf{X}|| \cdot ||\mathbf{Y}|| \tag{4.2.21}$$

with equality if and only if $\mathbf{X} = \mathbf{Y}$. This inequality follows immediately from figure 4-2, where $|\cos(\theta)| \leqslant 1$. Applying the Schwarz inequality to (4.2.20),

$$|k_{m+1}^b| \leqslant 1 . \tag{4.2.22}$$

The derivation of the order update for the backward prediction error, (4.2.8), is only slightly more complicated. Replacing m by $m+1$ in (4.2.2) (problem 4-13),

$$\begin{aligned} \mathbf{E}_b(m+1) &= z^{-m-1}\mathbf{Y} - \mathbf{P}_{M(0,m)}(z^{-m-1}\mathbf{Y}) \\ &= z^{-m-1}\mathbf{Y} - \mathbf{P}_{M(1,m)}(z^{-m-1}\mathbf{Y}) \\ &\quad - \mathbf{P}_{\{\mathbf{E}_f(m)\}}(z^{-m-1}\mathbf{Y}) \\ &= z^{-1}\mathbf{E}_b(m) - k_{m+1}^f \mathbf{E}_f(m) \end{aligned} \tag{4.2.23}$$

from (4.2.15). As before, (4.2.8) is verified except for the final step of calculating the constant in the last term of (4.2.23). It can be shown (problem 4-14) by the same method that k_{m+1}^f is also given by (4.2.20), which verifies that $k_m^f = k_m^b$. In the remainder of this section both these quantities will be replaced by just k_m.

A useful relationship for the length of the prediction error vector can also be derived. By the same reasoning (problem 4-15) as was used in (4.2.19),

$$
\begin{aligned}
||\mathbf{E}_f(m)||^2 &= <\mathbf{E}_f(m),\mathbf{Y}> \\
&= <\mathbf{E}_f(m),\mathbf{Y} - \mathbf{P}_{M(1,m-1)}\mathbf{Y}> \\
&= <\mathbf{E}_f(m),\mathbf{E}_f(m-1)> \qquad\qquad (4.2.24) \\
&= <\mathbf{E}_f(m-1) - k_m z^{-1}\mathbf{E}_b(m-1),\mathbf{E}_f(m-1)> \\
&= (1 - k_m^2)\ ||\mathbf{E}_f(m-1)||^2
\end{aligned}
$$

where (4.2.7) and (4.2.20) have also been used. The initialization for this recursion is of course $||\mathbf{E}_f(0)||^2 = ||\mathbf{Y}||^2$ from (4.1.45). This demonstrates that the prediction error decreases in squared-length by a factor of $(1 - k_m^2)$ from one lattice stage to the next (that is, as the predictor order is increased by one). Intuitively, one would expect the prediction error to decrease rapidly for the first few stages, and then begin to decrease more and more slowly as stages are added. When this is the case, as often happens in practice, (4.2.24) implies that the first few k_m coefficients will be near unity in magnitude, and then approach zero as the number of stages increases.

This completes the derivation of the lattice filter. It is useful to give concrete expressions for k_m for the particular cases of MMSE and LS criteria. For the MMSE criterion, applying definition (4.1.16) of inner product to (4.2.20),

$$k_m = \frac{E[e_f(T|m-1)e_b(T-1|m-1)]}{E[e_f^2(T|m-1)]}, \qquad (4.2.25)$$

and for the autocorrelation block LS criterion, applying definition (4.1.12),

$$k_m = \frac{\sum_T e_f(T|m-1)e_b(T-1|m-1)}{\sum_T e_f^2(T|m-1)}, \qquad (4.2.26)$$

where in each case the denominator can be replaced by a couple of equivalent forms using (4.2.6). There are a number of other options for calculating the k_m in a block LS context [2].

4.2.1. Advantages of the Lattice Filter

The lattice filter has a number of advantages over the transversal filter which will be discussed in this section.

The optimal PARCOR coefficients k_m, $1 \leqslant m \leqslant n$, are independent of the filter order n (this is why no filter order has been included in the notation). This can be seen from (4.2.20), where the coefficient depends only on the predictors of order m, independent of the final predictor order n. Successive lattice stages may therefore be added or existing stages subtracted without the necessity of recalculating already existing PARCOR coefficients. In contrast, the transversal predictor coefficients $f_m(n)$ all change when the order of the filter is changed.

The PARCOR coefficients can be calculated simply by applying the recursions of (4.2.25) or (4.2.26) to the input data. Once k_m is determined, knowing the two inputs to the m-th stage of the lattice, $e_f(T|m)$ and $e_b(T|m)$ can be determined from (4.2.9) and (4.2.10), and thereafter k_{m+1} can be determined from

(4.2.25-26). Note that unlike the transversal case, no solution of a system of linear equations is explicitly required. (In the next section it is shown that using the lattice formulation is in fact equivalent to performing a Levinson-Durbin solution to the system of linear equations (4.1.53)).

The fact that lower orders of prediction error are developed in this procedure is an advantage in some applications. For example, if it is not known in advance the order of predictor which is required to achieve a given error, successive orders of prediction errors can be calculated while monitoring the norm of the error vector using (4.2.24).

The lattice filter stages form an orthogonalization which speeds up the adaptation of subsequent stages when the stochastic gradient algorithm considered in section 4.5 is used. Since the output of each stage is a prediction error signal, this signal becomes increasingly white as the prediction order increases (see section 3.1.1). In a joint process estimation application, considered in section 4.4, the backward prediction errors for different filter orders are orthogonal after the PARCOR coefficients converge. When using the stochastic gradient algorithm, this results in a substantially faster convergence rate for the joint process coefficients than that achieved using the transversal gradient algorithm. The convergence properties of the adaptive filtering algorithms based on the lattice filter structure will be considered in chapter 7.

The lattice filter is minimum phase (all its zeros lie on or in the interior to the unit circle), if and only if (4.2.22) is satisfied with inequality for all the PARCOR coefficients. This will be established in chapter 5, and is important in speech applications, where the stability of the all-pole filter in the synthesis part of the system will be guaranteed if the PARCOR coefficients are less than unity in magnitude. There is no comparable simple condition in the case of transversal filter coefficients (one standard algorithm for determining whether a transversal filter is minimum phase is in fact based on calculating the PARCOR coefficients and seeing if they are less than unity in magnitude). There is therefore no simple way to ensure that the all-pole filter based on a transversal filter in a feedback loop is stable, with the result that finite precision effects occasionally result in an unstable filter.

If the PARCOR coefficients are to be quantized, the fact that they are bounded by unity is very helpful in establishing the quantizer overload point. No such rule exists for transversal tap coefficients, with the result that overflow of the quantization characteristic can result. These last two points will be developed in chapter 5.

The lattice filter performance using finite word length implementation is much superior to that exhibited by the transversal filter. The PARCOR coefficients have a natural ordering, and we will see in chapter 5 that the different PARCOR coefficients have different statistics, a property which can be exploited in their quantization. Further, incremental changes in PARCOR coefficients (due to quantization error) generally cause much less movement of filter zeroes than incremental changes in the corresponding transversal filter coefficients. This property is especially useful in the context of spectral estimation, and will be considered in detail in chapter 5.

These advantages have resulted in the lattice structure receiving much attention in the context of the applications previously mentioned, and particularly in speech applications.

4.2.2. All-Pole Lattice Filter

Thus far the transversal filter and the lattice filter which have been discussed have been all-zero (FIR) filters. In some applications it is also important to implement an all-pole filter. An example of such an application is in the synthesis of speech (section 2.1), which is often accomplished by exciting an all-pole filter by a pulse train or white noise excitation signal. Another example is the decoder for DPCM (section 2.2). An all-pole filter can be realized by putting a transversal filter in a feedback loop as shown in figure 4-10, where the transfer function will be (problem 4-17)

$$H(z) = \frac{1}{1 - \sum_{j=1}^{n} f_j(n) z^{-j}} . \tag{4.2.27}$$

It is possible to implement an all-pole filter using a similar modification to the lattice filter. To see this, if the transfer function of the lattice filter from the input to the $m-th$ order forward and backward prediction errors are respectively $F_m(z)$ and $B_m(z)$, then it follows directly from (4.2.9) and (4.2.10) that

$$F_m(z) = F_{m-1}(z) - k_m z^{-1} B_{m-1}(z), \quad F_0(z) = 1 \tag{4.2.28}$$

$$B_m(z) = z^{-1} B_{m-1}(z) - k_m F_{m-1}(z), \quad B_0(z) = 1 \tag{4.2.29}$$

where from (4.1.43) and (4.2.3)

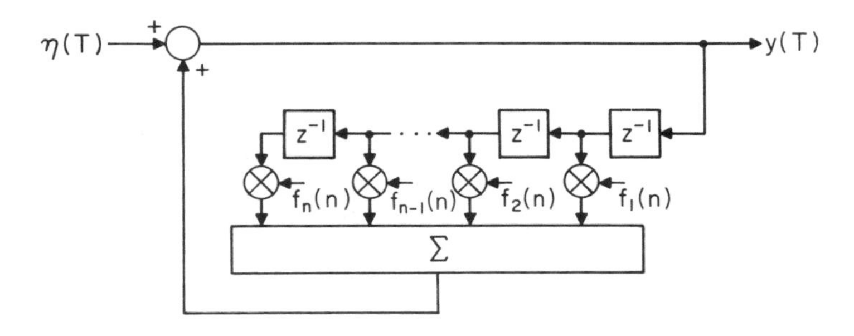

Figure 4-10. An all-pole filter implemented using a transversal filter.

$$F_m(z) = 1 - \sum_{j=1}^{m} f_j(m) z^{-j} \tag{4.2.30}$$

$$B_m(z) = z^{-m} - \sum_{j=1}^{m} b_j(m) z^{-j+1} . \tag{4.2.31}$$

In addition, for future reference it is useful to note (problem 4-18) that from (4.2.5),

$$B_m(z) = F_m(z^{-1}) z^{-m} . \tag{4.2.32}$$

Simply dividing (4.2.28) and (4.2.29) by $F_n(z)$ and rearranging,

$$\frac{F_{m-1}(z)}{F_n(z)} = \frac{F_m(z)}{F_n(z)} + k_m \frac{z^{-1} B_{m-1}(z)}{F_n(z)} \tag{4.2.33}$$

$$\frac{B_m(z)}{F_n(z)} = \frac{z^{-1} B_{m-1}(z)}{F_n(z)} - k_m \frac{F_{m-1}(z)}{F_n(z)} . \tag{4.2.34}$$

These equations are illustrated in figure 11a. In figure 11b it is then shown how these stages can be cascaded to yield an all-pole filter with transfer function $F_n^{-1}(z)$ (identical to (4.2.27)). Note how the output of the last stage is connected back to its input to achieve the desired transfer function.

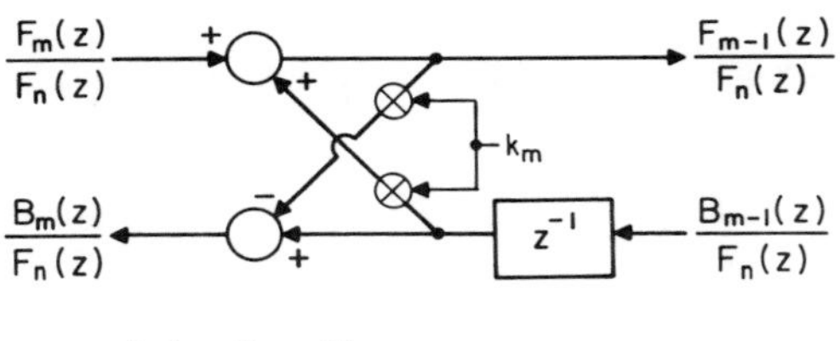

(a) One Stage

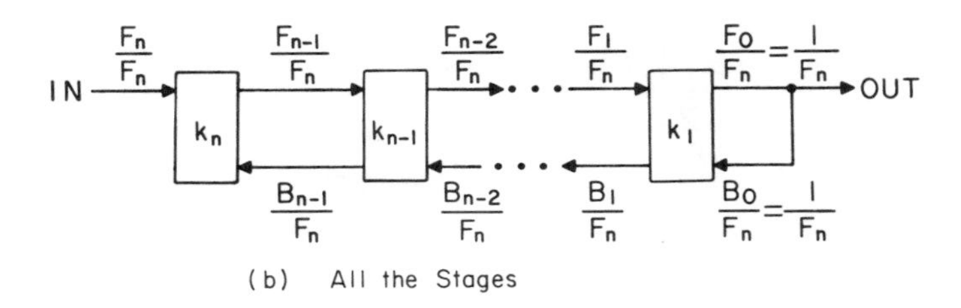

(b) All the Stages

Figure 4-11. Lattice realization of the all-pole filter.

In section 4.3 these results will be extended to yield a lattice filter structure with both poles and zeros.

4.3. CONVERTING AMONG REPRESENTATIONS

Thus far three different representations for the spectral information implicit in an optimum n-th order predictor have been defined. They are the autocorrelation coefficients ϕ_j , $0 \leqslant j \leqslant n$, the predictor coefficients $f_j(n)$, $1 \leqslant j \leqslant n$ (which are also the coefficients of a transversal filter realization of the predictor), and the PARCOR coefficients k_j , $1 \leqslant j \leqslant n$ (which are the coefficients of a lattice filter which generates the prediction error). It is often of interest to convert among these different representations.

The Levinson-Durbin algorithm, which is a computationally efficient method for solving system of equations (4.1.53), starts with the autocorrelation coefficients ϕ_j and generates the coefficients of both the transversal and lattice filter. It is derived in section 4.3.1. This algorithm is widely used for solving the linear equations in the block and recursive LS and the MMSE algorithms described in chapter 3. In section 4.3.2, methods are given for converting from the lattice filter coefficients to the transversal filter coefficients and vice versa. All of these conversions are a straightforward consequence of the lattice filter recursions which have been derived thus far.

4.3.1. Levinson-Durbin Algorithm

Suppose that the autocorrelation coefficients are given, and it is desired to find the transversal and lattice filter coefficients. The Levinson-Durbin algorithm accomplishes this. The key to this algorithm is relating k_m to the forward and backward prediction coefficients. Calculating the inner product in (4.2.20) in terms of the transversal filter coefficients,

$$\begin{aligned} <\mathbf{E}_f(m), z^{-1}\mathbf{E}_b(m)> &= <\mathbf{E}_f(m), z^{-m-1}\mathbf{Y} - \sum_{j=1}^{m} b_j(m) z^{-j}\mathbf{Y}> \\ &= <\mathbf{E}_f(m), z^{-m-1}\mathbf{Y}> \\ &= <\mathbf{Y} - \sum_{j=1}^{m} f_j(m) z^{-j}\mathbf{Y}, z^{-m-1}\mathbf{Y}> \\ &= \phi_{m+1} - \sum_{j=1}^{m} f_j(m)\phi_{m+1-j} \end{aligned} \tag{4.3.1}$$

where the orthogonality of $\mathbf{E}_f(m)$ to $M(1,m)$ has been used, as well as the definition of autocorrelation coefficients (4.1.54). Further, combining (4.2.28) and (4.2.32), and utilizing the mirror image symmetry of the forward and backward predictor coefficients of (4.2.5),

$$F_m(z) = F_{m-1}(z) - k_m z^{-m} F_{m-1}(z^{-1}) \ . \tag{4.3.2}$$

Equating the coefficients of z^{-j} on both sides of this equation (problem 4-19),

$$f_m(m) = k_m \tag{4.3.3}$$

$$f_j(m) = f_j(m-1) - k_m f_{m-j}(m-1) \ , \quad 1 \leqslant j \leqslant m-1 \ . \tag{4.3.4}$$

These equations relate the m-th order predictor coefficients to k_m and the $(m-1)$-st order prediction coefficients. The algorithm is easily completed when $E(m)$ is defined as the norm squared of the m-th order prediction error vector,

$$E(m) = ||\mathbf{E}_f(m)||^2 \ . \tag{4.3.5}$$

Then from (4.2.24)

$$E(m) = (1 - k_m^2)E(m-1), \ \ E(0) = \phi_0 \tag{4.3.6}$$

and from (4.3.1) and (4.2.20),

$$k_m = \frac{\phi_m - \sum_{j=1}^{m-1} f_j(m-1)\phi_{m-j}}{E(m-1)} \ , \tag{4.3.7}$$

which completes the algorithm.

In summary, the algorithm starts with knowledge of the autocorrelation coefficients ϕ_j , $0 \leqslant j \leqslant n$, and allows calculation of both the PARCOR coefficients and the predictor coefficients. It consists of (4.3.3-4), (4.3.6), and (4.3.7). At the first step, $E(0) = \phi_0$ is calculated from (4.3.6), and $k_1 = \phi_1/E(0)$ is determined from (4.3.7). Once k_1 is known, the transversal filter coefficient $f_1(1) = k_1$ follows from (4.3.3). This procedure is simply repeated, obtaining k_m, $E(m)$, and the transversal filter coefficients in that order, until all the PARCOR and transversal filter coefficients are obtained.

4.3.2. Conversion Between Lattice and Transversal Filter

The preceding shows how the lattice and transversal filters can be derived from knowledge of the autocorrelation coefficients. Also of interest in many applications is the ability to convert from a lattice filter to an equivalent transversal filter, or vice versa. This is required, for example, if the optimum predictor is known and it is desired to design a lattice filter equivalent. Another application is to the design of a lattice filter which implements a given all-zero or all-pole transfer function, as discussed in section 4.4.3.

Deriving a transversal filter equivalent to a given lattice filter (or equivalently finding the transfer function of a given lattice filter) is quite straightforward given the recursions (4.2.28-29). One simply iterates these recursions starting at $m=1$ to yield the successively higher order prediction error polynomials (forward and backward). Note that this recursion requires knowledge of the PARCOR coefficients.

Going in the direction of the lattice filter from the transversal filter is eased by (4.3.3), which gives a simple way to determine k_m from the m-th order predictor coefficients,

$$k_m = f_m(m) \ . \tag{4.3.8}$$

Thus, it only remains to find a way to start with the n-th order predictor, and work backwards to find the $(n-1)$-st order predictor, the $(n-2)$-nd order predictor, and so forth. This can be accomplished by solving (4.2.28) and (4.2.29) to find $F_{m-1}(z)$ in terms of $F_m(z)$ and $B_m(z)$,

$$F_{m-1}(z) = \frac{F_m(z) + k_m B_m(z)}{1 - k_m^2} \tag{4.3.9}$$

where $B_{m-1}(z)$ can be found from (4.2.35). Substituting for $F_m(z)$ and $B_m(z)$ from (4.2.30) and (4.2.31) and using (4.2.32), the following recursion similar to the Levinson-Durbin recursion can be derived,

$$f_j(m-1) = \frac{k_m f_{m-j}(m) + f_j(m)}{1 - k_m^2} , \quad 1 \leqslant j \leqslant m-1 . \tag{4.3.10}$$

Thus (4.3.8) and (4.3.9) (or equivalently (4.3.10)) give a complete procedure for finding the lattice filter given a transversal representation (or equivalently the transfer function).

4.4. LATTICE JOINT PROCESS ESTIMATOR

Thus far this chapter has emphasized the prediction application. The lattice filter is also applicable to the joint process estimator, as will be shown in this section. The lattice joint process estimator is obtained by adding additional coefficients and a summation to the lattice predictor. The resulting lattice structure for a joint process estimator is of interest primarily because of its ability to adapt faster than the transversal filter joint process estimator when the stochastic gradient algorithm is used. This faster adaptation is due to the orthogonalization performed by the lattice predictor.

In the case of the joint process estimator, rather than estimating $y(T)$ as a linear combination of $y(T-1)$ through $y(T-n)$, we are estimating another signal $d(T)$ in terms of $y(T)$ through $y(T-n+1)$.

Consistent with our earlier notation, let $\mathbf{D}$ denote the Hilbert space vector which is being estimated,

$$\mathbf{D} \longleftrightarrow d(T) \text{ or } (...,d(-1),d(0),d(1),...) . \tag{4.4.1}$$

As a consequence of (4.1.40), the m-th order joint process estimator of $\mathbf{D}$, the element of $M(0,m-1)$ which is closest to $\mathbf{D}$, is $\mathbf{P}_{M(0,m-1)}\mathbf{D}$, and the estimation error is

$$\mathbf{E}_c(m) = \mathbf{D} - \mathbf{P}_{M(0,m-1)}\mathbf{D} . \tag{4.4.2}$$

It is very simple to develop an order update relation for this estimation error in analogy to (4.2.7). For this purpose, a subspace decomposition analogous to (4.2.12-13) which is applicable is

$$M(0,m) = M(0,m-1) \oplus \{\mathbf{E}_b(m)\} \tag{4.4.3}$$

from which it follows that

$$\mathbf{P}_{M(0,m)}\mathbf{D} = \mathbf{P}_{M(0,m-1)}\mathbf{D} + \mathbf{P}_{\{\mathbf{E}_b(m)\}}\mathbf{D} . \tag{4.4.4}$$

Combining (4.4.2) and (4.4.4),

$$\begin{aligned}\mathbf{E}_c(m+1) &= \mathbf{D} - \mathbf{P}_{M(0,m)}\mathbf{D} \\ &= \mathbf{D} - \mathbf{P}_{M(0,m-1)}\mathbf{D} - \mathbf{P}_{\{\mathbf{E}_b(m)\}}\mathbf{D} \\ &= \mathbf{E}_c(m) - \mathbf{P}_{\{\mathbf{E}_b(m)\}}\mathbf{D}\end{aligned} \quad \textbf{(4.4.5)}$$

and again recognizing that the last projection is a constant, say k_{m+1}^c, times $\mathbf{E}_b(m)$, the order recursion

$$\mathbf{E}_c(m) = \mathbf{E}_c(m-1) - k_m^c \mathbf{E}_b(m-1) \quad (4.4.6)$$

follows, where the initial value is $\mathbf{E}_c(0) = \mathbf{D}$. Further, it is simple to verify by the method of (4.2.19) (problem 4-26) that

$$k_m^c = \frac{<\mathbf{D},\mathbf{E}_b(m-1)>}{||\mathbf{E}_b(m-1)||^2} \ . \quad (4.4.7)$$

By iterating (4.4.4), a different form for the n-th order joint process estimator can be developed,

$$\mathbf{E}_c(n) = \mathbf{D} - \sum_{j=1}^{n} k_j^c \mathbf{E}_b(j-1) \ , \quad (4.4.8)$$

which illustrates that the lattice joint process estimator expresses $\mathbf{E}_c(m)$ in terms of a linear combination of the orthogonal basis vectors for $M(0,m-1)$ given by (4.2.1-2).

The two forms of the joint process estimator given by (4.4.8) and (4.4.6) are shown in figure 4-12. In the "direct form" of figure 4-12a, the time domain equivalent of (4.4.8) is

$$e_c(T|n) = d(T) - \sum_{j=1}^{n} k_j^c e_b(T|j-1) \ . \quad (4.4.9)$$

In this form, the n-th order joint process estimate is expressed as a linear combination of successive higher order backward prediction errors rather than delayed samples of $y(T)$ as in the transversal case. One way to look on this form is as a transversal filter structure in which the delay elements have been replaced by lattice filter stages. In the equivalent recursive form of figure 4-12b, the time domain equivalent of (4.4.6) is

$$e_c(T|m) = e_c(T|m-1) - k_m^c e_b(T|m-1) \ . \quad (4.4.10)$$

The major distinction of this form is that all lower order joint process estimation errors are developed in the course of developing the n-th order error. On the other hand, unlike the direct form, the joint process estimator of $d(T)$ is never directly developed (only the error), which is a disadvantage in some applications.

It is important to note that in the structure of figure 4-12, the coefficients k_m of the lattice filter are dependent only on the properties of the input sequence $y(T)$ (in particular, the correlation function). These coefficients serve only to present a set of orthogonal vectors to the remainder of the joint process estimator. The coefficients k_m^c, on the other hand, depend on both the inputs $y(T)$ and $d(T)$.

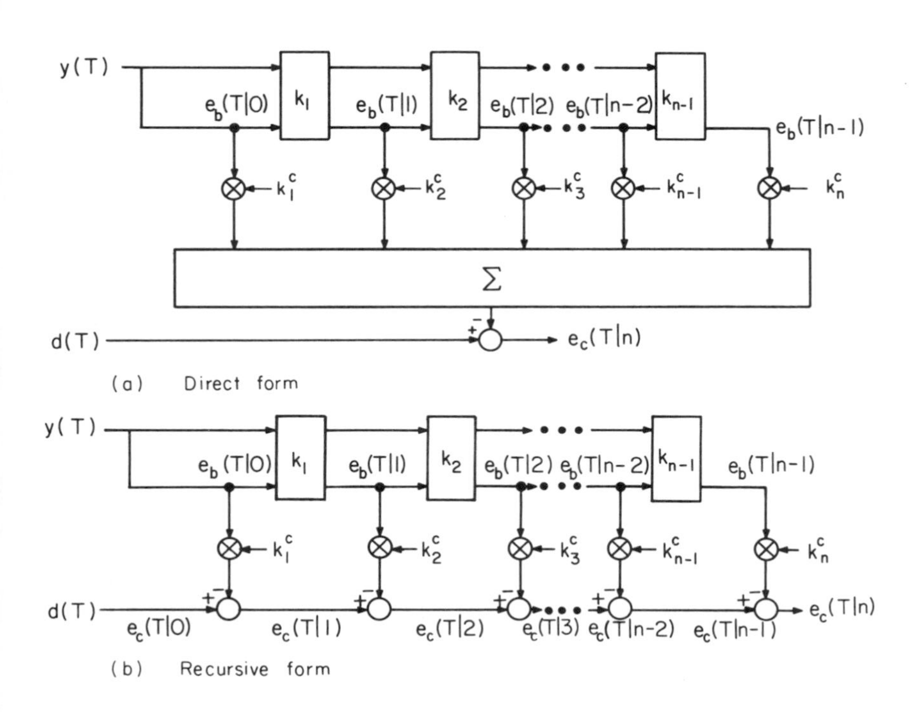

Figure 4-12. Lattice filter realization of a joint process estimator.

4.4.1. Extension of the Levinson-Durbin Algorithm

It is straightforward to extend the Levinson-Durbin algorithm to the joint process estimation case. In particular, a relation for the coefficients k_{m+1}^c in terms of the inner product of **D** and **Y** is easily developed. Define this cross-correlation as

$$\phi_m^c = <\mathbf{D}, z^{-m}\mathbf{Y}> \tag{4.4.11}$$

and note that

$$<\mathbf{D}, \mathbf{E}_b(m)> = <\mathbf{D}, z^{-m}\mathbf{Y} - \sum_{j=1}^{m} b_j(m) z^{-j+1}\mathbf{Y}> . \tag{4.4.12}$$

Applying (4.4.7), it follows directly that

$$k_{m+1}^c = \frac{\phi_m^c - \sum_{j=1}^{m} f_{m-j+1}(m)\phi_{j-1}^c}{E(m)} . \tag{4.4.13}$$

This relation when added to the Levinson-Durbin equations of (4.3.3-7) gives an extended Levinson-Durbin algorithm which is applicable to the joint process estimation case.

4.4.2. Conversion to Transversal Joint Process Estimator

The m-th order joint process estimation error can be written in the form

$$\mathbf{E}_c(m) = \mathbf{D} - \sum_{j=0}^{m-1} c_{j+1}(m) z^{-j} \mathbf{Y} \ , \tag{4.4.14}$$

where the $c_j(m)$, $1 \leqslant j \leqslant m$, are the coefficients of a transversal filter realization of the estimator. It is of interest to find relations for converting between the lattice and transversal forms of the joint process estimator. This is equivalent to converting between the coefficients k_j^c , $1 \leqslant j \leqslant m$, and $c_j(m)$, $1 \leqslant j \leqslant m$.

Defining the transfer function of the transversal joint process estimator as

$$C_m(z) = \sum_{j=0}^{m-1} c_{j+1}(m) z^{-j} \ , \tag{4.4.15}$$

then the equivalent to the order update of (4.4.10) is

$$C_m(z) = C_{m-1}(z) + k_m^c B_{m-1}(z) \ . \tag{4.4.16}$$

Equating the coefficients of each power of z,

$$c_m(m) = k_m^c \tag{4.4.17}$$

$$c_j(m) = c_j(m-1) - k_m^c b_j(m-1) \ , \ 1 \leqslant j \leqslant m-1 \ . \tag{4.4.18}$$

Thus, given the lattice representation, these recursive relations allow us to calculate the transversal representation for all orders of estimator, starting with $m=1$. Conversely, given an n-th order transversal filter joint process estimator, the lattice representation can be obtained by iterating the equations

$$k_m^c = c_m(m) \tag{4.4.19}$$

$$c_j(m-1) = c_j(m) + k_m^c b_j(m-1) \ , \ 1 \leqslant j \leqslant m-1 \tag{4.4.20}$$

starting at $m = n$.

4.4.3. Lattice Filter for Arbitrary Transfer Function

In the last section, it was shown that a lattice joint process estimator can be derived from a lattice predictor by adding additional coefficients and a summation. This result plus the fact that a lattice filter can also realize an all-pole filter suggest that perhaps the lattice filter can realize an arbitrary rational transfer function with both poles and zeros by adding coefficients and a summation to an all-pole lattice. This is the case, as was first pointed out in [3].

Suppose that the desired transfer function has the form

$$G(z) = \frac{\sum_{j=0}^{n} c_{j+1}(n+1)z^{-j}}{1 - \sum_{j=1}^{n} f_j(n)z^{-j}} , \qquad \textbf{(4.4.21)}$$

which is arbitrary except that the order of the numerator is no larger than n, the order of the denominator. From (4.2.30) and (4.4.15) this expression can be written in the form

$$G(z) = \frac{C_{n+1}(z)}{F_n(z)} , \qquad (4.4.22)$$

where $F_n(z)$ and $C_{n+1}(z)$ are respectively an n-th order predictor and $(n+1)$-st order joint process estimator. This identification of a predictor and joint process estimator is artificial, since the object was not to solve an estimation problem. However, it is useful to pretend that there is an estimation problem for which these particular given polynomials are the solution, in order to be able to apply the earlier results.

Given the denominator polynomial $F_n(z)$, by the procedure of section 4.3.2 the corresponding forward and backwards predictors of all lower orders can be generated, these being represented by the polynomials $F_m(z)$ and $B_m(z)$, $1 \leqslant m \leqslant n$. Further, this procedure gives the PARCOR coefficients for a lattice realization of the denominator transfer function of (4.4.21). This suggests using the all-pole lattice filter form to realize the poles of this transfer function.

The procedure of (4.4.19-20) also gives us a way to generate a set of coefficients k_m^c for a lattice realization of the transfer function $C_{n+1}(z)$. Furthermore, from (4.4.9) these coefficients satisfy the relation

$$C_m(z) = \sum_{j=1}^{m} k_j^c B_{j-1}(z) , \qquad (4.4.23)$$

and substituting into (4.4.22),

$$G(z) = \sum_{j=1}^{n+1} k_j^c \frac{B_{j-1}(z)}{F_n(z)} . \qquad (4.4.24)$$

The transfer functions in this relation are precisely those generated at intermediate stages of the all-pole lattice filter of figure 4-11. Thus, (4.4.24) can be realized by appending taps to these intermediate stages as shown in figure 4-13. One can think of the resulting structure as adding tap-weights and a summation, much in the style of a transversal filter, to an all-pole lattice in order to realize the desired zeros.

To summarize again the design procedure, a lattice filter realization of the given denominator polynomial $F_n(z)$ is designed using the procedure of section 4.3.2. Then (4.4.19-20) is used to generate a set of coefficients of a lattice joint process estimator from the numerator polynomial $C_{n+1}(z)$. Finally, these two sets of coefficients are combined in the structure of figure 4-13, in which an all-pole lattice filter is combined with taps and a summation to realize the given pole-zero transfer function. A normalized version of this filter has also been

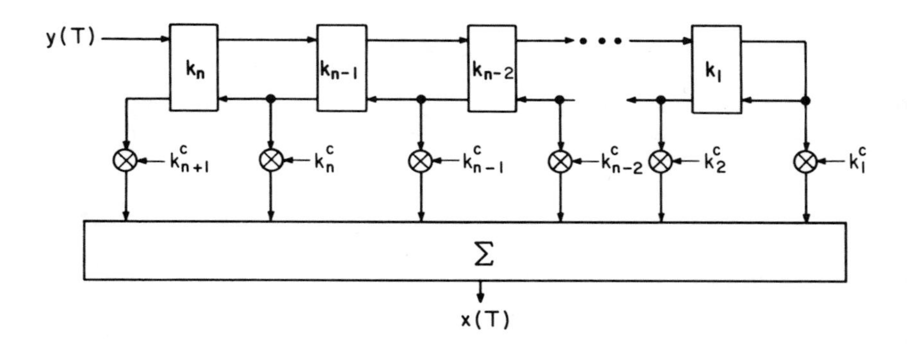

Figure 4-13. Arbitrary transfer function realized using lattice filter.

proposed [4] and a generalization of the structure derived here in which the delay elements are replaced by arbitrary discrete- or continuous-time all-pass filters has also been proposed [5]. The lattice filters have excellent sensitivity properties, and as such are members of a larger class of filters which includes the wave digital filter [6].

4.5. LATTICE STOCHASTIC GRADIENT ALGORITHMS

This section will describe the lattice SG adaptation algorithms for a predictor and joint process estimator which are analogous to the algorithms of section 3.2 but based on the lattice filter structures rather than the transversal [7, 2, 8].

In the transversal filter, the gradient of the mean-square n-th order prediction error was evaluated. In the case of the lattice filter, there is no simple expression for the prediction error at the output of the n-th stage as a function of $k_1, \cdots, k_n$. Since we cannot calculate the gradient of the squared error at the output of the n-th stage, we calculate the gradient of the squared error at the output of the m-th stage with respect to k_m instead. Fortunately, the previous sections have shown that minimizing the error at the output of the n-th stage is equivalent to minimizing the error at the output of each lower order stage with respect to the PARCOR coefficient at that stage, as long as all the previous stages have optimal PARCOR coefficients.

The lattice gradient algorithm which is most popular adapts $k_m(T)$ to minimize the sum of the forward and backward errors,

$$E\left[e_f^2(T|m) + e_b^2(T|m)\right] . \tag{4.5.1}$$

We know in the stationary case that minimizing one of these quantities is equivalent to minimizing the other or the sum, but for the SG algorithm the effect of minimizing the sum of both errors is to include data from both the forward and backward prediction errors in the adaptation algorithm. Following the method of section 3.2, the way to develop a SG algorithm is to take the derivative of the quantity

$$e_f^2(T|m) + e_b^2(T|m) \tag{4.5.2}$$

with respect to k_m, where from the order recursions of (4.2.9-10) this quantity can be expressed in terms of $e_f(T|m-1)$ and $e_b(T-1|m-1)$ (which are not a function of k_m) and k_m. This derivative, calculated using the current estimate for k_m, $k_m(T)$, is then subtracted from the current estimate to yield the new estimate $k_m(T)$. The resulting algorithm is (problem 4-28)

$$\begin{aligned} k_m(T+1) = \{1 - \beta_m(T)[e_b^2(T-1|m-1) + e_f^2(T|m-1)]\}k_m(T) \\ + 2\beta_m(T)e_f(T|m-1)e_b(T-1|m-1) \ . \end{aligned} \tag{4.5.3}$$

The current estimate of the PARCOR coefficient, $k_m(T)$ is used in (4.2.9-10) to obtain the prediction errors of order m at the current time T. It would be feasible to use the new estimate $k_m(T+1)$ from (4.5.3) to calculate the higher order prediction errors, but it has been established experimentally that the method described here gives better performance (smaller asymptotic MSE). The adaptation constant for the m-th stage, $\beta_m(T)$, has been made a function of both the stage m of the filter and T in order to use the normalization strategy described in section 3.2.2.3, thereby making the speed of adaptation relatively independent of input signal levels. Hence, normalize the step-size by an estimate of the sum of the $(m-1)$-st order prediction error variances (which are the input signals to the m-th stage),

$$\beta_m(T) = \frac{1}{E(T|m-1)} \ . \tag{4.5.4}$$

$E(T|m-1)$ is given by

$$\begin{aligned} E(T+1|m-1) = (1 - \beta)E(T|m-1) \\ + e_f^2(T|m-1) + e_b^2(T-1|m-1) \end{aligned} \tag{4.5.5}$$

and β is a constant which in conjunction with initial condition $E(0|m-1)$ controls the speed of this adaptation. As $T\rightarrow\infty$ (problem 4-29),

$$E[E(T|m-1)] \rightarrow \frac{1}{\beta}\{E[e_b^2(T|m-1)] + E[e_f^2(T|m-1)]\} \ , \tag{4.5.6}$$

and hence

$$E[\beta_m(T)] \rightarrow \frac{\beta}{E[e_b^2(T|m-1)] + E[e_f^2(T-1|m-1)]} \tag{4.5.7}$$

as long as the input signals are stationary. Hence, the constant β controls the speed of convergence of both estimates (4.5.3) and (4.5.5) simultaneously.

The preceding approach is similar to that used in chapter 3 for the transversal filter. An alternative approach is to find the optimum k_m which minimizes (4.5.1), and then find a recursive estimate for the quantities necessary to calculate this optimum. The value of k_m which minimizes (4.5.1) is (problem 4-30)

$$k_{m,\text{opt}} = \frac{2E[e_f(T|m-1)e_b(T-1|m-1)]}{E[e_f^2(T|m-1)] + E[e_b^2(T-1|m-1)]} \ . \tag{4.5.8}$$

This is of course equivalent to (4.2.25). Recursively estimating the numerator of (4.5.8) gives

$$K_m(T+1) = (1-\beta)K_m(T) + 2e_f(T|m-1)e_b(T-1|m-1) \tag{4.5.9}$$

where we have defined a new "unnormalized" version of the PARCOR coefficient as $K_m(T)$. An estimate of the denominator in (4.5.8) has already been developed in (4.5.5), and hence the final estimate of the PARCOR coefficient is

$$k_m(T) = \frac{K_m(T)}{E(T|m-1)} \ . \tag{4.5.10}$$

In fact, these two approaches to finding an adaptation algorithm are equivalent. It can be shown by algebraic manipulation (problem 4-31) that (4.5.3) and (4.5.5) are identical to (4.5.9) and (4.5.10).

Yet another approach to deriving a SG algorithm is to allow the two PARCOR coefficients in (4.2.9-10), k_m^b and k_m^f, to adapt independently. In this case, k_m^b is adapted to minimize $e_f^2(T|m)$, resulting in

$$\begin{aligned} k_m^b(T+1) &= [1 - \beta e_b^2(T-1|m-1)]k_m^b(T) \\ &\quad + \beta e_f(T|m-1)e_b(T-1|m-1) \ , \end{aligned} \tag{4.5.11}$$

and k_m^f can be adapted to minimize $e_b^2(T|m)$, as in

$$\begin{aligned} k_m^f(T) &= [1 - \beta e_f^2(T|m-1)]k_m^f(T-1) \\ &\quad + \beta e_f(T|m-1)e_b(T-1|m-1) \ . \end{aligned} \tag{4.5.12}$$

If the input is stationary, both (4.5.11) and (4.5.12) use the same error criterion and are in this sense equivalent. As before, the adaptation constant β in (4.5.11) can be normalized by an estimate of the mth-order forward residual energy, and the β in (4.5.12) can be normalized by an estimate of the backward residual energy.

Aside from the gradient algorithms given here which apply to the all-zero lattice filter, gradient algorithms have also been derived for pole-zero or autoregressive moving-average (ARMA) lattice structures [9, 10]. These algorithms are an extension of those given here.

The LMS gradient algorithm is easily extended to the lattice joint process estimator as well. In this case the lattice prediction filter is adapted in exactly the same fashion as before, taking into account the input signal $y(T)$ to maintain the orthogonality of the different orders of backward prediction errors. In addition to this adaptation, the joint process estimator coefficients which multiply the backward prediction errors can also be adapted by a gradient algorithm.

Referring to the non-order recursive form of the lattice joint process estimator shown in figure 12a, a gradient algorithm for the coefficient k_m^c is obtained by writing

$$
\begin{aligned}
k_m^c(T+1) &= k_m^c(T) - \frac{\beta}{2}\frac{\partial}{\partial k_m^c}[e_c^2(T|n)] \\
&= k_m^c(T) + \beta e_c(T|n)e_b(T|m-1) \,.
\end{aligned}
\tag{4.5.13}
$$

Similarly for the recursive form shown in figure 12b k_m^c can be adapted to minimize the mean square mth stage joint-process estimation error $E[e_c^2(T|m)]$. From (4.4.5), the value of k_m^c which minimizes $E[e_c^2(T|m)]$ is

$$
k_{m,opt}^c = \frac{E[d(T)e_b(T|m-1)]}{E[e_b^2(T|m-1)]} \,.
\tag{4.5.14}
$$

A gradient algorithm for k_m^c is given by

$$
\begin{aligned}
k_m^c(T+1) &= [1 - \beta_m(T)e_b^2(T|m-1)]k_m^c(T) \\
&\quad + \beta_m(T)e_c(T|m-1)e_b(T|m-1).
\end{aligned}
\tag{4.5.15}
$$

Notice the similarities between (4.5.13) and the transversal gradient algorithm (3.2.4) and between (4.5.14) and the lattice predictor gradient algorithm (4.5.3). Again, β can be normalized by an estimate of $E[e_b^2(T|m-1)]$, as in

$$
\beta_m(T) = \frac{1}{B(T|m-1)}
\tag{4.5.16}
$$

$$
B(T+1|m-1) = (1-\beta)B(T|m-1) + e_b^2(T|m-1) \,.
\tag{4.5.17}
$$

The primary advantage to using the lattice joint process estimator is its faster adaptation speed in comparison to the transversal filter joint process estimator. This faster convergence of the joint process coefficients is due to the fact that after convergence of the lattice predictor portion, the backward prediction errors of different orders used to form the joint process estimate are uncorrelated. Thus, the slowdown in convergence due to the correlated signal samples as discussed in chapter 3 is partially mitigated. On the other hand, the lattice predictor must converge first, a step which is not required in the transversal joint process estimator, and thus there remains a question of overall adaptation speed. This is considered in chapter 7, where it is shown that the lattice form of the joint process estimator does indeed adapt considerably faster than the transversal form.

4.6. APPLICATIONS OF LATTICE FILTERS

Lattice filters have found broad application, particularly in speech applications where speed of adaptation is critical. In this section, specific applications in LPC, ADPCM, and joint process estimation will be considered.

4.6.1. Lattice Linear Predictive Coder

It has become almost universal to use the lattice filter for the analysis and synthesis of speech using the autocorrelation method of linear predictive coding [11, 2, 12]. The lattice filter has several advantages for this application. One advantage is that the solution of a set of linear equations using the Levinson-

Durbin algorithm is eliminated, reducing the complexity of the software or hardware to perform the analysis. A more important advantage is the reduced precision required in the calculations, since the PARCOR coefficients require considerably less precision than the autocorrelation coefficients which are generated in the transversal filter autocorrelation method. In addition, as discussed in chapter 5, the PARCOR coefficients require fewer bits of precision than the transversal filter coefficients for a given subjective quality of synthesized speech, resulting in a lower bit rate for the transmission or storage of encoded speech. Finally, unlike the transversal filter coefficients, the PARCOR coefficients have a well defined maximum value of unity (in absolute value), making the scaling of the quantization a non-issue and ensuring the stability of the synthesis filter (see chapter 5).

The lattice filter LPC analyzer can be implemented as follows. First the given data sequence $y(T)$ is windowed for a given block, resulting in finite summations in all the subsequent calculations. Often the blocks are overlapped so that all the given data samples are represented with a large weight in one adjacent block. The next step is to use (4.2.26) to generate all the PARCOR coefficients. Using the given data, $e_f(T|0) = y(T)$ and $e_b(T|0) = y(T)$ are known for the duration of the block. From (4.2.26) the first PARCOR coefficient k_1 can be calculated, and using this value for k_1, $e_f(T|1)$ and $e_b(T|1)$ can be determined from (4.2.9-10). Again (4.2.26) allows us to determine k_2, and so forth. Note that in each stage of the calculation, the entire time sequence (vector) for the block corresponding to the successive higher orders of forward and backward prediction error are being calculated. After each PARCOR coefficient is calculated and the next higher order of prediction errors are calculated, the lower order prediction signals can be discarded from memory.

The synthesis of the speech proceeds as in figure 2-3a using the all-pole lattice filter structure illustrated in figure 4-11.

4.6.2. Lattice Adaptive Differential PCM

The predictor in an ADPCM encoder and decoder can be replaced by a lattice filter. For this case the substitution is not quite straightforward, since the lattice filter generates the forward prediction *error*, whereas in ADPCM what is desired is the prediction itself. To develop the prediction, iterate (4.2.7),

$$\mathbf{E}_f(n) = \mathbf{Y} - \sum_{j=1}^{n} k_j z^{-1} \mathbf{E}_b(j-1) \tag{4.6.1}$$

where it is evident that the summation term is the optimum forward predictor of order n. Therefore, the predictor in ADPCM in figure 2-4b can be written in the time domain as

$$\hat{y}(T) = \sum_{j=1}^{n} k_j e_b(T-1|j-1) \; . \tag{4.6.2}$$

The input to the lattice filter predictor in the ADPCM encoder and decoder is $\tilde{y}(T)$, the same as the decoder output. This signal is given by (2.2.2). This leads to the ADPCM predictor as shown in figure 4-14. This is actually similar to the lattice joint process estimator, except that the joint process coefficients are the same as the PARCOR coefficients and the output is taken from the output of the delays internal to the lattice stages rather than the input. Of course, the multiplication of the backward prediction errors by the PARCOR coefficients only need be done once. This illustrates how the lattice filter sometimes has to be modified to fit a particular application.

The performance improvement due to a lattice filter predictor has been estimated [13] using simulations. The results are disappointing (at least for the lattice filter advocate) in that no appreciable improvement in performance as measured by the mean-square error of the ADPCM encoder was observed. The reason for this result can be explained as follows. The overall mean-square error of the encoder is improved indirectly by the reduction in step-size in the quantizer which can be achieved when the predictor succeeds in reducing the quantizer input signal. But this mean-square error improvement is only realized to the extent that the adaptive quantizer can track the changing prediction error variance, and if the predictor adapts more rapidly than the quantizer there is no advantage. Further, speech tends to be characterized by periods of relatively slow changes in statistics, which even a transversal filter can track, and periods of such rapid change (such as the onset of a plosive) that even a lattice predictor cannot track it.

4.6.3. Lattice Joint Process Estimator

The lattice joint process estimator has an interesting property which reduces its value in some applications and increases it in others. Namely, the transfer function of the filter depends on both the lattice filter coefficients k_m, $1 \leqslant m \leqslant n$, and the joint process coefficients k_m^c, $1 \leqslant m \leqslant n$. The

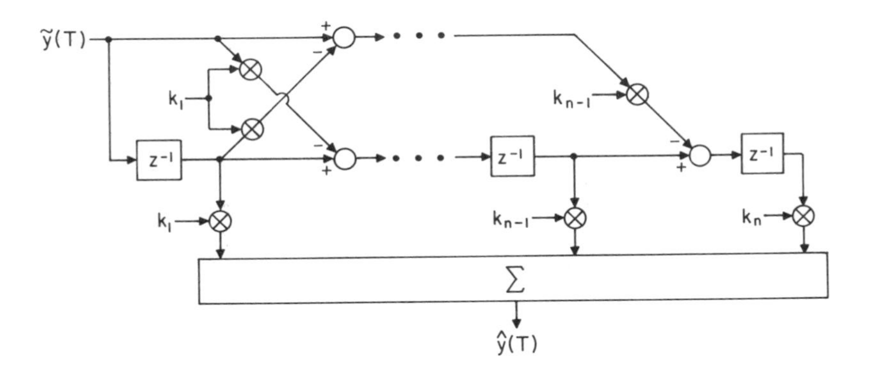

Figure 4-14. Lattice filter predictor for ADPCM.

former coefficients are adapted or chosen so that the successive orders of backward prediction error are orthogonal, and are thus chosen only on the basis of the reference input signal $y(T)$ or its statistics (depending on what criterion is chosen). The joint process coefficients are chosen to achieve the desired transfer function, or equivalently the desired joint process estimation. As the filter is adapting, there are two sets of coefficients adapting, one on the basis of the reference signal alone, and the other on the basis of both the reference and primary signals.

An example of where this is a consideration is in the echo canceler, where the object is to adapt the transfer function of the filter to be the same as the transfer function of the echo channel. Generally the echo channel is varying extremely slowly if at all, and hence the purpose of the adaptation is to adjust to an unknown transfer function, but not so much to track a changing transfer function. If the statistics of the reference signal are changing, as for a voice canceler, the lattice filter coefficients will be changing to track those statistics, and the joint process coefficients will also have to be changing to keep the overall transfer function fixed.

This illustrates a disadvantage of the lattice filter joint process estimator for this application. The transversal filter when used for the same application will require changes in the filter coefficients only at the speed that the echo channel is changing, since those coefficients directly reflect the echo channel. Putting it another way, the transversal filter coefficients only have to track the slowly varying echo channel, while both the lattice and joint process coefficients of the lattice filter joint process estimator have to track the much more rapidly varying reference signal input statistics. It should be emphasized, however, that this disadvantage applies only to the lattice joint process estimator using a gradient algorithm, and not the lattice joint process estimator using an LS criterion as described in chapter 6.

The primary benefit of using a lattice filter joint process estimator is that the successive orders of backward prediction error are orthogonal after convergence of the lattice coefficients, and therefore the joint process coefficients adapt more quickly since the interaction of the coefficient adaptations is eliminated. This advantage is lost when the successive delays of the input signal are orthogonal, or in other words the input samples are uncorrelated. For this case the optimum lattice coefficients are zero, and the lattice filter is equivalent to the transversal filter. The presence of the lattice filter in this situation not only unnecessarily increases the complexity of the filter, but also increases the excess error after adaptation due to the fluctuation of the lattice coefficients around zero.

An example where the lattice filter joint process estimator would not be beneficial would be the echo canceler for data transmission. In this application, particularly where scrambling of the data signal is included, the reference signal samples are essentially uncorrelated.

There are other applications of noise canceling, where the input signal samples are highly correlated, and a lattice filter joint process estimator will improve the convergence speed.

Another example of an application favorable to the lattice filter approach is the equalizer for data transmission [14]. Even though the data symbols may be uncorrelated (as in the data echo canceler), the intersymbol interference at the receiver causes the equalizer input samples to be correlated. From (2.3.1), assuming that the data symbols $d(T)$ and noise samples $n(T)$ are uncorrelated, the correlation of equalizer input samples is given by

$$E[y(T)y(T+i)] = \sigma_d^2 \sum_j h_j h_{j-i} + \sigma_n^2 \delta_i \tag{4.6.3}$$

where σ_d^2 is the variance of the data symbols $d(T)$ and σ_n^2 is the variance of the noise. The summation in (4.6.3) is the autocorrelation function of the samples of the pulse waveform. This illustrates that as the intersymbol interference gets more severe (as measured by the autocorrelation function of the pulse waveform), the correlation of successive samples gets larger. Where there is intersymbol interference, the lattice filter equalizer will result in faster convergence.

4.7. CONCLUSIONS

This chapter has derived an alternative structure to the transversal filter, namely the lattice filter. The lattice filter is used in many of the same applications as the transversal filter, but has many advantageous properties. Not the least of these is the faster adaptation which can be obtained in an adaptive filtering context.

In chapter 5, the guaranteed stability properties of the lattice structure for realizing an all-pole filter will be established. In addition, the sensitivity of the transversal and lattice filter transfer functions to the filter coefficients will be characterized. Then in chapter 6, recursive least-square algorithms will be derived for both the transversal and lattice structures.

4.8. FURTHER READING

This chapter has emphasized the application of lattice filters to estimating the parameters of an autoregressive model, as in linear predictive coding of speech. The lattice filter can also be applied, although with greater difficulty, to the estimation of moving average and autoregressive moving average spectra [15]. An extensive treatise on the lattice filter is given in [16] and an excellent perspective on the history of linear filtering is given in [17].

PROBLEMS

4-1 Verify that the definitions of linear spaces given by (4.1.16-20) and by (4.1.21-26) satisfy the axioms of (4.1.7-15).

4-2 Verify that the definitions of inner product of (4.1.33) and (4.1.34) satisfy the axioms of (4.1.29-32).

4-3 Prove (4.1.39) based on the statement of the projection theorem.

4-4 Prove (4.1.40) based on the statement of the projection theorem.

4-5 Show that for both of the definitions of the delay operator z^{-i} of (4.1.41) and (4.1.50) that

$$z^{-i}(z^{-j}\mathbf{Y}) = z^{-(i+j)}\mathbf{Y} .$$

4-6 Show that (4.1.45) is valid for both definitions of vector $\mathbf{Y}$ and z^{-i}.

4-7 Show that for both spaces of interest, the definition of autocorrelation given by (4.1.54) leads to the following properties:

a. $\phi_{-j} = \phi_j$.

b. $<z^{-i}\mathbf{Y},z^{-j}\mathbf{Y}> = \phi_{j-i}$.

4-8 Establish (4.2.5); namely, that the optimal forward and backward predictors have coefficients which are mirror images of one another.

4-9 Establish (4.2.6); namely, that the optimal forward and backward prediction error vectors have the same norm.

4-10 Show that, as claimed in (4.2.13), $z^{-1}\mathbf{E}_b(m)$ is orthogonal to subspace $M(1,m)$.

4-11 In the following parts state whether the condition is TRUE or FALSE:

a. $\mathbf{P}_{M(1,10)}\mathbf{Y}$ is orthogonal to $z^{-7}\mathbf{Y}$.

b. $\mathbf{P}_{M(0,5)}\mathbf{Y}$ is a vector with nonzero length.

c. $z^{-1}\mathbf{Y}$ is in the subspace $M(1,10)$.

d. $\mathbf{E}_b(5)$ is orthogonal to $\mathbf{E}_b(6)$.

e. $\mathbf{E}_f(5)$ is orthogonal to $\mathbf{E}_f(6)$.

f. $\mathbf{E}_b(5)$ is orthogonal to $\mathbf{E}_f(5)$.

g. $\mathbf{E}_b(5)$ is orthogonal to $z^{-1}\mathbf{E}_f(4)$.

h. $\mathbf{E}_f(5) = \mathbf{E}_b(5)$.

i. $<\mathbf{E}_b(5),\mathbf{E}_f(3)) = <\mathbf{E}_b(5),\mathbf{P}_{M(1,3)}\mathbf{Y}>$

4-12 Prove the Schwarz inequality, (4.2.21), using the definition of an inner product.

4-13 Carefully justify the steps in the derivation of (4.2.23).

4-14 Establish that k^f_{m+1} is given by (4.2.20).

4-15 Carefully justify the steps in the derivation of (4.2.24).

4-16 For a given data signal $y(T)$, specify the steps in calculating the PARCOR coefficients for the autocorrelation LS method.

4-17 Show that the transfer function of the filter shown in figure 4-10 is given by (4.2.27).

4-18 Establish (4.2.32).

4-19 Establish that (4.3.3-4) follows from (4.3.2).

4-20 Design a two-pole lattice filter with poles at $Re^{j\theta}$ and $Re^{-j\theta}$.

4-21 An input process is autoregressive with complex conjugate poles at $z = 0.5j$ and $z = -0.5j$. Find a lattice filter realization of the optimum third order forward prediction error $e_f(T|3)$.

4-22 Find the PARCOR coefficients for the predictor derived in Problem 3-18.

4-23 Show that the one multiplier version of the all-pole lattice filter drawn in figure P4-1 has transfer function

$$\frac{\prod_{i=1}^{n}(1 - k_i)}{A_n(z)} .$$

Hint: Define

$$\pi_m = \begin{cases} 1, & m = n \\ \prod_{i=m+1}^{n} (1 - k_i), & m<n \end{cases}$$

$$\hat{A}_m = \pi_m A_m$$

$$\hat{B}_m = \pi_m B_m$$

and then derive a recursion for $\hat{A}_m$ in terms of $\hat{A}_{m+1}$ and $\hat{B}_m$; derive a recursion for $\hat{B}_{m+1}$ in terms of $\hat{B}_m$ and $\hat{A}_m$; substitute from the first recursion to obtain $\hat{B}_{m+1}$ in terms of $\hat{B}_m$ and $\hat{A}_{m+1}$.

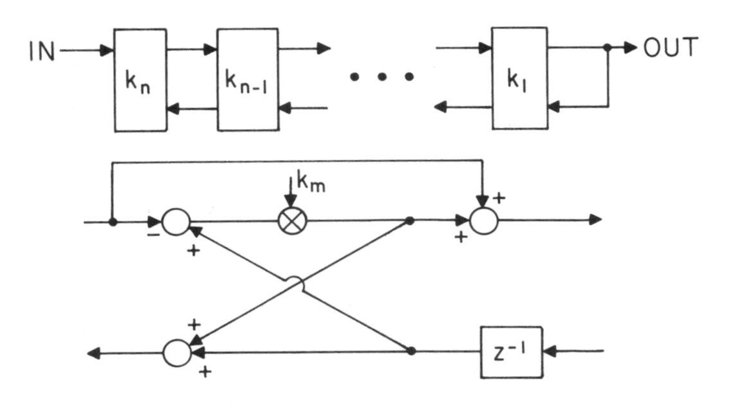

Figure P4-1. One multiplier lattice filter.

4-24 A wide-sense stationary signal has autocorrelation

$$\Phi(z) = \sum_k \phi_k z^{-k} = \frac{A}{(1 - \alpha z)(1 - \alpha z^{-1})} ,$$

where α and A are real, $1 < \alpha < 1$, and the region of convergence includes the unit circle. Find

a. The autocorrelation ϕ_k,

b. The predictor coefficients $f_j(n)$ of an n-th order predictor (hint: the results of a. are not needed), and

c. The PARCOR coefficients $k_j,\ 1 \leqslant j \leqslant n$.

4-25 Write out the equations necessary to solve the Levinson-Durbin algorithm up through a third order predictor. Be certain that the equations are listed in the proper order so that all unknowns have been previously computed.

4-26 Verify (4.4.7), the equation for the joint process estimation coefficients.

4-27 Equations (4.4.18) and (4.4.20) are written in terms of backward predictor coefficients. Rewrite them in terms of forward predictor coefficients.

4-28 Verify (4.5.3), one stochastic gradient algorithm for a lattice predictor.

4-29 Take the expected value of (4.5.5) and verify that (4.5.6) is valid.

4-30 Verify (4.5.8). Why is this relation equivalent to (4.2.25) when all the previous lattice stages have optimal PARCOR coefficients? If the previous coefficients are not optimal, are (4.5.8) and (4.2.25) equivalent?

4-31 Show that (4.5.3) and (4.5.5) are equivalent to (4.5.9-10).

4-32 Derive a stochastic gradient lattice algorithm which uses the minimization of $E[e_f^2(T|n)]$, $1 \leqslant n \leqslant N$, as the optimality criterion. Assume that $k_n^b(T) = k_n^f(T)$ for all n and T.

REFERENCES

1. A.W. Naylor and G.R. Sell, *Linear Operator Theory in Engineering and Science,* Holt, Rinehart and Winston, Inc., New York (1971).

2. J. Makhoul, "Stable and Efficient Lattice Methods for Linear Prediction," *IEEE Trans. on ASSP* **ASSP-25** pp. 423-428 (Oct. 1977).

3. A.H. Gray, Jr. and J.D. Markel, "Digital Lattice and Ladder Filter Synthesis," *IEEE Trans. Audio Electroacoust.* **AU-21** pp. 491-500 (Dec. 1973).

4. A.H. Gray, Jr. and J.D. Markel, "A Normalized Digital Filter Structure," *IEEE Trans. ASSP* **ASSP-23** pp. 268-277 (June 1975).

5. D. G. Messerschmitt, "A Class of Generalized Lattice Filters," *IEEE Trans. on ASSP* **ASSP-28** pp. 198-204 (April 1980).

6. P.P. Vaidyanathan and S.K. Mitra, "Low Passband Sensitivity Digital Filters: A Generalized Viewpoint and Synthesis Procedures," *IEEE Proceedings* **72**(4) p. 401 (April 1984).

7. L. J. Griffiths, "A Continuously-Adaptive Filter Implemented as a Lattice Structure ," *Proc. 1977 IEEE ICASSP*, pp. 683-686 (May 1977).

8. J. Makhoul, "A Class of All-Zero Lattice Digital Filters: Properties and Applications ," *IEEE Trans. on ASSP* **ASSP-26** pp. 304-314 (Aug. 1978).

9. S. Horvath, Jr., "Lattice Form Adaptive Recursive Digital Filters: Algorithms and Applications ," *IEEE Conf. on Circuits and Systems*, pp. 128-133 (1980).

10. D. Parikh, N. Ahmed, and S. D. Stearns, "An Adaptive Lattice Algorithm for Recursive Filters," *IEEE Trans. on ASSP* **ASSP-28** pp. 110-111 (Feb. 1980).

11. L.R. Rabiner and R.W. Schafer, *Digital Processing of Speech Signals,* Prentice-Hall, Englewood Cliffs, N.J. (1978).

12. J. Makhoul and L.K. Cosell, "Adaptive Lattice Analysis of Speech," *IEEE Trans. on ASSP* **ASSP-29**(3) p. 654 (June 1981).

13. M.L. Honig and D.G. Messerschmitt, "Comparison of adaptive linear prediction algorithms in ADPCM," *IEEE Trans. on Communications* **COM-30**(7) pp. 1775-1785 (July 1982).

14. E.H. Satorius and S.T. Alexander, "Channel Equalization Using Adaptive Lattice Algorithms," *IEEE Trans. Communications* **COM-27** p. 899 (June 1979).

15. B. Friedlander, "Lattice Methods for Spectral Estimation," *Proc. IEEE* **70**(9) p. 990 (Sept. 1982).

16. B. Friedlander, "Lattice Filters for Adaptive Processing," *Proceedings of the IEEE* **70**(8) pp. 829-867 (August 1982).

17. T. Kailath, "A View of Three Decades of Linear Filtering Theory," *IEEE Trans. on Inform. Theory* **IT-20** pp. 146-181 (March 1974).

5

SENSITIVITY AND STABILITY

In this chapter, two related issues which both relate to the transversal and lattice filter structures are discussed. The first is the condition under which the filters can be guaranteed to be minimum phase, and the second is the sensitivity of the filter transfer functions to small perturbations of the filter coefficients.

These two issues are of special interest in speech encoding. It was seen in chapter 2 that a common method of coding the spectral information in LPC speech vocoding is to quantize the coefficients of an adaptive predictor represented either as a transversal filter or a lattice filter. Since these encoded coefficients make up most of the bit rate of a vocoder parameter set (the other parameters being the voiced/unvoiced decision and the pitch), it is important to minimize the number of bits used in the coefficient quantization. The sensitivity of the model spectrum to the filter coefficients is thus of importance, since less sensitivity will allow a coarser quantization and thereby a lower bit rate. We choose to study the sensitivity through the movement of the poles of the all-pole model spectrum.

More generally, there is fluctuation in the coefficients of an adaptive filter after convergence, and the sensitivity of the transfer function to perturbations in filter coefficients is important in characterizing the effect of this inaccuracy on the filter transfer function.

In situations where an all-pole filter is implemented, as in the decoder for an LPC vocoder or an ADPCM decoder, the stability of that all-pole filter is also of concern. This filter is stable if and only if the poles are in the interior of the unit circle. When the poles and zeros of a rational filter are interior to the unit circle, the filter is said to be *minimum phase* [1]. Thus, the conditions under which the transversal or lattice filter is minimum phase are important in these applications. We are free to study either the all-zero or all-pole versions of these filters, since the zeros of the former are equal to the poles of the latter, and if one is minimum phase then so is the other.

The conclusions of this chapter can be summarized succinctly as follows. The lattice filter is minimum phase if and only if all the PARCOR coefficients are less than unity in magnitude. Thus, it is possible to ensure stability of an all-pole modeling filter by encoding the reflection coefficients with a maximum value which is slightly less than unity. There is no similar simple condition for the transversal filter, and hence it is more difficult to guarantee stability.

With respect to the sensitivity of the filter zeros or poles to perturbations of the filter coefficients, all the transversal filter coefficients have a similar effect on zero movement. The direction of the zero movement for the deviation in a single filter coefficient is not predictable, and may well be in a direction which results in the zero moving outside the unit circle.

For the lattice filter structure, the sensitivity of the zero locations to the first couple of PARCOR coefficients (k_1 and k_2) is actually greater than that of the transversal filter coefficients. However, the direction of movement of the zeros is predictably approximately tangent to the unit circle, particularly as the coefficients get near the unity, so that these large perturbations nevertheless tend not to move the zeros outside the unit circle (unless the perturbation makes the PARCOR coefficient larger than unity). The sensitivity of the zero locations to the later PARCOR coefficients is less than that of the transversal filter coefficients, although the direction of zero movement is again unpredictable.

The material in this chapter is of interest in speech applications. This chapter can safely be skipped by those readers desiring a broader introduction to the topic of adaptive filters.

5.1. MINIMUM PHASE LATTICE FILTERS

The transfer function of an n-th order predictor, given by (4.2.30), is a polynomial in z with real coefficients of order n. This polynomial therefore has n roots, the zeros of the transfer function, which are either real-valued or occur in complex conjugate pairs. Denote these n zeros by $Z_{n,1}, \ldots, Z_{n,n}$. A very important property of the lattice prediction filter is that all the zeros are interior to the unit circle,

$$|Z_{n,i}| < 1, \quad 1 \leqslant i \leqslant n \; , \tag{5.1.1}$$

if and only if the magnitude of the PARCOR coefficients are less than unity in magnitude,

$$|k_i| < 1, \quad 1 \leqslant i \leqslant n \; . \tag{5.1.2}$$

An elementary proof of this result is given in appendix 5-A.

It was shown in chapter 4, (4.2.22), that when the PARCOR coefficients are calculated using either the MMSE or autocorrelation method of block LS, the PARCOR coefficients satisfy

$$|k_i| \leqslant 1, \quad 1 \leqslant i \leqslant n \; . \tag{5.1.3}$$

It is therefore important to investigate the conditions under which a PARCOR coefficient is unity in magnitude, since according to the results of appendix 5-A in that instance one of the zeros of the predictor filter lies on the unit circle. From (4.2.24), when $|k_m| = 1$ the corresponding prediction error is zero,

$$||\mathbf{E}_f(m)|| = 0 \; , \tag{5.1.4}$$

which implies that

$$\begin{aligned} \mathbf{E}_f(m) &= 0 \\ &= \mathbf{Y} - \sum_{j=1}^{m} f_j(m) z^{-j} \mathbf{Y} \,. \end{aligned} \tag{5.1.5}$$

Examining the recursive solution of section 4.2, we see that once one of the orders of prediction error is zero, the remaining PARCOR coefficients are obtained by dividing zero by zero; that is, the solution is indeterminate. This suggests that the solution for the remaining PARCOR coefficients is not unique, and this is in fact the case.

Recall from section 3.2.2.4 that the solution to the MMSE problem is not unique when the correlation matrix $\mathbf{\Phi}$ is singular. We can generalize that condition if the correlation matrix is defined in terms of the inner product of (4.1.54). From this relation, the inner product can be calculated for a set of vectors which satisfys (5.1.5), namely

$$\begin{aligned} \phi_i &= <\mathbf{Y}, z^{-i}\mathbf{Y}> \\ &= <\sum_{j=1}^{m} f_j(m) z^{-j}\mathbf{Y}, z^{-i}\mathbf{Y}> \\ &= \sum_{j=1}^{m} f_j(m)\phi_{i-j} \,. \end{aligned} \tag{5.1.6}$$

Since (5.1.6) is of the form

$$\mathbf{f}\mathbf{\Phi} = 0 \tag{5.1.7}$$

for a non-zero vector $\mathbf{f}$, it follows that $\mathbf{\Phi}$ is indeed singular. Conversely, (5.1.6) could never be satisfied for a nonsingular $\mathbf{\Phi}$. It follows that whenever the PARCOR coefficients are the solution of the set of linear equations given by (4.1.53), then (5.1.2) is satisfied if and only if the matrix $\mathbf{\Phi}$ is nonsingular and the solution is unique.

Of course, these results apply only when the PARCOR coefficients are calculated in accordance with the recursive procedure of chapter 4 using infinite precision arithmetic. In practice, if finite precision arithmetic is used a PARCOR coefficient can be larger than unity in magnitude. To ensure stability of the synthesis filter in this instance, the offending PARCOR coefficient would be set to a value slightly less than unity.

These results apply to the transversal filter as well. Although there is no simple criterion for the transversal filter being minimum phase, when the transversal filter coefficients are calculated using the Levinson-Durbin algorithm using infinite precision arithmetic, then these results establish that the filter is minimum phase.

5.2. FILTER ZERO SENSITIVITY

Where the coefficients of a transversal or lattice filter are encoded and used to represent the spectrum of the input signal, as in LPC encoding of speech, the relationship of the spectrum and the coefficients is of importance. In particular, since the encoding process will result in some quantization error in the representation of the coefficients, it is important to understand how small perturbations of the coefficients affect the resulting transfer function of the filter (which is the inverse spectrum of the reconstructed speech).

The speech signal can be represented approximately as an all-pole spectrum, as discussed in section 2.1. Each of the resonances, or poles, is called a *formant.* A great deal is understood about how changes in the center frequencies and bandwidths of the formants affect the subjective quality of the synthesized speech. Therefore, the question we would like to answer is how the formants change in center frequency and bandwidth as the filter coefficients are perturbed. This question is answered in this section.

Let $F_n(z)$ given by (4.2.30) be the transfer function of the predictor used to analyze the speech signal. As in appendix 5-A, define the zeros of this all-zero transfer function as $Z_1, \ldots, Z_n$ (the order notation of the appendix is not needed and has therefore been eliminated). Because the polynomial $F_n(z)$ has real-valued coefficients, these zeros must either be real-valued or come in complex conjugate pairs. Each complex conjugate pair corresponds to one formant of speech, where the center frequency of the formant is related to the angle of the zero, and the bandwidth of the formant is related to how close the zero is to the unit circle (the closer to the unit circle the narrower the bandwidth). Let Z_1 be the zero under consideration, assume it is complex, and assume that Z_2 is its complex conjugate. Further, assume that all the zeros are distinct (which is empirically true in speech applications).

Since we are concerned with how Z_1 moves as the filter coefficients are perturbed, it is appropriate to calculate the partial derivative of Z_1 with respect to each of the filter coefficients. This will indicate the direction and magnitude of a shift in the zero location as the filter coefficient is perturbed by a small amount. This derivative is calculated in the next two sections, first for the simpler transversal filter and then for the lattice filter.

5.2.1. Zero Sensitivity of the Transversal Filter

For the transversal filter, the transfer function is given by (4.2.30), where we eliminate the order notation on the filter coefficients because it is not needed,

$$F_n(z) = 1 - \sum_{j=1}^{n} f_j z^{-j} . \tag{5.2.1}$$

The goal is to calculate the partial derivative $\dfrac{\partial Z_1}{\partial f_m}$ for each coefficient f_m, and the approach is to use the implicit function theorem [2],

$$\frac{\partial Z_1}{\partial f_m} = - \left. \frac{\dfrac{\partial F_n(z)}{\partial f_m}}{\dfrac{\partial F_n(z)}{\partial z}} \right|_{z = Z_1} . \tag{5.2.2}$$

In particular, the numerator of (5.2.2) is trivial to evaluate for the transversal structure,

$$\left.\frac{\partial F_n(z)}{\partial f_m}\right|_{z = Z_1} = -Z_1^{-m} \,. \tag{5.2.3}$$

The denominator is a little more challenging, but writing the transfer function in terms of its zeros,

$$F_n(z) = \prod_{i=1}^{n} (1 - Z_i z^{-1}) \tag{5.2.4}$$

as in (5.A.9), then utilizing the assumption of distinct zeros it is easy to evaluate the derivative (problem 5-1) as

$$\left.\frac{\partial F_n(z)}{\partial z}\right|_{z = Z_1} = -Z_1^{-1} \prod_{i=2}^{n} (1 - Z_i Z_1^{-1}) \,. \tag{5.2.5}$$

Finally, evaluating (5.2.2) using (5.2.3) and (5.2.5),

$$\frac{\partial Z_1}{\partial f_m} = \frac{Z_1^{-m+1}}{\prod_{i=2}^{n} (1 - Z_i Z_1^{-1})} \,. \tag{5.2.6}$$

No general statement can be made about the angle of $\frac{\partial Z_1}{\partial f_m}$ based on (5.2.6). There is a good chance that for some m the angle will be such that the direction of zero movement will be towards the unit circle, resulting in instability of the synthesis filter for zeros near the unit circle.

The magnitude of (5.2.6) can be expressed as

$$\left|\frac{\partial Z_1}{\partial f_m}\right| = \frac{|Z_1|^{n-m}}{\prod_{i=2}^{n} |Z_1 - Z_i|} \,. \tag{5.2.7}$$

The denominator is the product of the distances from each of the other zeros to the zero in question. If there are zeros that are very close to Z_1 then the denominator will be small, resulting in a large derivative and a large sensitivity. For a zero with angle close to 0 or π radians, the complex conjugate zero in particular will be very near to the zero in question. This implies that low frequency zeros and zeros near half the sampling rate are very sensitive to perturbations in the filter coefficients regardless of the location of the other zeros. The numerator of (5.2.7) is largest (unity) for $m = n$ and gets progressively smaller as m gets smaller. Thus, the sensitivity is greatest for the later coefficients of the transversal filter, and smallest for the first coefficients, although for zeros near the unit circle this effect is not very strong.

For speech spectra, since low frequency zeros close to the unit circle occur quite frequently, the preceding results demonstrate that a fine quantization of all the transversal coefficients is necessary just to maintain stability of the filter.

5.2.2. Zero Sensitivity of the Lattice Filter

The zero sensitivity of the lattice filter is a little more difficult to obtain. It is in fact apparently intractable to obtain this sensitivity analytically for other than the first few and last few coefficients of the lattice filter [3]. We will restrict attention here to only the sensitivity to the first and last PARCOR coefficient, and the reader is referred to [3] for a more complete exposition. The first coefficient is representative of the first few PARCOR coefficients and the last is representative of the last few. What will be striking is that the sensitivity properties of the first and last PARCOR coefficients are quite different.

The sensitivity of the last coefficient is easiest to obtain, so consider that case first. Note that (4.2.28) gives a relation for $F_n(z)$ in terms of k_n and $F_{n-1}(z)$, where the latter quantity is not a function of k_n. Therefore,

$$\begin{aligned} \frac{\partial F_n(z)}{\partial k_n} &= -z^{-1}B_{n-1}(z) \\ &= z^{-n}F_{n-1}(z^{-1}) \\ &= \frac{-z^{-n}F_n(z^{-1}) - k_nF_n(z)}{1-k_n^2}, \end{aligned} \tag{5.2.8}$$

and finally using the fact that $F_n(Z_1) = 0$,

$$\begin{aligned} \left.\frac{\partial F_n(z)}{\partial k_n}\right|_{z=Z_1} &= \frac{-Z_1^{-n}F_n(Z_1^{-1})}{1-k_n^2} \\ &= \frac{-\prod_{i=1}^{n}(Z_1^{-1}-Z_i)}{1-k_n^2}. \end{aligned} \tag{5.2.9}$$

Finally using (5.2.5) and substituting into (5.2.2),

$$\frac{\partial Z_1}{\partial k_n} = \frac{Z_1\prod_{i=1}^{n}(Z_1^{-1}-Z_i)}{(1-k_n^2)\prod_{i=2}^{n}(1-Z_iZ_1^{-1})}. \tag{5.2.10}$$

This rather formidable expression can be simplified by recalling that by assumption Z_2 is the complex conjugate of Z_1, and hence

$$\frac{\partial Z_1}{\partial k_n} = \frac{(1-Z_1^2)(1-|Z_1|^2)}{(1-k_n^2)(Z_1-Z_1^*)}\prod_{i=3}^{n}\left(\frac{Z_1^{-1}-Z_i^*}{1-Z_1^{-1}Z_i}\right). \tag{5.2.11}$$

In the product term the zeros and their complex conjugates have been grouped together in the numerator and denominator.

The main interest is in the sensitivity of zeros which are close to the unit circle, $|Z_1|\approx 1$, for two reasons. First, those are the zeros which are candidates to go outside the unit circle, and second, the formants with the narrowest bandwidth are the most critical in terms of the subjective effects of changes in center frequency and bandwidth. Each of the terms in the product has a magnitude close to unity for this case (see the lemma in appendix 5-A), and the first term can also be approximated for this case. The result is (problem 5-2),

$$\left| \frac{\partial Z_1}{\partial k_n} \right| \approx \frac{2(1 - |Z_1|)}{1 - k_n^2} . \tag{5.2.12}$$

For speech applications, k_n is typically no larger than 0.3 in magnitude, in which case the denominator can be neglected.

No general statement can be made about the angle of $\frac{\partial Z_1}{\partial k_n}$. Thus the direction of movement of the zero is unpredictable as in the transversal filter, but what is interesting is that as the zero approaches the unit circle, the sensitivity of the distance of movement to the PARCOR coefficient approaches zero. This is consistent with the stability property of section 5.1, since as long as the PARCOR coefficients are kept less than unity in magnitude, there is no concern about the zeros migrating outside the unit circle. In fact as the zero approaches the unit circle the zeros don't migrate at all as the later PARCOR coefficients are perturbed.

The sensitivity of zero movement to the first PARCOR coefficient k_1 is a little more difficult to obtain, but the method is basically the same. This case is considered in appendix 5-B, where it is shown that

$$\frac{\partial Z_1}{\partial k_1} \approx - \frac{e^{j\theta} \prod_{i=2}^{N} (1 - k_i^2)}{2j \sin(\theta)\ (1 - |Z_1|^2)} \cdot \frac{1}{\prod_{i=3}^{n} |1 - Z_1 Z_i|^2} , \tag{5.2.13}$$

where θ is the angle of Z_1 (that is, $Z_1 = |Z_1| e^{j\theta}$). Since all the terms in this expression are positive real-valued except for

$$-\frac{e^{j\theta}}{j} = e^{j(\theta - \frac{3\pi}{4})} , \tag{5.2.14}$$

the angle of $\frac{\partial Z_1}{\partial k_1}$ is perpendicular to θ, or in other words the direction of movement of Z_1 for small perturbations of k_1 is tangent to the unit circle. In fact, the movement is counterclockwise in the upper half plane, resulting in a higher formant center frequency, when k_1 increases.

How large is the movement of the zero? This requires examining the other terms in (5.2.13). The $\prod_{i=2}^{n} (1 - k_i^2)$ term is the ratio of the prediction error energy of the n-th stage to that of the first stage. Experimentally, this product term ranges from .1 to .8 in value for speech prediction. Each of the product terms can be written in the form

$$|1 - Z_1 Z_i|^2 = |Z_1|^2 |Z_1^{-1} - Z_i|^2 . \tag{5.2.15}$$

This term will be very small only if Z_1 and Z_i are both near $z=1$; that is, there are two low frequency zeros. If the angle of Z_1 is close to 0 or π radians (

corresponding to "low" and "high" frequencies), then the $\sin(\theta)$ term of (5.2.13) will be small, and $\left|\frac{\partial Z_1}{\partial k_1}\right|$ will be large. Finally, as $|Z_1| \to 1$, the $1 - |Z_1|^2$ term will become small and $\left|\frac{\partial Z_1}{\partial k_1}\right|$ will be large.

The conclusion is that zeros which are very sensitive to perturbation of k_1 are those close to the unit circle or those at low frequencies. Even though these zeros move substantially as k_1 is perturbed, there is no danger of the zero moving outside the unit circle because the direction of motion is approximately tangent to the unit circle. Thus, it is the center frequency of the formant which is changed, and not the bandwidth. For voiced speech, the occurrence of narrow bandwidth formants at low frequencies is common, and the center frequency of these formants is quite sensitive to perturbations in k_1. For this reason it is necessary to have a fine quantization of k_1 (and usually k_2 as well) in vocoding applications.

In conclusion, we have derived zero sensitivity for the first and last PARCOR coefficients, and they are quite different in nature. For a zero close to the unit circle, the zero sensitivity of the last coefficient will be small,

$$\frac{\left|\frac{\partial Z_1}{\partial k_n}\right|}{1 - |Z_1|} \approx 2 , \tag{5.2.16}$$

and random in direction. The zero sensitivity of the first coefficient will be large compared to that of the last coefficient, with an angle of zero movement tangent to the unit circle. As $|Z_1|$ approaches unity, the sensitivity of k_1 increases without bound, and in contrast, the sensitivity of k_N decreases to zero. For coefficients in between the first and last coefficients, roughly a transitional behavior occurs between these two extremes.

5.2.3. Comparison of Transversal and Lattice Filters

In order to verify the analytical results which have just been derived, it is appropriate to give a numerical example of the zero sensitivity of the transversal and lattice filter coefficients. This example is shown in figure 5-1a and 5-1b. The complex values of $\frac{\partial Z_1}{\partial k_l}$ and $\frac{\partial Z_1}{\partial f_l}$ are plotted for the zero closest to the unit circle of a 10 zero inverse speech spectrum. These values are normalized by a factor $(1 - |Z_1|)$, so that the magnitude of the sensitivity vectors must be multiplied by an increasingly large number as the zero approaches the unit circle. In addition, the angle of these vectors are relative to a radial vector through the zero, so that the positive real axis in the figure is in the direction of the unit circle and the imaginary axis is in the direction tangent to the unit circle. The spectrum itself ($|F_n^{-1}(z)|^2$ for z on the unit circle), plotted in figure 5-2, is representative of voiced speech. The largest formant (peak in the spectrum) corresponds to the pole whose sensitivity is plotted in figure 5-1.

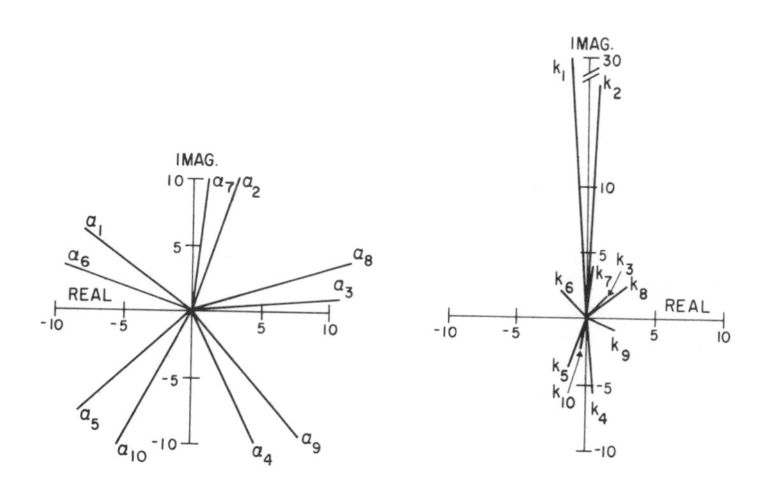

Figure 5-1. Sensitivity of a zero location to transversal and lattice coefficients.

In figure 5-1a the transversal zero sensitivity vector is scattered at all angles with approximately constant magnitude, as we predicted. In figure 5-1b the lattice zero sensitivity vectors for the first few coefficients are close to being tangent to the unit circle, and are large in magnitude. The normalized magnitude for the final lattice coefficients is roughly two, as predicted, with random angles. Observe that for this example the first two PARCOR coefficients are more sensitive than the transversal coefficients, while the later PARCOR coefficients are less sensitive.

5.3. CONCLUSIONS

One striking difference between the transversal and lattice filter structures is the quite different effects of perturbations of the filter coefficients. For the lattice filter it is easy to guarantee stability of the synthesis filter in an LPC or ADPCM speech encoding system by quantizing the PARCOR coefficient with a maximum magnitude slightly less than unity. There is no comparable simple criterion for the transversal filter. The sensitivity with respect to the different transversal filter coefficients is roughly comparable, particularly for zeros near the unit circle. On the other hand, there is a natural ordering of the PARCOR coefficients, where perturbations of the first coefficients have a marked effect on the center frequency of zeros near the unit circle while the later PARCOR coefficients have very little effect on those zeros.

In LPC encoding of speech, quantization and encoding of the PARCOR coefficients generally requires fewer total bits per frame than the quantization and encoding of the transversal filter coefficients. This is because of the natural scaling of the PARCOR coefficients, and because the third and remaining

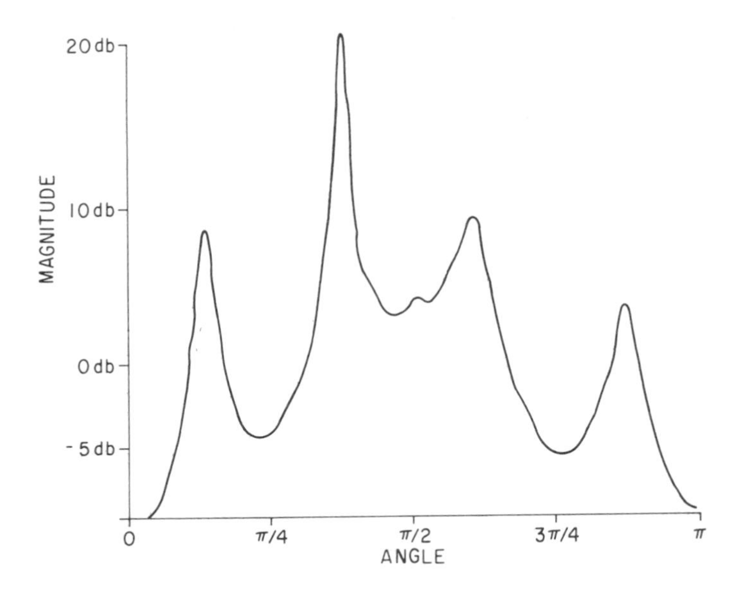

Figure 5-2. Typical all-pole spectrum for speech.

PARCOR coefficients can be quantized with less precision for a given subjective impairment.

All these properties work to the advantage of the lattice filter in speech encoding applications, and help to explain its popularity.

5.4. FURTHER READING

More details on the properties of speech can be found in [4]. Two papers in particular outline in detail the quantization properties of the PARCOR coefficients in LPC encoding of speech [5, 6].

PROBLEMS

5-1 Verify the derivative given in (5.2.5).

5-2 Verify the approximation of (5.2.11) given in (5.2.12).

5-3 For the two-pole lattice filter derived in problem 4-20:

a. Derive the transfer function as a function of the PARCOR coefficients.

b. Find an exact relation for the zero sensitivity with respect to perturbations of one of the PARCOR coefficients, and then verify the validity of (5.2.12) for this case.

5-4 Derive equation (5.A.3) of appendix 5-A.

REFERENCES

1. A.V. Oppenheim and R.W. Schafer, *Digital Signal Processing,* Prentice-Hall, Englewood Cliffs, N.J. (1975).
2. W. Rudin, *Principles of Mathematical Analysis*, McGraw-Hill, (1976).
3. P.L. Chu and D.G. Messerschmitt, "Zero Sensitivity Properties of the Digital Lattice Filter," *IEEE Trans. on Acoustics, Speech, and Signal Processing* **ASSP-31**(3) p. 685 (June 1983).
4. L.R. Rabiner and R.W. Schafer, *Digital Processing of Speech Signals,* Prentice-Hall, Englewood Cliffs, N.J. (1978).
5. A.H. Gray and J.D. Markel, "Quantization and Bit Allocation in Speech Processing," *IEEE Trans. on ASSP* **ASSP-24** pp. 459-473 (Dec. 1976).
6. R. Viswanathan and J. Makhoul, "Quantization Properties of Transmission Parameters in Linear Predictive Systems," *IEEE Trans. on ASSP* **ASSP-23** pp. 309-321 (June 1975).

APPENDIX 5-A

PROOF OF MINIMUM PHASE

This appendix will show that the lattice filter is minimum phase (zeros or poles fall inside the unit circle) if and only if the PARCOR coefficients are less than unity in magnitude. The proof will apply directly to the all-zero lattice filter (predictor), but is also applicable to the all-pole lattice filter since the poles of the latter are the zeros of the former.

The proof is based on the following simple lemma:

(Lemma) Define the complex function

$$A(z) = \frac{1 - Z_0 z^{-1}}{z^{-1} - Z_0^*}, \tag{5.A.1}$$

where Z_0 is a complex number with modulus less than unity, $|Z_0| < 1$, and Z_0^* is its conjugate. Then the function $A(z)$ has the property

$$|A(z)| \begin{cases} < 1, & |z| < 1 \\ = 1, & |z| = 1 \\ > 1, & |z| > 1 \end{cases} . \tag{5.A.2}$$

The proof of the lemma is simple, since by calculating the modulus squared of $A(z)$ directly (problem 5-4),

$$|A(z)|^2 = \frac{|z|^2 - 2\mathrm{Re}(Z_0 z^*) + |Z_0|^2}{1 - 2\mathrm{Re}(Z_0 z^*) + |Z_0|^2 |z|^2} , \tag{5.A.3}$$

and taking the difference between the numerator and denominator, we obtain

$$(|z|^2 - 1)(1 - |Z_0|^2) . \tag{5.A.4}$$

Since by assumption the second quantity satisfies

$$0 < (1 - |Z_0|^2) < 1 , \tag{5.A.5}$$

the numerator is less than the denominator for $|z| < 1$, the numerator is equal to the denominator for $|z| = 1$, and the numerator is greater than the denominator for $|z| > 1$.

Proceeding with the proof of the main result, we first prove that if $|k_m| < 1$ for $1 \leqslant m \leqslant n$ then $F_n(z)$ given by (4.2.28) is minimum phase. Proof is by induction: first show that $F_1(z)$ is minimum phase, and then show that if $F_{m-1}(z)$ is minimum phase then $F_m(z)$ is minimum phase.

Since

$$\begin{aligned} F_1(z) &= 1 - f_1(1) z^{-1} \\ &= 1 - k_1 z^{-1} \end{aligned} \tag{5.A.6}$$

by (4.3.3), the single zero of $F_1(z)$ is at $z = k_1$ which by assumption is interior to the unit circle.

Let the zeros of $F_m(z)$ be denoted by $Z_{m\,j}$, $1 \leqslant j \leqslant m$, and assume that $F_{m-1}(z)$ is minimum phase so that the $Z_{m-1\,j}$ all have modulus less than unity. Then $F_m(z)$ can be calculated from (4.2.28), and evaluating it at a zero,

$$\begin{aligned} F_m(Z_{m\,j}) &= 0 \\ &= F_{m-1}(Z_{m\,j}) - k_m Z_{m\,j}^{-1} B_{m-1}(Z_{m\,j}), \quad 1 \leqslant j \leqslant m , \end{aligned} \tag{5.A.7}$$

which implies that

$$|k_m|^2 = \left| \frac{F_{m-1}(Z_{m\,j})}{Z_{m\,j}^{-1} B_{m-1}(Z_{m\,j})} \right|^2, \quad 1 \leqslant j \leqslant m . \tag{5.A.8}$$

Writing $F_{m-1}(z)$ in terms of its zeros,

$$F_{m-1}(z) = \prod_{i=1}^{m-1} (1 - Z_{m-1,i} z^{-1}) , \tag{5.A.9}$$

and similarly from (4.2.32)

$$
\begin{aligned}
B_{m-1}(z) &= z^{-m+1}F_{m-1}(z^{-1}) \\
&= \prod_{i=1}^{m-1}(z^{-1} - Z_{m-1,i}) .
\end{aligned}
\tag{5.A.10}
$$

Since $F_{m-1}(z)$ is a polynomial with real coefficients, its zeros are either real or come in complex conjugate pairs. Thus, a zero and its complex conjugate can always be grouped in the numerator and denominator of (5.A.8), as in

$$
|k_m|^2 = \prod_{i=1}^{m-1}\left|\frac{1 - Z_{m-1,i}Z_{m,j}^{-1}}{Z_{m,j}^{-1} - Z_{m-1,i}^{*}}\right|^2 |Z_{m,j}|^2, \quad 1 \leqslant j \leqslant m . \tag{5.A.11}
$$

Applying the lemma to each of these factors, it follows that

$$
|k_m| \begin{cases} < 1, & |Z_{m,j}| < 1 \\ = 1, & |Z_{m,j}| = 1 \\ > 1, & |Z_{m,j}| > 1 \end{cases} , \tag{5.A.12}
$$

from which it follows that $|Z_{m,j}| < 1, 1 \leqslant j \leqslant m$.

Now we prove that if $F_n(z)$ is minimum phase, then $|k_m| < 1, 1 \leqslant m \leqslant n$. Proof is again by induction, showing that $F_m(z)$ minimum phase implies that $|k_m| < 1$ and $F_{m-1}(z)$ is minimum phase. From (4.3.3), $|k_m|$ can be calculated directly from the zeros of $F_m(z)$,

$$
\begin{aligned}
k_m &= f_m(m) \\
&= \prod_{i=1}^{m} Z_{m,i} ,
\end{aligned}
\tag{5.A.13}
$$

and hence

$$
|k_m| = \prod_{i=1}^{m}|Z_{m,i}| < 1 . \tag{5.A.14}
$$

To show that $F_{m-1}(z)$ is minimum phase, from (4.3.9) we can find a condition for $Z_{m-1,j}$ to be a zero of $F_{m-1}(z)$,

$$
F_{m-1}(Z_{m-1,j}) = \frac{F_m(Z_{m-1,j}) + k_m B_m(Z_{m-1,j})}{1 - k_m^2} = 0 \tag{5.A.15}
$$

which implies that

$$
|k_m|^2 = \left|\frac{F_m(Z_{m-1,j})}{B_m(Z_{m-1,j})}\right|^2 . \tag{5.A.16}
$$

By using the lemma as before, since $|k_m| < 1$ it follows that $|Z_{m-1,j}| < 1$ for $1 \leqslant j \leqslant m-1$ and $F_{m-1}(z)$ is minimum phase.

APPENDIX 5-B

ZERO SENSITIVITY OF FIRST PARCOR COEFFICIENT

In this appendix we calculate the sensitivity of the zero locations to perturbations in the first PARCOR coefficient k_1.

For this purpose, write the transfer function of the lattice filter in the form

$$\begin{bmatrix} F_n(z) \\ B_n(z) \end{bmatrix} = \mathbf{M}_n(z)\mathbf{M}_{n-1}(z) \cdots \mathbf{M}_1(z) \begin{bmatrix} 1 \\ 1 \end{bmatrix}, \tag{5.B.1}$$

where $\mathbf{M}_m(z)$ is the matrix transfer function of the m-th stage of the lattice filter,

$$\mathbf{M}_m(z) = \begin{bmatrix} 1 & -k_m z^{-1} \\ -k_m & z^{-1} \end{bmatrix}, \tag{5.B.2}$$

and its inverse is

$$\mathbf{M}_m^{-1}(z) = \frac{1}{1 - k_m^2} \begin{bmatrix} 1 & k_m \\ k_m z & z \end{bmatrix}. \tag{5.B.3}$$

Defining the matrix $\mathbf{K}(z)$ as

$$\mathbf{K}(z) = \mathbf{M}_n(z)\mathbf{M}_{n-1}(z) \cdots \mathbf{M}_1(z), \tag{5.B.4}$$

then the determinant of $\mathbf{K}(z)$ is the product of the determinants of $\mathbf{M}_m(z)$, $1 \leqslant m \leqslant n$, or

$$\det \mathbf{K}(z) = z^{-n} \prod_{m=1}^{n} (1 - k_m^2). \tag{5.B.5}$$

Further, defining the i,j element of $\mathbf{K}(z)$ as $K_{i,j}(z)$ then we can solve for these elements at the zero location as follows. Inverting (5.B.1),

$$\begin{bmatrix} 1 \\ 1 \end{bmatrix} = \mathbf{K}^{-1}(z) \begin{bmatrix} F_n(z) \\ B_n(z) \end{bmatrix}, \tag{5.B.6}$$

and evaluating this equation at the zero Z_1 of $F_n(z)$,

$$\begin{bmatrix} 1 \\ 1 \end{bmatrix} = \frac{1}{\det \mathbf{K}(Z_1)} \cdot \begin{bmatrix} K_{2,2}(Z_1) & -K_{1,2}(Z_1) \\ -K_{2,1}(Z_1) & K_{1,1}(Z_1) \end{bmatrix} \begin{bmatrix} 0 \\ B_n(Z_1) \end{bmatrix} \tag{5.B.7}$$

from which it follows that

$$K_{1,1}(Z_1) = \frac{\det \mathbf{K}(Z_1)}{B_n(Z_1)} \tag{5.B.8}$$

$$K_{1,2}(Z_1) = -\frac{\det \mathbf{K}(Z_1)}{B_n(Z_1)}. \tag{5.B.9}$$

Now we are prepared to calculate the partial derivative. Noting in (5.B.1) that the only term which is a function of k_1 is $\mathbf{M}_1(z)$, the derivative of the transfer function can be evaluated as

$$\begin{bmatrix} \dfrac{\partial F_n(z)}{\partial k_1} \\ \dfrac{\partial B_n(z)}{\partial k_1} \end{bmatrix} = \mathbf{M}_n(z)\mathbf{M}_{n-1}(z) \cdots \mathbf{M}_2(z) \frac{\partial}{\partial k_1} \mathbf{M}_1(z) \begin{bmatrix} 1 \\ 1 \end{bmatrix}$$

$$= \mathbf{K}(z)\mathbf{M}_1^{-1}(z) \frac{\partial}{\partial k_1} \mathbf{M}_1(z) \begin{bmatrix} 1 \\ 1 \end{bmatrix}, \tag{5.B.10}$$

where the derivative is evaluated from (5.B.2) as

$$\frac{\partial}{\partial k_1} \mathbf{M}_1(z) = \begin{bmatrix} 0 & -z^{-1} \\ -1 & 0 \end{bmatrix}. \tag{5.B.11}$$

Finally, evaluating (5.B.10) at the zero Z_1,

$$\left. \frac{\partial F_n(z)}{\partial k_1} \right|_{z=Z_1} = \frac{Z_1^{-n}(Z_1 - Z_1^{-1}) \prod_{m=2}^{n} (1 - k_m^2)}{B_n(Z_1)}. \tag{5.B.12}$$

Substituting for $B_n(Z_1)$ from (5.A.10),

$$\left. \frac{\partial F_n(z)}{\partial k_1} \right|_{z=Z_1} = -\frac{\prod_{m=2}^{n} (1 - k_m^2)}{Z_1 \prod_{m=2}^{n} (1 - Z_m Z_1)}, \tag{5.B.13}$$

and evaluating the derivative of Z_1,

$$\frac{\partial Z_1}{\partial k_1} = -\frac{\prod_{m=2}^{n} (1 - k_m^2)}{\prod_{m=2}^{n} (1 - Z_m Z_1^{-1})(1 - Z_m Z_1)}. \tag{5.B.14}$$

Finally, pulling out the $m = 2$ term in the denominator, and recalling that $Z_2 = Z_1^*$,

$$\frac{\partial Z_1}{\partial k_1} = -\frac{\prod_{m=2}^{n} (1 - k_m^2)}{(1-Z_1^* Z_1^{-1})(1-|Z_1|^2) \prod_{m=3}^{n} (1-Z_m^* Z_1^{-1})(1-Z_m Z_1)} \tag{5.B.15}$$

The one approximation made in obtaining (5.2.13) is that for $|Z_1| \approx 1$, $Z_1^{-1} \approx Z_1^*$. In addition, the first term in the denominator is written in the form

$$\frac{1}{1 - Z_1^* Z_1^{-1}} = \frac{1}{1 - e^{-j2\theta}} = \frac{e^{j\theta}}{2j\sin\theta} \; . \tag{5.B.16}$$

6

RECURSIVE LEAST SQUARES

The transversal and lattice algorithms of chapters 3 and 4 start with a filter structure derived using a MMSE criterion, with adaptation to nonstationary statistics provided by way of stochastic gradient (SG) algorithms. In order to handle nonstationary or unknown statistics, the filter coefficients were adapted using gradient techniques, where a noisy gradient is substituted for the true LMS gradient. A second approach discussed in those chapters was the least squares (LS) criterion, which when properly formulated is capable of handling the nonstationary case directly.

In this chapter the class of recursive LS algorithms will be developed further. In particular, adaptation algorithms based on both the transversal and lattice structures and a recursive LS criterion will be derived.

The distinction between transversal and lattice algorithms in this chapter is really a distinction between "fixed-order" and "order-recursive" algorithms. The primary characteristic of the transversal algorithms is that they calculate estimates for a single order N, whereas the lattice algorithms calculate estimates for all orders $n \leqslant N$. A time-varying order-recursive (lattice) filter structure, which is analogous to the lattice SG algorithms presented in chapter 4, can be used to recursively minimize the sum of the squared estimation errors.

Recent empirical results suggest that for some (but not all) applications the LS algorithms do not offer an adaptation speed significantly faster than that of an appropriately defined SG lattice algorithm [1, 2, 3, 4]. However, this does not really diminish the importance of the LS algorithms, since they offer a solid theoretical foundation for adaptive filtering algorithms, whereas the gradient algorithms are derived by an approximation to an exact method. Thus, the LS algorithms offer a solid basis of comparison with which to judge the gradient algorithms, and have suggested some desirable modifications to the gradient algorithms to improve their performance.

In the case of SG algorithms, it was stated in chapter 4 that the lattice algorithms have faster convergence than the transversal algorithms. This is not true for adaptation algorithms based on the LS criterion. The LS lattice and transversal algorithms minimize identical criteria, and hence have identical performance when finite precision effects are ignored. Further, in this chapter the transversal and lattice algorithms will be derived together, with each structure resulting from simply choosing an alternate set of recursions. The choice of the filter structure is thus governed solely by factors such as required arithmetic precision and implementation complexity.

The recursive LS algorithms derived in this chapter originated from a computationally efficient technique for inverting the sample covariance matrix arising in LS prediction [5, 6]. This led to the derivation of a computationally efficient recursive prewindowed LS adaptation algorithm for the transversal filter structure [5, 7]. This algorithm has been referred to as the "fast Kalman algorithm", since it is a computationally efficient version of an algorithm originally derived using Kalman filtering theory [8]. Order-recursive LS algorithms based on a lattice structure were subsequently derived [9, 10]. More recently, "normalized" LS lattice algorithms have been derived that involve fewer recursions than the original unnormalized versions, and which have the important advantage that nearly all internal variables are less than or equal to unity in magnitude [11].

Since the original work on computationally efficient recursive LS algorithms first appeared, numerous papers have derived computationally efficient algorithms that solve different types of autoregressive LS estimation problems [4, 12, 13, 14, 15]. These recursive algorithms use three windowing techniques described in chapter 3 in the context of nonrecursive, or block, LS estimation: (1) prewindowed recursive LS, (2) sliding window recursive LS, and (3) growing memory covariance LS. For each type of windowing scheme, both fixed-order (transversal) and order-recursive (lattice) algorithms have been derived. In addition, there are both normalized and unnormalized versions of each of these algorithms, giving a total of at least twelve different recursive LS algorithms so far! This is not counting the computationally efficient *nonrecursive*, or block, LS algorithms that have also been presented [6, 16, 17], along with other associated algorithms which, for example, compute LS prediction coefficients at time T given the parameters which enter a LS lattice algorithm at time T (so called "whitening filter algorithms"). Other variations on the recursive LS theme include LS joint process estimation algorithms [18], instrumental variables LS algorithms [19], and extended LS algorithms for ARMA, or pole-zero, models [20, 4]. Fortunately, all these algorithms can be derived in a similar manner, so that once one knows how to derive one or two of these algorithms, derivation of the remainder becomes significantly easier.

In this chapter both transversal (fixed-order) and lattice (order-recursive) algorithms which recursively solve the prewindowed and covariance LS problems described in chapter 2 are derived. As in chapter 4, a Hilbert space, or geometric, approach is developed by appropriately defining quantities such as norm, inner product, and projection operator. This approach was originally used to derive the prewindowed LS lattice algorithm [11, 21].

Throughout this chapter and book, the given data is assumed to be scalar-valued. These algorithms can be extended to the vector-valued given data case.

The LS recursions in the following sections can be derived in a number of different ways. In particular, the updates for the LS projection operator in section 6.2 can be equivalently derived via a series of matrix manipulations in which the shifting property of the data (discussed in chapter 3) is exploited. Other techniques which have been used are a Hilbert space array method [22] and a generalized update formula for the LS projection operator [4]. The Hilbert space approach used here has the advantage that complicated recursions involving the LS projection operator can be translated into geometric relations among vectors in Euclidean space. Once one understands the "pictures" associated with the LS projection updates, derivation of recursive LS algorithms becomes a straightforward (although somewhat tedious) task.

This chapter is organized as follows. Section 6.1 develops the necessary mathematical background needed to cast the LS problem into a geometric setting. This development is very similar to that given in chapter 4. This geometric formalism is then used in section 6.2 to derive order and time updates for the LS projection operator. The recursive prewindowed and covariance LS prediction algorithms in sections 6.3 and 6.4, respectively, are obtained by simply substituting the appropriate vectors and subspaces into the projection updates in section 6.2. In section 6.5 recursive LS algorithms for joint process estimation are derived. Derivations of normalized lattice algorithms are given in section 6.6. Finally, section 6.7 discusses the application of recursive LS algorithms to pitch detection.

6.1. HILBERT SPACE BACKGROUND

The geometric or Hilbert space interpretation of linear LS estimation is reviewed in this section.

The discussion of Hilbert space in chapter 4 defined the concepts of an inner product and Hilbert space. In that discussion, two examples of linear spaces and their associated inner products were given. One was the linear space of random variables with zero mean and finite second moments, where the inner product of two vectors was the expected value of the product of the two corresponding random variables. The second linear space included as vectors all infinite sequences of given data samples such that the sum of the samples squared is finite. This is essentially Euclidean space with an infinite number of components in each vector. The inner product for this linear space was the standard Euclidean inner product, which is the sum of the products of the individual components of each vector.

For the recursive LS criterion, the algorithms are defined in terms of the given data samples from some initial time T_0 to the current time T. In other words, the vector of given data samples has a finite number of components, although that number of components is growing as T increases. For this case the appropriate inner product space is the familiar Euclidean space consisting of vectors with $T-T_0+1$ components. In this chapter that space will be denoted for convenience as $\mathbf{R}^{T-T_0+1}$. This notation, which displays the dimension of the vectors in the linear space, is important since as T increases the dimension is also increasing. This inner product space is a Hilbert space, and therefore the all-important projection theorem applies and will be the basis for all the algorithms that will be derived.

If the standard Euclidean inner product is used in the recursive LS algorithms, since the dimensionality is growing with T, the weighting in the criterion given to the most recent samples will be decreasing with time. Therefore, it is appropriate to modify the definition of the inner product to allow a greater weighting to be given to more recent samples. Accordingly, given two n-dimensional vectors **X** and **Y**, whose components are real numbers, the inner product of **X** and **Y** is defined as

$$<\mathbf{X},\mathbf{Y}> \equiv \mathbf{X}'\mathbf{W}\mathbf{Y} , \tag{6.1.1}$$

where **W** is some prespecified and fixed $n \times n$ weighting matrix. Using (6.1.1) as the definition of inner product, all n dimensional vectors have a finite norm, and hence the discussion in section 4.1.2 indicates that the space consisting of n dimensional real-valued vectors is a Hilbert space. A weighting matrix which will be particularly useful is the exponential weighting matrix

$$\mathbf{W}_n = [1 \; w \; w^2 \cdots w^{n-1}] \, \mathbf{I} , \tag{6.1.2}$$

where **I** is the $n \times n$ identity matrix. Expanding (6.1.1) for this case gives

$$<\mathbf{X},\mathbf{Y}> = \sum_{j=1}^{n} w^{j-1} x_j y_j , \tag{6.1.3}$$

where x_j and y_j are respectively the jth components of **X** and **Y**. This definition has the desired property that the most recent samples are more heavily weighted (if j is interpreted as time) as long as $w < 1$. For the time being, we will assume that **W** is the identity matrix. Modification of the results in this chapter to the case of arbitrary **W** is straightforward.

Before proceeding with the LS criterion further, let us first review inner product spaces, the projection theorem, and also introduce some new notation which will be used in this chapter. The distance between two vectors **X** and **Y** with the same dimension is the norm (which for this chapter will be the regular Euclidean distance),

$$d(\mathbf{X},\mathbf{Y}) = ||\mathbf{Y}-\mathbf{X}|| \equiv <\mathbf{Y}-\mathbf{X},\mathbf{Y}-\mathbf{X}>^{1/2} . \tag{6.1.4}$$

as in (4.1.28).

In this chapter it will be common to calculate the projection of a vector **Y** onto the subspace spanned by another vector **X**. In chapter 4, the notation for that projection was $\mathbf{P}_{\{\mathbf{X}\}}\mathbf{Y}$, but since this vector is used so often here we will use the streamlined notation $\mathbf{P}_\mathbf{X}\,\mathbf{Y}$. Recall that this vector must equal $a\mathbf{X}$ where a is some scalar. Also, a special notation will be defined for the error vector between **Y** and its projection $\mathbf{P}_\mathbf{X}\mathbf{Y}$, which we will call the *orthogonal projection* of **Y** onto **X** and write as

$$\mathbf{P}_\mathbf{X}^\perp \, \mathbf{Y} \equiv \mathbf{Y} - \mathbf{P}_\mathbf{X} \, \mathbf{Y} . \tag{6.1.5}$$

From the projection theorem, the orthogonal projection will be orthogonal to **X**, so that

$$<\mathbf{X},\mathbf{P}_\mathbf{X}^\perp \, \mathbf{Y}> = <\mathbf{X},\mathbf{Y}-a\mathbf{X}> = 0, \tag{6.1.6}$$

which implies that

$$\mathbf{P}_\mathbf{X}\,\mathbf{Y} = a\mathbf{X} = \frac{<\mathbf{X},\mathbf{Y}>}{||\mathbf{X}||^2}\,\mathbf{X}\,. \tag{6.1.7}$$

The proportionality constant a is called the *regression coefficient.*

Similarly, the projection of a vector $\mathbf{Y}$ onto a subspace M, which is spanned by the vectors $\{\mathbf{X}_1,\mathbf{X}_2, \ldots, \mathbf{X}_n\}$, is denoted as $\mathbf{P}_M\,\mathbf{Y}$ and must satisfy

$$<\mathbf{X}_j,\mathbf{Y} - \mathbf{P}_M\,\mathbf{Y}> = 0, \quad \text{for } j=1, \ldots, n\,. \tag{6.1.8}$$

Since $\mathbf{P}_M\,\mathbf{Y}$ is in M, there exist constants $f_1, f_2, \ldots, f_n$ such that

$$\mathbf{P}_M\,\mathbf{Y} = \sum_{j=1}^{n} f_j \mathbf{X}_j = \mathbf{S}\,\mathbf{f}\,, \tag{6.1.9}$$

where $\mathbf{S} \equiv [\mathbf{X}_1 \;\cdots\; \mathbf{X}_n]$ and $\mathbf{f} \equiv [f_1 \;\cdots\; f_n]'$. The single regression coefficient in (6.1.7) is now replaced by the *vector* of regression coefficients $\mathbf{f}$. Combining (6.1.8) and (6.1.9) gives

$$\begin{aligned} <\mathbf{X}_j,\mathbf{Y}> &= <\mathbf{X}_j,\mathbf{P}_M\,\mathbf{Y}> = <\mathbf{X}_j,\mathbf{S}\,\mathbf{f}> \\ &= \mathbf{X}_j{}'\mathbf{S}\,\mathbf{f} \quad \text{for } j=1, \ldots, n\,, \end{aligned} \tag{6.1.10}$$

or equivalently,

$$\mathbf{S}'\mathbf{Y} = (\mathbf{S}'\mathbf{S})\,\mathbf{f}\,. \tag{6.1.11}$$

Assuming that the matrix $\mathbf{S}'\mathbf{S}$ is nonsingular, we can solve for $\mathbf{f}$ and substitute into (6.1.9) to get

$$\mathbf{P}_M\,\mathbf{Y} = \mathbf{S}\,(\mathbf{S}'\mathbf{S})^{-1}\mathbf{S}'\mathbf{Y}\,. \tag{6.1.12}$$

This matrix relation for the projection arises so often in this chapter that it is appropriate to define some suggestive notation. Specifically, if for matrices $\mathbf{A}$ and $\mathbf{B}$ we define a vector valued "inner product",

$$<\mathbf{A},\mathbf{Y}> \equiv \mathbf{A}'\mathbf{Y}, \quad <\mathbf{Y},\mathbf{A}> \equiv \mathbf{Y}'\mathbf{A}\,, \tag{6.1.13}$$

and a matrix valued inner product,

$$<\mathbf{A},\mathbf{B}> \equiv \mathbf{A}'\mathbf{B}\,, \tag{6.1.14}$$

then (6.1.12) can be rewritten as

$$\mathbf{P}_M\,\mathbf{Y} = \mathbf{S}<\mathbf{S},\mathbf{S}>^{-1}<\mathbf{S},\mathbf{Y}>\,. \tag{6.1.15}$$

There is a striking analogy between the projection on a single vector given by (6.1.7) and the projection on a subspace spanned by a set of vectors given by (6.1.15).

The linear LS estimate of $\mathbf{Y}$ given the vectors $\mathbf{X}_1, \ldots, \mathbf{X}_n$ is defined as the linear combination of those vectors which is closest to $\mathbf{Y}$. This LS estimate can therefore be defined in terms of the projection relations just derived. Let $f_1, \ldots, f_n$ be the set of coefficients which minimize the length of the error vector $\boldsymbol{\epsilon}$,

$$||\epsilon||^2 \equiv ||\mathbf{Y} - \sum_{i=1}^{n} f_i \mathbf{X}_i||^2 . \qquad \mathbf{(6.1.16)}$$

From the projection theorem the optimum estimate $\hat{\mathbf{Y}}$ according to this criterion is the projection on the subspace spanned by $\mathbf{X}_1, \ldots, \mathbf{X}_n$,

$$\hat{\mathbf{Y}} \equiv \sum_{i=1}^{n} f_i \mathbf{X}_i = \mathbf{P}_M \mathbf{Y} , \qquad (6.1.17)$$

and the vector of estimation errors is the orthogonal projection

$$\epsilon = \mathbf{P}_M^{\perp} \mathbf{Y} . \qquad (6.1.18)$$

We will call the operator $\mathbf{P}$ an *LS projection.*

Returning to the problem of recursive LS estimation, let T_0 be the starting time for the recursive estimation, and let $T > T_0$ be the current time for which an estimate is desired. The sequence of given data samples to be used in the estimate is therefore $\{y(j)\}$, $j=T_0, T_0+1, T_0+2, \ldots, T$ where $y(T)$ is the current input sample. The "data vector", or "vector of given data samples", is defined as

$$\mathbf{Y}(T) \equiv [y(T) \;\; y(T-1) \;\; \cdots \;\; y(T_0)]' \in \mathbf{R}^{T-T_0+1} , \qquad (6.1.19)$$

where T_0 to T is the time "window" containing the data values of interest. A shift operator z^{-j} is defined by

$$z^{-j}\mathbf{Y}(T) \equiv [y(T-j) \;\; y(T-j-1) \;\; \cdots \;\; y(T_0-j)]' \in \mathbf{R}^{T-T_0+1} . \qquad (6.1.20)$$

When T is near T_0 (we are just starting up the algorithm) this vector includes some given data samples prior to T_0. The way in which we handle this is different for the the prewindowed and covariance LS methods used in sections 6.3 and 6.4. In the covariance case the data values $y(j)$ for $j<T_0$ are never used, whereas in the prewindowed case these data values are used but assumed to be zero.

As in chapter 4 define a special notation for the subspace spanned by $\{z^{-l}\mathbf{Y}(T), z^{-l-1}\mathbf{Y}(T), \ldots, z^{-n}\mathbf{Y}(T)\}$, where $n>l$, as $M(T|l,n)$. We will call this subspace the *subspace of past observations.* Particularly note that as time increases this subspace changes, and hence the need to include T in the notation for this subspace (in contrast to chapter 4). It is also convenient to define a matrix of vectors which span $M(T|l,n)$ as

$$\mathbf{S}(T|l,n) \equiv [z^{-l}\mathbf{Y}(T) \;\; \cdots \;\; z^{-n}\mathbf{Y}(T)] \in \mathbf{R}^{T-T_0+1} \times \mathbf{R}^{n-l+1} . \qquad (6.1.21)$$

From the previous discussion, if $f_1, \ldots, f_{n-l+1}$ are selected such that

$$||\epsilon||^2 \equiv ||\mathbf{Y}(T) - \sum_{j=0}^{n-l} f_{j+1}[z^{-(l+j)}\mathbf{Y}(T)]||^2 \qquad (6.1.22)$$

is minimized, then the vector of estimation errors can be written as the orthogonal projection,

$$\epsilon = \mathbf{P}_{M(T|l,n)}^{\perp}\mathbf{Y}(T) . \qquad (6.1.23)$$

Projecting the current data vector onto the subspace $M(T|l,n)$ therefore results in an LS estimate of the current data vector based on a linear combination of the shifted data vectors

$$z^{-l}\mathbf{Y}(T),\ z^{-(l+1)}\mathbf{Y}(T),\ \ldots,\ z^{-n}\mathbf{Y}(T)\ .$$

The LS algorithms in this chapter are derived by first obtaining fundamental order and time updates for the LS projection operator, and then substituting appropriately shifted data vectors and subspaces into these updates. The linear space used throughout is the space spanned by the data values $\{y(j)\}$, $j=T_0, T_0+1, \ldots, T$, or $\mathbf{R}^{T-T_0+1}$. In contrast to the discussion in chapter 4, vectors in this linear space have an added time argument T. This argument, which designates the time at which the LS estimation is being performed, distinguishes the time varying estimation problem considered here from the stationary estimation problem considered in chapter 4. Assuming T_0 is fixed, as T increases, more data is used to obtain the LS estimate, and hence the dimension of the underlying linear space increases.

The forthcoming derivation of the LS lattice order recursions is nearly identical to the derivation of the order recursions for the MMSE lattice in chapter 4. To illustrate the close similarity between the LS estimation problem just formulated and solved and the MMSE solution, recall that for the latter the appropriate linear space is the space of random variables with zero mean and finite second moment, with an inner product given by

$$<\mathbf{X},\mathbf{Y}> \equiv E[XY]\ . \tag{6.1.24}$$

The linear MMSE estimate of a random variable Y, given the random variables $X_1, \ldots, X_n$, which span a space M, is obtained by minimizing

$$||\mathbf{Y} - \sum_{j=1}^{n} f_j \mathbf{X}_j||^2 = E[(\mathbf{Y} - \mathbf{s}'\mathbf{f})^2]\ , \tag{6.1.25}$$

where $\mathbf{s}=[X_1 \ \cdots \ X_n]'$. The resulting estimate of Y is

$$\begin{aligned}\hat{\mathbf{Y}} &= \mathbf{s}'\mathbf{f} = \mathbf{P}_M\ \mathbf{Y} \\ &= \mathbf{s}'\{E[\mathbf{s}\,\mathbf{s}']\}^{-1}E[Y\mathbf{s}]\ ,\end{aligned} \tag{6.1.26}$$

which is analogous to (6.1.15) and (6.1.17).

6.2. PROJECTION OPERATOR UPDATE FORMULAS

In this section some fundamental relationships satisfied by the LS projection operator are derived. These projection updates fall into two main categories: order updates and time updates. In this sense the LS algorithms derived here will be very similar to the SG lattice recursions of chapter 4. For example, the order updates for the SG lattice are given by (4.2.9-10) and the time updates for the PARCOR coefficients are given by (4.5.3), (4.5.9), or (4.5.15). This chapter will uncover interesting similarities between the LS and SG algorithms.

The derivation of time updates for the LS projection operator is more difficult than the derivation of order updates, especially since the order update derivation is so similar to the MMSE case of chapter 4. Therefore, the order updates are derived first, followed by the time updates.

6.2.1. Order Updates

To enhance understanding of order updates, consider first a simple example of an order update. Given three vectors $\mathbf{Y}$, $\mathbf{X}_1$, and $\mathbf{X}_2$ in $\mathbf{R}^n$, suppose the LS estimate of $\mathbf{Y}$ based on a linear combination of vectors $\mathbf{X}_1$ and $\mathbf{X}_2$ is desired. We therefore wish to find the two coefficients a_1 and a_2 such that $||\mathbf{Y} - (a_1\mathbf{X}_1 + a_2\mathbf{X}_2)||^2$ is minimized. From the discussion in the last section, the linear LS estimate of $\mathbf{Y}$ is known to be

$$a_1\mathbf{X}_1 + a_2\mathbf{X}_2 = \mathbf{P}_{\{\mathbf{X}_1\}\oplus\{\mathbf{X}_2\}}\mathbf{Y} , \tag{6.2.1}$$

where $\{\mathbf{X}_1\}\oplus\{\mathbf{X}_2\}$ denotes the space spanned by $\mathbf{X}_1$ and $\mathbf{X}_2$. Figure 6-1 illustrates this projection as the sum of two projections onto vectors orthogonal to each other. In particular, the space spanned by $\mathbf{X}_1$ and $\mathbf{X}_2$ can be decomposed into the orthogonal components $\mathbf{X}_1$ and $\mathbf{P}_{\mathbf{X}_1}^{\perp}\mathbf{X}_2$. That is, any vector which can be written as a linear combination of $\mathbf{X}_1$ and $\mathbf{X}_2$ can also be written as a linear combination of the orthogonal vectors $\mathbf{X}_1$ and $\mathbf{P}_{\mathbf{X}_1}^{\perp}\mathbf{X}_2$. In mathematical terms, we have the orthogonal subspace decomposition,

$$\{\mathbf{X}_1\}\oplus\{\mathbf{X}_2\} = \{\mathbf{X}_1\} \oplus \{\mathbf{P}_{\mathbf{X}_1}^{\perp}\mathbf{X}_2\} . \tag{6.2.2}$$

By the projection theorem in chapter 4, (4.1.10), for any vector $\mathbf{Y} \in \mathbf{R}^n$,

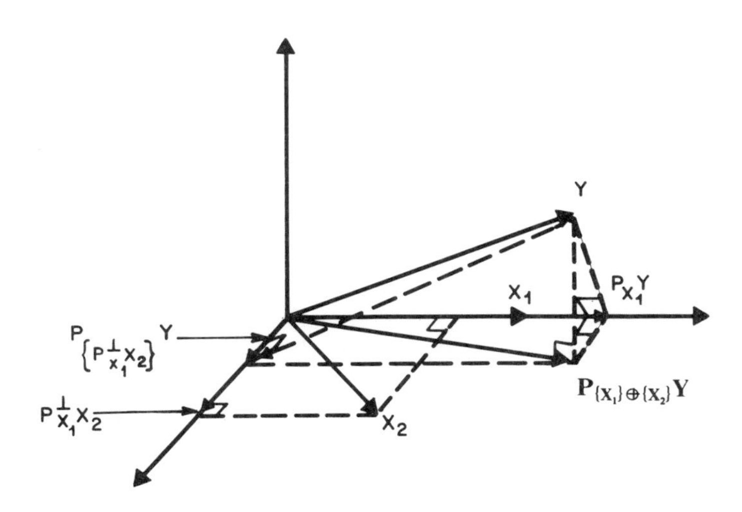

Figure 6-1. Decomposition of the projection $\mathbf{P}_{\{\mathbf{X}_1\}\oplus\{\mathbf{X}_2\}}\mathbf{Y}$.

$$\mathbf{P}_{\{\mathbf{X}_1\}\oplus\{\mathbf{X}_2\}}\mathbf{Y} = \mathbf{P}_{\{\mathbf{X}_1\}}\mathbf{Y} + \mathbf{P}_{\{\mathbf{P}^{\perp}_{\mathbf{X}_1}\mathbf{X}_2\}}\mathbf{Y} . \tag{6.2.3}$$

If the "order" of a projection is defined as the number of basis vectors which span the space onto which a given vector (in this case $\mathbf{Y}$) is being projected, then this equation is an elementary projection order update. The first-order projections on the right hand side of (6.2.3) are added to form the second-order projection $\mathbf{P}_{\{\mathbf{X}_1\}\oplus\{\mathbf{X}_2\}}\mathbf{Y}$.

This result can now be extended to an arbitrary order. Replace the vector $\mathbf{X}_1$ by a subspace M spanned by the vectors $\mathbf{X}_1, \mathbf{X}_2, \ldots, \mathbf{X}_n$, all in $\mathbf{R}^n$, and replace $\mathbf{X}_2$ by the vector $\mathbf{X} \in \mathbf{R}^n$. The LS estimate of $\mathbf{Y}$ given $\mathbf{X}_1, \ldots, \mathbf{X}_n$, and $\mathbf{X}$ can be related to the LS estimate without $\mathbf{X}$. In analogy with the simpler example, write the orthogonal subspace decomposition for the space spanned by $\mathbf{X}_1, \mathbf{X}_2, \ldots, \mathbf{X}_n$, and $\mathbf{X}$, which is denoted as $M\oplus\{\mathbf{X}\}$ (problem 6-1),

$$M\oplus\{\mathbf{X}\} = M \oplus \{\mathbf{P}^{\perp}_M\ \mathbf{X}\} . \tag{6.2.4}$$

By the projection theorem of (4.1.10), for any vector $\mathbf{Y} \in \mathbf{R}^n$, since the two subspaces on the right side of (6.2.4) are orthogonal

$$\mathbf{P}_{M\oplus\{\mathbf{X}\}}\mathbf{Y} = \mathbf{P}_M\mathbf{Y} + \mathbf{P}_{\{\mathbf{P}^{\perp}_M\mathbf{X}\}}\mathbf{Y} . \tag{6.2.5}$$

This equation constitutes a fundamental order update for the LS projection operator. The $(n+1)$st order projection $\mathbf{P}_{M\oplus\{\mathbf{X}\}}$ is expressed as the sum of the nth order projection, $\mathbf{P}_M$, and the first order projection, $\mathbf{P}_{\{\mathbf{P}^{\perp}_M\mathbf{X}\}}$. By subtracting both sides of (6.2.5) from $\mathbf{Y}$, an order update for the orthogonal projection operator $\mathbf{P}^{\perp}$ is obtained,

$$\mathbf{P}^{\perp}_{M\oplus\{\mathbf{X}\}}\mathbf{Y} = \mathbf{P}^{\perp}_M\mathbf{Y} - \mathbf{P}_{\{\mathbf{P}^{\perp}_M\mathbf{X}\}}\mathbf{Y} . \tag{6.2.6}$$

This recursion can be used to derive all of the LS order updates in this chapter by simply redefining the space M and the vectors $\mathbf{X}$ and $\mathbf{Y}$.

6.2.2. Forward Time Updates

The forward time updates derived in this section compute an LS projection at time T given the same LS projection computed from data collected at time $T-1$. These recursions, when combined with the order recursions in the last section, can be used to derive prewindowed LS algorithms. Analogous "backward" time updates, which are presented in the next section, are also needed to derive covariance LS algorithms.

Consider the vectors $\mathbf{X}(T)$ and $\mathbf{Y}(T)$, which are composed of data samples from time T_0 to T, i.e.,

$$\mathbf{X}(T) \equiv [x(T)\ x(T-1)\ \cdots\ x(T_0)]' , \tag{6.2.7}$$

and $\mathbf{Y}(T)$ is defined by (6.1.19). Our objective is to compute the linear LS estimate of $\mathbf{Y}(T)$ given $\mathbf{X}(T)$ in terms of a LS estimate which does not make use of the most recent samples $y(T)$ and $x(T)$. A unit vector is first defined,

$$\mathbf{u}(T) \equiv [1\ 0\ \cdots\ 0\ 0]'. \tag{6.2.8}$$

This vector has the same dimension as $\mathbf{Y}(T)$, so $\mathbf{u}(T)$ is a vector in the same space $\mathbf{R}^{T-T_0+1}$. Denote by $U(T)$ the subspace of $\mathbf{R}^{T-T_0+1}$ which is spanned by $\mathbf{u}(T)$. This is of course precisely the set of vectors in $\mathbf{R}^{T-T_0+1}$ whose components other than the first component are all zero. We therefore call the subspace $U(T)$ the *subspace of most recent data samples*. The projection of the current data vector onto this subspace is

$$\begin{aligned}\mathbf{P}_{U(T)}\mathbf{Y}(T) &= [y(T)\ 0\ \cdots\ 0\ 0]\\ &= y(T)\mathbf{u}(T)\end{aligned}$$

and is a vector made up of the current data sample plus all zeros in the other components.

The subspace $U(T)$ is associated with the current given data sample, and it is appropriate to also characterize the subspace associated with the past data samples. To this end, it is useful to define the *orthogonal complement* of the subspace $U(T)$ as $U^{\perp}(T)$. This is the subspace of $\mathbf{R}^{T-T_0+1}$ consisting of all vectors orthogonal to the subspace $U(T)$. It is easy to see that this is the subspace consisting of vectors with a zero first component, and therefore we call it the *subspace of past data values*. By definition the inner product of any vector in $U(T)$ with any vector in $U^{\perp}(T)$ is zero.

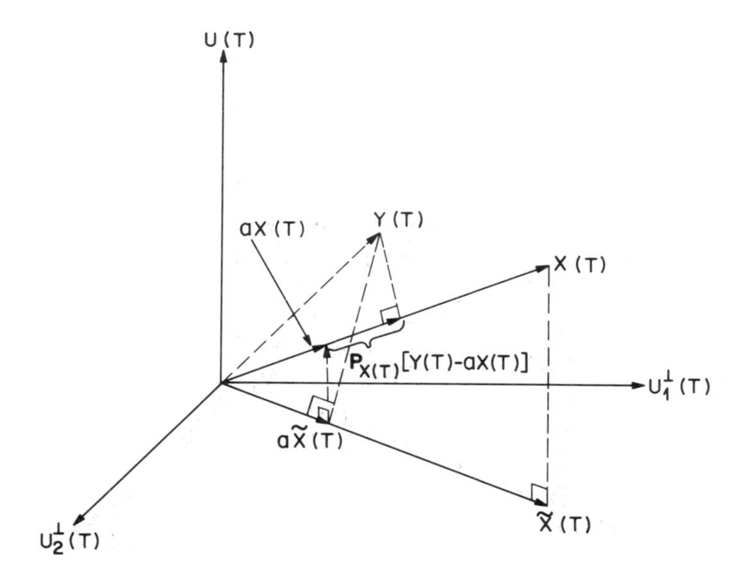

Figure 6-2. Geometric illustration of the projection operator forward time update.

For notational convenience a "tilde" operator is defined as,

$$\tilde{\mathbf{Y}}(T) \equiv \mathbf{P}^{\perp}_{U(T)}\mathbf{Y}(T) = [0\; y(T-1)\; y(T-2) \cdots y(T_0)]' . \tag{6.2.9}$$

The vector $\tilde{\mathbf{Y}}(T)$ is therefore the data vector at time $T-1$ with an added zero component to make it the same dimension as the data vector at time T. Notice that $\tilde{\mathbf{Y}}(T) = \mathbf{P}_{U^{\perp}(T)}\,\mathbf{Y}(T)$, i.e. $\tilde{\mathbf{Y}}(T)$ is the projection of $\mathbf{Y}(T)$ onto the subspace of past data values.

The basic prediction problem is illustrated in figure 6-2 where $\mathbf{Y}(T)$ is a vector having its endpoint in back of the plane of the paper and $\mathbf{X}(T)$ has its endpoint in front of the plane of the paper. Consider the case where a LS estimate of $\mathbf{Y}(T)$ is desired given the vector $\mathbf{X}(T)$. For example, in linear prediction the vector $\mathbf{X}(T)$ is a shifted version of $\mathbf{Y}(T)$, i.e., $\mathbf{X}(T) = z^{-j}\mathbf{Y}(T)$. The LS estimate $\mathbf{P}_{\mathbf{X}(T)}\mathbf{Y}(T)$ is to be decomposed as the sum of a projection onto a subspace of past data values, which has been computed at time $T-1$, and a term which uses the data values $x(T)$ and $y(T)$. This latter term thus represents the change due to the current data sample, and is given the special name *innovation.*

The regression coefficient a computed at time $T-1$ is defined by $\mathbf{P}_{\mathbf{X}(T-1)}\mathbf{Y}(T-1) = a\mathbf{X}(T-1)$. This regression coefficient can equivalently be defined in terms of the "tilde" vectors at time T, since $\mathbf{P}_{\tilde{\mathbf{X}}(T)}\tilde{\mathbf{Y}}(T) = a\tilde{\mathbf{X}}(T)$, which gives the regression coefficient at time $T-1$ in terms of the data vector at time T. Figure 6-2 shows $\mathbf{P}_{\mathbf{X}(T)}\mathbf{Y}(T)$ decomposed into the two vectors $a\mathbf{X}(T)$ and $\mathbf{P}_{\mathbf{X}(T)}[\mathbf{Y}(T)-a\mathbf{X}(T)]$. Figure 6-3 illustrates the plane spanned by the vectors $\mathbf{X}(T)$, $\tilde{\mathbf{X}}(T)$, and $U(T)$. Referring to figure 6-3, since ABC and ADE are similar triangles,

$$\frac{AB}{AD} = \frac{AC}{AE} = a , \tag{6.2.10}$$

and therefore[1] $\overline{AC} = a\mathbf{X}(T)$. Figure 6-4 again illustrates the decomposition of the projection $\mathbf{P}_{\mathbf{X}(T)}\mathbf{Y}(T)$, and attempts to include vectors not shown in figure 6-2. (Note that only the endpoint of $\mathbf{Y}(T)$ is present.)

The identity

$$\mathbf{P}_{\mathbf{X}(T)}\mathbf{Y}(T) = a\mathbf{X}(T) + \mathbf{P}_{\mathbf{X}(T)}\mathbf{P}_{U(T)}[\mathbf{Y}(T)-a\mathbf{X}(T)] . \tag{6.2.11}$$

is now illustrated. Referring to figure 6-3, let G denote the endpoint of $\mathbf{P}_{\mathbf{X}(T)}\mathbf{Y}(T)$ and H the endpoint of $\mathbf{P}_{U(T)}\mathbf{Y}(T)$. Construct a plane N orthogonal to $\mathbf{X}(T)$ at point G and a plane V orthogonal to $U(T)$ at point H. Since $\mathbf{P}^{\perp}_{\mathbf{X}(T)}\mathbf{Y}(T)$ must lie in N and $\mathbf{P}^{\perp}_{U(T)}\mathbf{Y}(T)$ must lie in V, it follows that the endpoint of $\mathbf{Y}(T)$ must lie in $N \cap V$ which is a line orthogonal to the plane of the paper at point F. The plane spanned by $N \cap V$ and $[\mathbf{Y}(T)-a\tilde{\mathbf{X}}(T)]$ is designated as W. Since both $N \cap V$ and $[\mathbf{Y}(T)-a\tilde{\mathbf{X}}(T)]$ are orthogonal to $\tilde{\mathbf{X}}(T)$, W must be orthogonal to $\tilde{\mathbf{X}}(T)$. Figure 6-5 illustrates the vectors in the plane W. The vectors $\mathbf{P}_{U(T)}[\mathbf{Y}(T)-a\mathbf{X}(T)]$ and $\overline{BC}$ are colinear by construction,

---

[1] The following notational conventions are adhered to. A "bar" above the two endpoints of a segment denotes the vector from the first endpoint to the second endpoint. The length of a segment is denoted by the two endpoints with no bar.

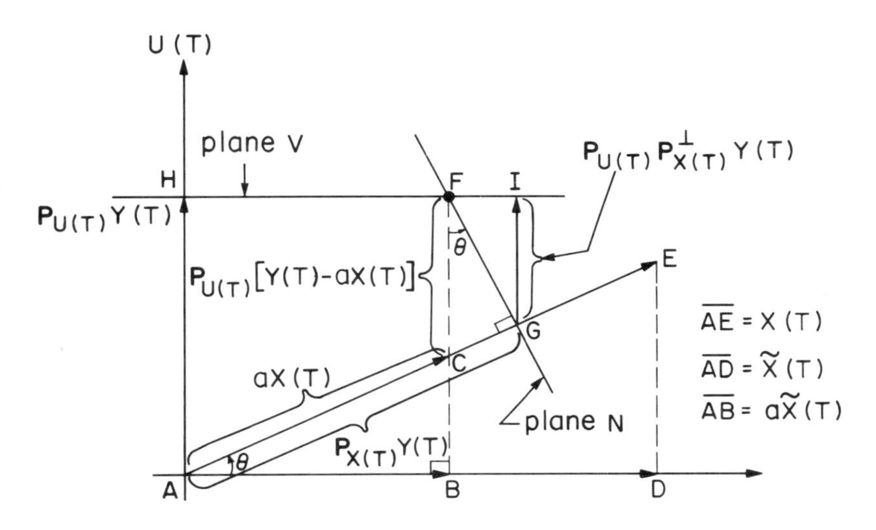

Figure 6-3. Plane spanned by the vectors $\mathbf{X}(T)$, $\tilde{\mathbf{X}}(T)$, and $\mathbf{u}(T)$ shown in figure 6-2.

and since $\overline{BC}$ is orthogonal to $\tilde{\mathbf{X}}(T)$ and intersects $[\mathbf{Y}(T)-a\tilde{\mathbf{X}}(T)]$, it follows that $\mathbf{P}_{U(T)}[\mathbf{Y}(T)-a\mathbf{X}(T)]$ is in W. Since $N \cap V$ is in W, and since it can be assumed that V and W are not coincident, it follows that $W \cap V = V \cap N$. The endpoint of $\mathbf{P}_{U(T)}[\mathbf{Y}(T)-a\mathbf{X}(T)]$ must lie in both W and V and therefore lies in N. The projection of any point in N onto $\mathbf{X}(T)$ is G so that we get the desired result

$$\mathbf{P}_{\mathbf{X}(T)}\mathbf{P}_{U(T)}[\mathbf{Y}(T)-a\mathbf{X}(T)] = \overline{CG} = \mathbf{P}_{\mathbf{X}(T)}[\mathbf{Y}(T) - a\mathbf{X}(T)]. \quad (6.2.12)$$

This time update formula for the LS projection operator forms the basis for the LS time update equations derived in this chapter. Other decompositions of $\mathbf{P}_{\mathbf{X}(T)}\mathbf{Y}(T)$ are possible. In particular, given an arbitrary constant b, one has the trivial identity

$$\mathbf{P}_{\mathbf{X}(T)}\mathbf{Y}(T) = b\mathbf{X}(T) + \mathbf{P}_{\mathbf{X}(T)}[\mathbf{Y}(T) - b\mathbf{X}(T)]\,. \quad (6.2.13)$$

Equation (6.2.12) states that if $b = a = \dfrac{<\tilde{\mathbf{X}}(T),\tilde{\mathbf{Y}}(T)>}{<\tilde{\mathbf{X}}(T),\tilde{\mathbf{X}}(T)>}$, then

$$\mathbf{P}_{\mathbf{X}(T)}[\mathbf{Y}(T)-b\mathbf{X}(T)] = \mathbf{P}_{\mathbf{X}(T)}\mathbf{P}_{U(T)}[\mathbf{Y}(T)-b\mathbf{X}(T)]\,. \quad (6.2.14)$$

Given a, only the most recent value of $\mathbf{Y}(T)$ is needed to compute the projection update represented by the second term on the right-hand side of (6.2.11).

Introducing the notation,

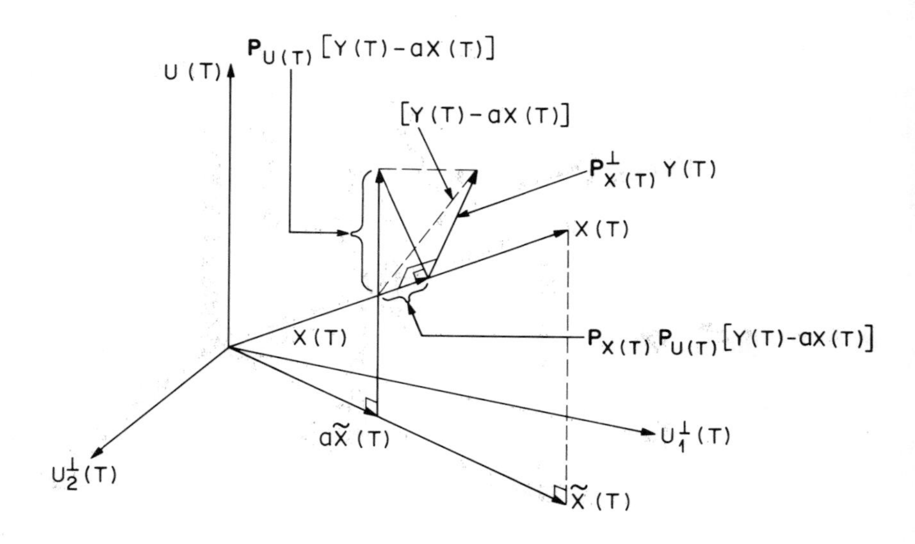

Figure 6-4. Decomposition of the projection $\mathbf{P}_{\mathbf{X}(T)}\mathbf{Y}(T)$. This figure is a rotated version of figure 6-2.

$$\hat{\mathbf{Y}}_{T-1}(T) \equiv \mathbf{P}_{\mathbf{X}_{T-1}(T)}\mathbf{Y}(T) \equiv a\,\mathbf{X}(T) = \frac{<\tilde{\mathbf{X}}(T),\tilde{\mathbf{Y}}(T)>}{<\tilde{\mathbf{X}}(T),\tilde{\mathbf{X}}(T)>}\mathbf{X}(T) \quad \mathbf{(6.2.15)}$$

and

$$\mathbf{P}^{\perp}_{\mathbf{X}_{T-1}(T)}\mathbf{Y}(T) = \mathbf{Y}(T) - \hat{\mathbf{Y}}_{T-1}(T) \ , \qquad (6.2.16)$$

(6.2.11) can be rewritten as

$$\mathbf{P}_{\mathbf{X}(T)}\mathbf{Y}(T) = \hat{\mathbf{Y}}_{T-1}(T) + \mathbf{P}_{\mathbf{X}(T)}\mathbf{P}_{U(T)}\mathbf{P}^{\perp}_{\mathbf{X}_{T-1}(T)}\mathbf{Y}(T) \ . \qquad (6.2.17)$$

The innovation due to the new data sample $y(T)$ is represented by the second term on the right hand side. Although this result has only been shown to be true when $\mathbf{X}(T)$ is a vector, it is shown in the appendix that it is also true when $\mathbf{X}(T)$ is replaced by some subspace $M(T)$ spanned by vectors $\{\mathbf{X}_1(T), \mathbf{X}_2(T), \ldots, \mathbf{X}_n(T)\}$. For this case, letting

$$\mathbf{S}(T) = [\mathbf{X}_1(T) \ \ \mathbf{X}_2(T) \ \ \cdots \ \ \mathbf{X}_n(T)] \qquad (6.2.18)$$

and

$$\tilde{\mathbf{S}}(T) = [\tilde{\mathbf{X}}_1(T) \ \ \tilde{\mathbf{X}}_2(T) \ \ \cdots \ \ \tilde{\mathbf{X}}_n(T)] \ , \qquad (6.2.19)$$

the "projection" is defined as,

$$\mathbf{P}_{M_{T-1}(T)}\mathbf{Y}(T) = \mathbf{S}(T)\ \mathbf{f} = \mathbf{S}(T)\ [\tilde{\mathbf{S}}^T(T)\tilde{\mathbf{S}}(T)]^{-1}\tilde{\mathbf{S}}^T(T)\tilde{\mathbf{Y}}(T), \quad (6.2.20)$$

i.e., $\mathbf{P}_{M_{T-1}(T)}\mathbf{Y}(T)$ lies in $M(T)$, but uses regression coefficients computed at time $T-1$. Replacing the vector $\mathbf{X}(T)$ in (6.2.17) by the subspace $M(T)$, (6.2.17) becomes

$$\mathbf{P}_{M(T)}\mathbf{Y}(T) = \mathbf{P}_{M_{T-1}(T)}\mathbf{Y}(T) + \mathbf{P}_{M(T)}\mathbf{P}_{U(T)}\mathbf{P}^{\perp}_{M_{T-1}(T)}\mathbf{Y}(T) \quad . \tag{6.2.21}$$

It will be convenient in the subsequent sections to replace the far right hand term by a term involving projections only on $M(T)$. We therefore wish to specify the relation between $\mathbf{P}_{U_T}\mathbf{P}_{M_{T-1}(T)}$ and $\mathbf{P}_{M(T)}$. Referring to figure 6-3, the angle θ is given by

$$\cos^2\theta = \frac{||\tilde{\mathbf{X}}(T)||^2}{||\mathbf{X}(T)||^2} = 1 - \frac{x^2(T)}{||\mathbf{X}(T)||^2} , \tag{6.2.22}$$

and hence indicates the amount of new information received at time T. If $\cos^2\theta$ is close to zero, then the new value $x(T)$ is relatively "unexpected". It is easily shown that (6.2.22) can be rewritten in a more useful form,

$$1 - \cos^2\theta = \sin^2\theta = ||\mathbf{P}_{\mathbf{X}(T)}\mathbf{u}(T)||^2 . \tag{6.2.23}$$

Figure 6-3 also indicates that

$$\theta = \angle EAD = \angle CFG = \angle FGI , \tag{6.2.24}$$

and hence

$$\cos\theta = \frac{FG}{CF} = \frac{GI}{FG} . \tag{6.2.25}$$

Since $\overline{GI} = \mathbf{P}_{U(T)}\mathbf{P}^{\perp}_{\mathbf{X}(T)}\mathbf{Y}(T)$ and $\overline{CF} = \mathbf{P}_{U(T)}\mathbf{P}^{\perp}_{\mathbf{X}_{T-1}(T)}\mathbf{Y}(T)$, it follows that

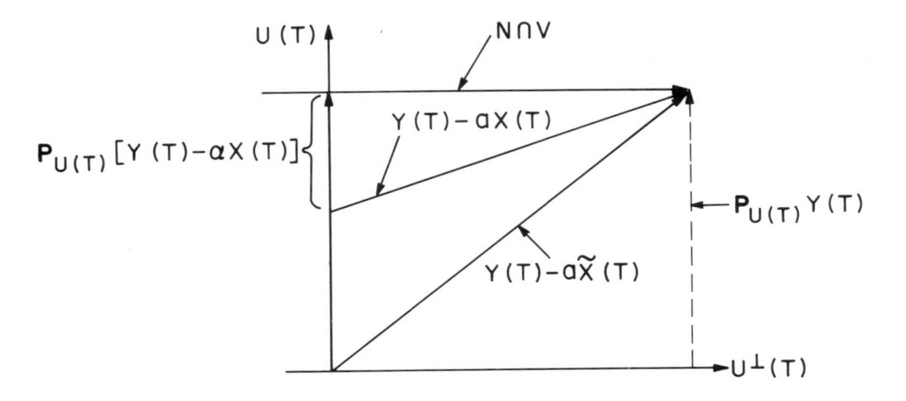

Figure 6-5. Plane containing $\mathbf{Y}(T) - a\tilde{\mathbf{X}}(T)$ which is perpendicular to $\tilde{\mathbf{X}}(T)$.

$$\frac{||\mathbf{P}_{U(T)}\mathbf{P}^{\perp}_{\mathbf{X}(T)}\mathbf{Y}(T)||}{||\mathbf{P}_{U(T)}\mathbf{P}^{\perp}_{\mathbf{X}_{T-1}(T)}\mathbf{Y}(T)||} = \frac{GI}{CF} = \frac{GI}{FG}\,\frac{FG}{CF} = \cos^2\theta\ . \qquad \textbf{(6.2.26)}$$

Since $\mathbf{P}_{U(T)}\mathbf{P}^{\perp}_{\mathbf{X}(T)}\mathbf{Y}(T)$ and $\mathbf{P}_{U(T)}\mathbf{P}^{\perp}_{\mathbf{X}_{T-1}(T)}\mathbf{Y}(T)$ are both parallel to $\mathbf{u}(T)$ (i.e. only the first element of the vectors are nonzero), it has been shown that

$$\mathbf{P}_{U(T)}\mathbf{P}^{\perp}_{\mathbf{X}_{T-1}(T)}\mathbf{Y}(T) = [\mathbf{P}_{U(T)}\mathbf{P}^{\perp}_{\mathbf{X}(T)}\mathbf{Y}(T)]\sec^2\theta\ , \qquad (6.2.27)$$

where $\sec\theta = (\cos\theta)^{-1}$. In the appendix this result is shown to remain valid when $\mathbf{X}(T)$ is replaced by a subspace $M(T)$. The variable θ then becomes the "angle" between the spaces spanned by the matrices of basis vectors $\mathbf{S}(T)$ and $\tilde{\mathbf{S}}(T)$. This notion is illustrated in figure 6-6 where $\theta(T)$ is the angle between the two planes $M(T)$ and $\tilde{M}(T)$. In analogy with (6.2.23) the following relation can be used to define the angle $\theta(T)$,

$$\sin^2\theta(T) = ||\mathbf{P}_{M(T)}\mathbf{u}(T)||^2\ . \qquad (6.2.28)$$

Noting that

$$\mathbf{P}_{M(T)}\mathbf{u}(T) = \mathbf{S}(T)[\mathbf{S}'(T)\mathbf{S}(T)]^{-1}\mathbf{S}'(T)\mathbf{u}(T)\ , \qquad (6.2.29)$$

it follows that

$$\begin{aligned}\sin^2\theta(T) &= [\mathbf{u}'(T)\mathbf{S}(T)][\mathbf{S}'(T)\mathbf{S}(T)]^{-1}[\mathbf{S}'(T)\mathbf{u}(T)]\ . \\ &= <\mathbf{u}(T)\ ,\ \mathbf{P}_{M(T)}\mathbf{u}(T)>\ .\end{aligned} \qquad (6.2.30)$$

Equation (6.2.27) therefore has the generalization,

$$\mathbf{P}_{U(T)}\mathbf{P}^{\perp}_{M_{T-1}(T)}\mathbf{Y}(T) = [\mathbf{P}_{U(T)}\mathbf{P}^{\perp}_{M(T)}\mathbf{Y}(T)]\sec^2\theta(T) \qquad (6.2.31)$$

where $\sec^2\theta(T) = \dfrac{1}{1-\sin^2\theta(T)}$ and $\theta(T)$ is defined by (6.2.30).

Using (6.2.31), the projection update (6.2.21) can be rewritten as

$$\mathbf{P}_{M(T)}\mathbf{Y}(T) = \mathbf{P}_{M_{T-1}(T)}\ \mathbf{Y}(T) + \mathbf{P}_{M(T)}\mathbf{P}_{U(T)}\mathbf{P}^{\perp}_{M(T)}\mathbf{Y}(T)\ \sec^2\theta(T)\ . \qquad (6.2.32)$$

This can be simplified as follows,

$$\begin{aligned}\mathbf{P}_{M(T)}\ \mathbf{Y}(T) &= \mathbf{P}_{M_{T-1}(T)}\mathbf{Y}(T) + \mathbf{P}_{M(T)}\Big[<\mathbf{u}(T),\mathbf{P}^{\perp}_{M(T)}\mathbf{Y}(T)>\mathbf{u}(T)\Big]\ \sec^2\theta(T) \\ &= \mathbf{P}_{M_{T-1}(T)}\mathbf{Y}(T) + \mathbf{P}_{M(T)}\mathbf{u}(T)\ <\mathbf{u}(T),\mathbf{P}^{\perp}_{M(T)}\mathbf{Y}(T)>\sec^2\theta(T)\ .\end{aligned} \qquad (6.2.33)$$

An update equation for the orthogonal projection operator can be obtained by subtracting both sides of (6.2.33) from $\mathbf{Y}(T)$,

$$\mathbf{P}^{\perp}_{M(T)}\mathbf{Y}(T) = \mathbf{P}^{\perp}_{M_{T-1}(T)}\mathbf{Y}(T) - \mathbf{P}_{M(T)}\mathbf{u}(T)\ <\mathbf{u}(T),\mathbf{P}^{\perp}_{M(T)}\mathbf{Y}(T)>\sec^2\theta(T)\ . \qquad (6.2.34)$$

One more relation which will be useful in the subsequent sections is easily derived from the time update recursion (6.2.34). This is a recursive equation for the inner product $<\mathbf{v}(T),\mathbf{P}^{\perp}_{M(T)}\mathbf{Y}(T)>$, where $\mathbf{v}(T)$ is an arbitrary vector in $\mathbf{R}^{T-T_0+1}$. It will be helpful to first rewrite (6.2.34) as

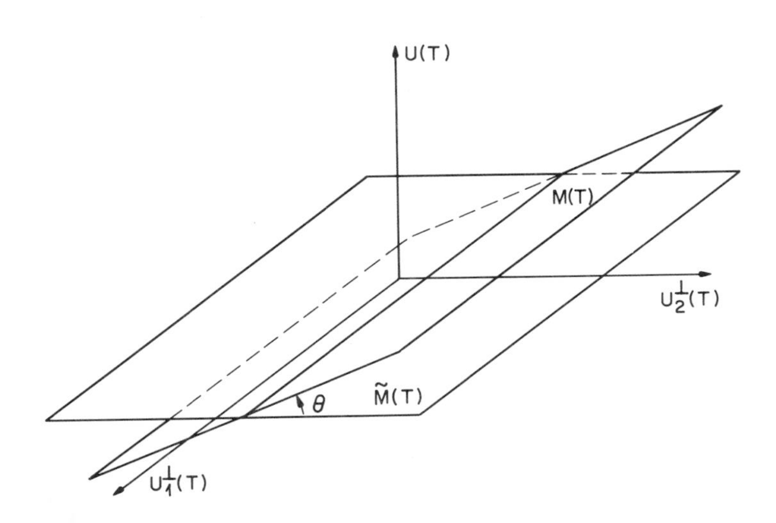

Figure 6-6. Illustration of the angle $\theta(T)$ between two planes $M(T)$ and $\tilde{M}(T)$.

$$\begin{aligned}
\mathbf{P}^{\perp}_{M(T)}\mathbf{Y}(T) &= \mathbf{P}^{\perp}_{M_{T-1}(T)}\mathbf{Y}(T) - \mathbf{P}_{U(T)}\mathbf{P}^{\perp}_{M_{T-1}(T)}\mathbf{Y}(T) + \mathbf{P}_{U(T)}\mathbf{P}^{\perp}_{M_{T-1}(T)}\mathbf{Y}(T) \\
&\quad - \mathbf{P}_{M(T)}\mathbf{u}(T) \ <\mathbf{u}(T),\mathbf{P}^{\perp}_{M(T)}\mathbf{Y}(T)>\sec^2\theta(T) \\
&= \mathbf{P}^{\perp}_{U(T)}\mathbf{P}^{\perp}_{M_{T-1}(T)}\mathbf{Y}(T) + \mathbf{u}(T)<\mathbf{u}(T),\ \mathbf{P}^{\perp}_{M(T)}\mathbf{Y}(T)>\sec^2\theta(T) \qquad \textbf{(6.2.35)} \\
&\quad - \mathbf{P}_{M(T)}\mathbf{u}(T) \ <\mathbf{u}(T),\mathbf{P}^{\perp}_{M(T)}\mathbf{Y}(T)>\sec^2\theta(T) \\
&= \mathbf{P}^{\perp}_{U(T)}\mathbf{P}^{\perp}_{M_{T-1}(T)}\mathbf{Y}(T) + \mathbf{P}^{\perp}_{M(T)}\mathbf{u}(T) \ <\mathbf{u}(T),\mathbf{P}^{\perp}_{M(T)}\mathbf{Y}(T)>\sec^2\theta(T) .
\end{aligned}$$

The term $\mathbf{P}_{M_{T-1}(T)}\mathbf{Y}(T)$ in (6.2.34) has now been replaced by the term $\mathbf{P}^{\perp}_{U(T)}\mathbf{P}^{\perp}_{M_{T-1}(T)}\mathbf{Y}(T)$. Taking the inner product of both sides of (6.2.35) with $\mathbf{v}(T)$ gives

$$\begin{aligned}
<\mathbf{v}(T),\mathbf{P}^{\perp}_{M(T)}\mathbf{Y}(T)> &= <\mathbf{v}(T),\mathbf{P}^{\perp}_{U(T)}\mathbf{P}^{\perp}_{M_{T-1}(T)}\mathbf{Y}(T)> \\
&\quad + <\mathbf{v}(T),\mathbf{P}^{\perp}_{M(T)}\mathbf{u}(T)><\mathbf{u}(T),\mathbf{P}^{\perp}_{M(T)}\mathbf{Y}(T)>\sec^2\theta(T) .
\end{aligned} \qquad (6.2.36)$$

It is easily verified that for arbitrary vectors $\mathbf{v}(T)$ and $\mathbf{w}(T)$,

$$<\mathbf{v}(T),\mathbf{P}^{\perp}_{M(T)}\mathbf{w}(T)> = <\mathbf{w}(T),\mathbf{P}^{\perp}_{M(T)}\mathbf{v}(T)> , \qquad (6.2.37)$$

and hence (6.2.36) can be rewritten in the more useful form

$$\begin{aligned}
<\mathbf{v}(T),\mathbf{P}^{\perp}_{M(T)}\mathbf{Y}(T)> &= <\tilde{\mathbf{v}}(T),\mathbf{P}^{\perp}_{\tilde{M}(T)}\tilde{\mathbf{Y}}(T)> \\
&\quad + <\mathbf{u}(T),\mathbf{P}^{\perp}_{M(T)}\mathbf{v}(T)><\mathbf{u}(T),\mathbf{P}^{\perp}_{M(T)}\mathbf{Y}(T)>\sec^2\theta(T) ,
\end{aligned} \qquad (6.2.38)$$

where $\tilde{M}(T)$ is the space spanned by $\tilde{\mathbf{S}}(T)$.

If the unweighted inner product defined by (6.1.1) where $\mathbf{W} = \mathbf{I}$ is used, then the first term on the right hand side of (6.2.38) is the value of the left hand side at time $T-1$. However, if the exponentially weighted inner product (6.1.3) is used, for any two vectors $\mathbf{v}(T)$ and $\mathbf{X}(T) \in \mathbf{R}^{T-T_0+1}$,

$$\begin{aligned} <\tilde{\mathbf{v}}(T),\tilde{\mathbf{X}}(T)> &= \tilde{\mathbf{v}}'(T)\mathbf{W}(T)\tilde{\mathbf{X}}(T) \\ &= \mathbf{v}'(T-1)[w\mathbf{W}(T-1)]\mathbf{X}(T-1) \\ &= w<\mathbf{v}(T-1),\mathbf{X}(T-1)> , \end{aligned} \tag{6.2.39}$$

and hence (6.2.38) becomes

$$\begin{aligned} <\mathbf{v}(T),\mathbf{P}^{\perp}_{M(T)}\mathbf{Y}(T)> &= w<\mathbf{v}(T-1),\mathbf{P}^{\perp}_{M(T-1)}\mathbf{Y}(T-1)> \\ &\quad + <\mathbf{u}(T),\mathbf{P}^{\perp}_{M(T)}\mathbf{v}(T)><\mathbf{u}(T),\mathbf{P}^{\perp}_{M(T)}\mathbf{Y}(T)>\sec^2\theta(T) . \end{aligned} \tag{6.2.40}$$

The prewindowed algorithms in section 6.3 can be derived by using only the order update (6.2.6), the time update (6.2.34), and the inner product update (6.2.40).

6.2.3. Backward Time Updates

In order to derive the covariance algorithms in section 6.4, an additional set of recursions closely related to (6.2.21-40) are needed. Consider again the data vectors $\mathbf{X}(T)$ and $\mathbf{Y}(T)$ specified by (6.1.19) and (6.2.7). Suppose the linear LS estimate of $\mathbf{Y}(T)$ given $\mathbf{X}(T)$ is desired in terms of an LS estimate which does not make use of the most *distant* or *past* values $y(T_0)$ and $x(T_0)$. Clearly, this problem can be solved in exactly the same fashion as the time update problem stated at the beginning of the last section. By turning the vectors $\mathbf{Y}(T)$ and $\mathbf{X}(T)$ "upside down" and assuming that $y(T_0)$ and $x(T_0)$ are the most recent samples, one can solve this LS problem by using time updates already derived. The same argument holds when $\mathbf{X}(T)$ is replaced by the subspace $M(T)$ spanned by vectors $\mathbf{X}_1(T)$, $\mathbf{X}_2(T)$, . . . , $\mathbf{X}_n(T)$. In this case we want to decompose the projection $\mathbf{P}_{M(T)}\mathbf{Y}(T)$ into the sum of two vectors, the first of which is a projection onto the space spanned by the matrix of basis vectors $\mathbf{S}(T)$ in which the *bottom row* has been replaced by zeroes. This is in contrast to the previous forward time updates that expressed $\mathbf{P}_{M(T)}\mathbf{Y}(T)$ as the sum of two vectors, the first of which was a projection onto the space spanned by $\mathbf{S}(T)$ in which the *top* row has been replaced by zeroes (i.e., $\tilde{M}(T)$).

The backward time updates in this section are derived by first defining analogous notation to the notation used in the last section. The derivation of forward time updates in the last section can then be repeated word for word, where the notation in the last section is replaced with the analogous notation defined here. A unit vector, which is analogous to $\mathbf{u}(T)$, is first defined,

$$\mathbf{u}(T_0) \equiv [0\ 0\ \cdots\ 0\ 1]' \in \mathbf{R}^{T-T_0+1} , \tag{6.2.41}$$

and the space spanned by $\mathbf{u}(T_0)$ is denoted as $U(T_0)$. An "asterisk" operator is defined in analogy with the previous "tilde" operator,

$$\mathbf{Y}^*(T) \equiv \mathbf{P}_{U^\perp(T_0)}\mathbf{Y}(T) = \mathbf{P}_{U^\perp(T_0)}\mathbf{Y}(T) \\ = [y(T)\ y(T-1)\ \cdots\ y(T_0+1)\ 0]' , \tag{6.2.42}$$

where $U^\perp(T_0)$ is the orthogonal complement of $U(T_0)$. The vector $\mathbf{Y}^*(T)$ is therefore the same as $\mathbf{Y}(T)$, except that the bottom element is set to zero. Similarly,

$$\mathbf{S}^*(T) \equiv [\mathbf{X}_1^*(T)\ \mathbf{X}_2^*(T)\ \cdots\ \mathbf{X}_n^*(T)] . \tag{6.2.43}$$

The projection of $\mathbf{Y}(T)$ onto $M(T)$ using regression coefficients computed from $\mathbf{S}^*(T)$, which contains the data values $y(T_0+1), y(T_0+2), \ldots, y(T)$, is defined as

$$\mathbf{P}_{M_{T_0+1}(T_0,T)}\mathbf{Y}(T) \equiv \mathbf{S}(T)[\mathbf{S}^{*\prime}(T)\mathbf{S}^*(T)]^{-1}[\mathbf{S}^{*\prime}(T)\mathbf{Y}(T)] . \tag{6.2.44}$$

The regression coefficients which multiply the basis vectors of $M(T)$ are in this case elements of the vector

$$[\mathbf{S}^{*\prime}(T)\mathbf{S}^*(T)]^{-1}[\mathbf{S}^{*\prime}(T)\mathbf{Y}(T)] .$$

This definition is analogous to $\mathbf{P}_{M_{T-1}(T)}\mathbf{Y}(T)$, which is the projection of $\mathbf{Y}(T)$ onto $M(T)$ using regression coefficients computed from $\tilde{\mathbf{S}}(T)$.

The derivation of (6.2.33) can be repeated with $\mathbf{u}(T)$ replaced by $\mathbf{u}(T_0)$, "tildes" replaced by "asterisks", and $\mathbf{P}_{M_{T-1}(T)}$ replaced by $\mathbf{P}_{M_{T_0+1}(T_0,T)}$ to give a projection decomposition,

$$\mathbf{P}_{M(T)}\mathbf{Y}(T) = \mathbf{P}_{M_{T_0+1}(T_0,T)}\mathbf{Y}(T) \\ + \mathbf{P}_{M(T)}\mathbf{u}(T_0)\ \langle\mathbf{u}(T_0),\mathbf{P}^\perp_{M(T)}\mathbf{Y}(T)\rangle \sec^2\theta^*(T) \tag{6.2.45}$$

where

$$\sin^2\theta^*(T) = ||\mathbf{P}_{M(T)}\mathbf{u}(T_0)||^2 \\ = [\mathbf{u}'(T_0)\mathbf{S}(T)][\mathbf{S}'(T)\mathbf{S}(T)]^{-1}[\mathbf{S}'(T)\mathbf{u}(T_0)] \\ = \langle\mathbf{u}(T_0),\mathbf{P}_{M(T)}\mathbf{u}(T_0)\rangle \tag{6.2.46}$$

and

$$\sec^2\theta^*(T) = \frac{1}{1-\sin^2\theta^*(T)} . \tag{6.2.47}$$

Subtracting both sides of (6.2.45) from $\mathbf{Y}(T)$ gives

$$\mathbf{P}^\perp_{M(T)}\mathbf{Y}(T) = \mathbf{P}^\perp_{M_{T_0+1}(T_0,T)}\mathbf{Y}(T) \\ - \mathbf{P}_{M(T)}\mathbf{u}(T_0)\ \langle\mathbf{u}(T_0),\mathbf{P}^\perp_{M(T)}\mathbf{Y}(T)\rangle \sec^2\theta^*(T) . \tag{6.2.48}$$

Finally, the following inner product update equation is analogous to (6.2.38),

$$<\mathbf{v}(T),\mathbf{P}^{\perp}_{M(T)}\mathbf{Y}(T)> \;=\; <\mathbf{v}^{*}(T),\mathbf{P}^{\perp}_{M^{*}(T)}\mathbf{Y}^{*}(T)> \\ +\; <\mathbf{u}(T_0),\mathbf{P}^{\perp}_{M(T)}\mathbf{v}(T)><\mathbf{u}(T_0),\mathbf{P}^{\perp}_{M(T)}\mathbf{Y}(T)>\sec^2\theta^{*}(T)\;. \tag{6.2.49}$$

This completes the derivation of projection operator recursions needed to derive the algorithms in the following sections. The following five updates can be used to derive any of the existing computationally efficient (unnormalized) LS algorithms: projection order update (6.2.6), projection forward time update (6.2.34), projection backward time update (6.2.48), inner product forward time update (6.2.40), and inner product backward time update (6.2.49).

6.2.4. Generalized Projection Update Formula

The LS projection operator order and time updates are summarized by the recursions (6.2.6), (6.2.34), and (6.2.45). These update equations can be considered special cases of the following generalized problem. Given the matrix $\mathbf{S}(T) = [\mathbf{X}_1(T), \mathbf{X}_2(T), \ldots, \mathbf{X}_n(T)]$, whose columns span the space $M(T)$, decompose the projection of a vector $\mathbf{Y}(T)$ onto $M(T)$ into the sum of two vectors, the first of which is a projection of $\mathbf{Y}(T)$ onto a *reduced* space $M^{-}(T)$ spanned by a reduced matrix $\mathbf{S}^{-}(T)$ in which a single row or column from $\mathbf{S}(T)$ has been replaced by zeroes. Replacing the first or last *column* of $\mathbf{S}(T)$ by zeroes leads to the order updates in section 6.2.1, whereas replacing the first or last *row* of $\mathbf{S}(T)$ by zeroes leads to the time updates in section 6.2.2. This observation has led to the interpretation of (6.2.6) as a very general projection operator update equation from which both order and time update equations can be derived [4].

Writing (6.2.6) out explicitly gives

$$\mathbf{P}^{\perp}_{\{M+\mathbf{X}\}}\mathbf{Y} = \mathbf{P}^{\perp}_{M}\mathbf{Y} + \mathbf{P}^{\perp}_{M}\mathbf{X}\Big[(\mathbf{P}^{\perp}_{M}\mathbf{X})'(\mathbf{P}^{\perp}_{M}\mathbf{X})\Big]^{-1}(\mathbf{P}^{\perp}_{M}\mathbf{X})'\mathbf{Y}\;. \tag{6.2.50}$$

Noting that

$$\begin{aligned}(\mathbf{P}^{\perp}_{M}\mathbf{X})'(\mathbf{P}^{\perp}_{M}\mathbf{X}) &= <\mathbf{X}-\mathbf{P}_{M}\mathbf{X},\mathbf{P}^{\perp}_{M}\mathbf{X}> \\ &= <\mathbf{X},\mathbf{P}^{\perp}_{M}\mathbf{X}> - <\mathbf{P}_{M}\mathbf{X},\mathbf{P}^{\perp}_{M}\mathbf{X}> \\ &= <\mathbf{X},\mathbf{P}^{\perp}_{M}\mathbf{X}>\;,\end{aligned} \tag{6.2.51}$$

and premultiplying (6.2.49) by an arbitrary vector $\mathbf{v}'$ gives,

$$\mathbf{v}'\mathbf{P}^{\perp}_{M\oplus\{\mathbf{X}\}}\mathbf{Y} = \mathbf{v}'\mathbf{P}^{\perp}_{M}\mathbf{Y} + (\mathbf{v}'\mathbf{P}^{\perp}_{M}\mathbf{X})[\mathbf{X}'\mathbf{P}^{\perp}_{M}\mathbf{X}]^{-1}(\mathbf{P}^{\perp}_{M}\mathbf{X})'\mathbf{Y}\;. \tag{6.2.52}$$

It has been shown that by selecting the vectors $\mathbf{v}$, $\mathbf{Y}$, $\mathbf{X}$, and the space M in an appropriate fashion, both order and time updates can be derived [23]. While this approach is mathematically elegant, it loses some of the geometric interpretations associated with the Hilbert space approach used here.

We proceed now to the derivation of recursive LS transversal and lattice algorithms. The prewindowed algorithms are easier to derive than the covariance algorithms, and are therefore considered first.

6.3. PREWINDOWED LS ALGORITHMS

The first step needed to derive any of the LS algorithms in this chapter is to specify the data vector $\mathbf{Y}(T)$ in (6.1.1) and the effect of the shift operator z^{-j}. Prewindowed LS estimation was discussed in chapter 3 in the context of block processing. Recall that for this case a sequence of data values originating from time zero is assumed, i.e. $y(0), y(1), \ldots, y(T)$ where $y(T)$ is the current data value. The value of T_0 in (6.1.19) is therefore zero. In order to calculate prediction residuals for times $j<N$, where N is the order of the LS predictor, the data $y(T)$ for $T<0$ is assumed to be zero. The prewindowed data vector and shifted data vector (by j samples) are therefore, respectively,

$$\mathbf{Y}(T) = [y(T)\ y(T-1)\ \cdots\ y(0)]' \in \mathbf{R}^{T+1} \tag{6.3.1}$$

and

$$z^{-j}\mathbf{Y}(T) = [y(T-j)\ \cdots\ y(0)\ 0\ \cdots\ 0\,]' \in \mathbf{R}^{T+1}. \tag{6.3.2}$$

Using these definitions, the subspace spanned by

$$\{z^{-l}\mathbf{Y}(T)\ z^{-l-1}\mathbf{Y}(T)\ \cdots\ z^{-n}\mathbf{Y}(T)\},$$

where $n>l$, is defined as $M(T|l,n)$. The matrix of basis vectors which span $M(T|l,n)$ is

$$\mathbf{S}(T|l,n) = [z^{-l}\mathbf{Y}(T)\ z^{-l-1}\mathbf{Y}(T)\ \cdots\ z^{-n}\mathbf{Y}(T)] \in \mathbf{R}^{T+1}\times\mathbf{R}^{n-l+1}, \tag{6.3.3}$$

and in particular,

$$\mathbf{S}(T|0,n) = \begin{bmatrix} y(T) & y(T-1) & \cdots & y(T-n+1) & y(T-n) \\ y(T-1) & y(T-2) & \cdots & \vdots & \vdots \\ y(T-2) & & & \vdots & y(0) \\ \vdots & \vdots & & y(0) & 0 \\ \vdots & \vdots & \ddots & 0 & 0 \\ \vdots & & \ddots & \vdots & \vdots \\ & y(0) & & \vdots & \vdots \\ y(0) & 0 & \cdots & 0 & 0 \end{bmatrix}. \tag{6.3.4}$$

The nth order forward and backward residuals are defined, respectively, as

$$e_f(T|n) = y(T) - \sum_{j=1}^{n} f_j(T|n)y(T-j) \tag{6.3.5a}$$

and

$$e_b(T|n) = y(T-n) - \sum_{j=0}^{n-1} b_{j+1}(T|n)y(T-j), \tag{6.3.5b}$$

where $f_j(T|n)$ and $b_j(T|n)$, $1\le j\le n$, are respectively the forward and backward prediction coefficients at time T. Similarly, the nth order forward and backward residual *vectors* are defined as

$$\mathbf{E}_f(T|n) \equiv \mathbf{Y}(T) - \sum_{j=1}^{n} f_j(T|n)[z^{-j}\mathbf{Y}(T)]$$
$$= \mathbf{Y}(T) - \mathbf{S}(T|1,n)\mathbf{f}(T|n) \tag{6.3.6a}$$

and

$$\mathbf{E}_b(T|n) \equiv z^{-n}\mathbf{Y}(T) - \sum_{j=0}^{n-1} b_j(T|n)[z^{-j}\mathbf{Y}(T)]$$
$$= z^{-n}\mathbf{Y}(T) - \mathbf{S}(T|0,n-1)\mathbf{b}(T|n) , \tag{6.3.6b}$$

where $\mathbf{f}(T|n)$ and $\mathbf{b}(T|n)$ are the n-dimensional forward and backward prediction vectors at time T. Notice that

$$\mathbf{E}_f(T|n) = [e_f(T|n)\; e_f(T-1|n)\; \cdots \; e_f(0|n)]' \tag{6.3.7a}$$

and

$$\mathbf{E}_b(T|n) = [e_b(T|n)\; e_b(T-1|n)\; \cdots \; e_b(0|n)]' . \tag{6.3.7b}$$

The residual $e_f(j|n)$ in (6.3.7a) refers to the forward residual obtained at time j using the prediction coefficients computed at time T.

Our objective is to choose the forward prediction coefficient vector $\mathbf{f}(T|n)$ to minimize the forward LS cost function

$$\epsilon_f(T|n) \equiv \sum_{j=0}^{T} e_f^2(j|n) = ||\mathbf{E}_f(T|n)||^2 . \tag{6.3.8}$$

From the discussion in section 6.1, this implies that

$$\mathbf{E}_f(T|n) = \mathbf{P}^{\perp}_{M(T|1,n)}\mathbf{Y}(T) . \tag{6.3.9}$$

Similarly, minimization of the backward LS cost function

$$\epsilon_b(T|n) \equiv \sum_{j=0}^{T} e_b^2(j|n) = ||\mathbf{E}_b(T|n)||^2 \tag{6.3.10}$$

implies that

$$\mathbf{E}_b(T|n) = \mathbf{P}^{\perp}_{M(T|0,n-1)}[z^{-n}\mathbf{Y}(T)] . \tag{6.3.11}$$

The projection updates in section 6.2 can now be used to derive order and time updates for the LS variables $\mathbf{E}_f$, $\mathbf{E}_b$, $\mathbf{f}$, and $\mathbf{b}$. Referring to the projection order update equation (6.2.6), M can be replaced by $M(T|1,n)$, $\mathbf{X}$ can be replaced by $z^{-n-1}\mathbf{Y}(T)$, and $\mathbf{Y}$ can be replaced by the data vector $\mathbf{Y}(T)$ to obtain

$$\mathbf{P}^{\perp}_{M(T|1,n+1)}\mathbf{Y}(T) = \mathbf{P}^{\perp}_{M(T|1,n)}\mathbf{Y}(T) - \mathbf{P}_{\{z^{-1}\mathbf{E}_b(T|n)\}}\mathbf{Y}(T) , \tag{6.3.12}$$

where the "shifted" backward residual is defined as

$$z^{-1}\mathbf{E}_b(T|n) \equiv \mathbf{P}^{\perp}_{M(T|1,n)}[z^{-n-1}\mathbf{Y}(T)] . \tag{6.3.13}$$

Writing (6.3.12) out explicitly gives

$$\mathbf{E}_f(T|n+1) = \mathbf{E}_f(T|n) - \frac{<\mathbf{Y}(T),z^{-1}\mathbf{E}_b(T|n)>}{||z^{-1}\mathbf{E}_b(T|n)||^2}[z^{-1}\mathbf{E}_b(T|n)] . \tag{6.3.14}$$

Similarly, replacing $\mathbf{Y}(T)$ in (6.2.6) by $z^{-n-1}\mathbf{Y}(T)$, M by $M(T|1,n)$, and $\mathbf{X}$ by the data vector gives

$$\mathbf{P}^{\perp}_{M(T|0,n)}[z^{-n-1}\mathbf{Y}(T)] = \mathbf{P}^{\perp}_{M(T|1,n)}[z^{-n-1}\mathbf{Y}(T)] - \mathbf{P}_{\{\mathbf{E}_f(T|n)\}}[z^{-n-1}\mathbf{Y}(T)] . \tag{6.3.15}$$

Writing this out explicitly gives

$$\mathbf{E}_b(T|n+1) = z^{-1}\mathbf{E}_b(T|n) - \frac{<z^{-n-1}\mathbf{Y}(T),\mathbf{E}_f(T|n)>}{||\mathbf{E}_f(T|n)||^2}\mathbf{E}_f(T|n) . \tag{6.3.16}$$

For any subspace M and vectors $\mathbf{Y}$ and $\mathbf{X}$, $\mathbf{P}^{\perp}_M\mathbf{X}$ is orthogonal to any vector contained in M and hence

$$<\mathbf{P}^{\perp}_M\mathbf{X},\mathbf{P}_M\mathbf{Y}> = 0 . \tag{6.3.17}$$

It follows that

$$\begin{aligned} <\mathbf{Y}(T),z^{-1}\mathbf{E}_b(T|n)> &= <\mathbf{P}^{\perp}_{M(T|1,n)}\mathbf{Y}(T),\mathbf{P}^{\perp}_{M(T|1,n)}[z^{-n-1}\mathbf{Y}(T)]> \\ &= <\mathbf{E}_f(T|n),z^{-n-1}\mathbf{Y}(T)> \\ &= <\mathbf{E}_f(T|n),z^{-1}\mathbf{E}_b(T|n)> . \end{aligned} \tag{6.3.18}$$

Denoting this inner product as

$$K_{n+1}(T) \equiv <\mathbf{E}_f(T|n),z^{-1}\mathbf{E}_b(T|n)> , \tag{6.3.19}$$

(6.3.14) and (6.3.16) can be rewritten as

$$\mathbf{E}_f(T|n+1) = \mathbf{E}_f(T|n) - \frac{K_{n+1}(T)}{\epsilon_b(T-1|n)}[z^{-1}\mathbf{E}_b(T|n)] \tag{6.3.20a}$$

and

$$\mathbf{E}_b(T|n+1) = [z^{-1}\mathbf{E}_b(T|n)] - \frac{K_{n+1}(T)}{\epsilon_f(T|n)}\mathbf{E}_f(T|n) . \tag{6.3.20b}$$

Taking norms of both sides of (6.3.20) gives the order recursions for $\epsilon_f(T|n)$ and $\epsilon_b(T|n)$,

$$\epsilon_f(T|n+1) = \epsilon_f(T|n) - \frac{K^2_{n+1}(T)}{\epsilon_b(T-1|n)} \tag{6.3.21a}$$

and

$$\epsilon_b(T|n+1) = \epsilon_b(T-1|n) - \frac{K^2_{n+1}(T)}{\epsilon_f(T|n)} . \tag{6.3.21b}$$

If the LS inner product is replaced by the MMSE inner product (expectation), recursions (6.3.20) become the order recursions for the MMSE lattice where

$$K_{n+1}(T) = E\,[e_f(T|n)e_b(T-1|n)]\ , \tag{6.3.22}$$

$$\epsilon_f(T|n) = E\,[e_f^2(T|n)]\ , \tag{6.3.23}$$

and

$$\epsilon_b(T|n) = E\,[e_b^2(T|n)]\ . \tag{6.3.24}$$

Assuming the input is stationary, then the mean squared forward residual equals the mean squared backward residual, so that using (6.3.22-23) implies that

$$\frac{K_{n+1}(T)}{\epsilon_b(T-1|n)} = \frac{K_{n+1}(T)}{\epsilon_f(T|n)} = k_{n,opt}\ ,$$

where $k_{n,opt}$ is the optimal MMSE PARCOR coefficient given by (4.2.26) and is constant. In addition, (6.3.21) translates to

$$\begin{aligned} E\,[e_f^2(T|n+1)] &= E\,[e_f^2(T|n)] - \frac{\{E\,[e_f(T|n)e_b(T-1|n)]\}^2}{E\,[e_b^2(T-1|n)]} \\ &= (1 - k_{n+1,opt}^2)E\,[e_f^2(T|n)]\ . \end{aligned} \tag{6.3.25}$$

Using the definitions (6.3.6a) and (6.3.6b), equations (6.3.20a) and (6.3.20b) can be used to obtain order updates for the prediction vectors $\mathbf{f}\,(T|n)$ and $\mathbf{b}\,(T|n)$. Equation (6.3.20a) can be rewritten as

$$\begin{aligned} \mathbf{Y}(T) - \mathbf{S}(T|1,n+1)\mathbf{f}\,(T|n+1) &= \mathbf{Y}(T) - \mathbf{S}(T|1,n)\mathbf{f}\,(T|n) \\ &- \frac{K_{n+1}(T)}{\epsilon_b(T-1|n)}\left[z^{-n-1}\mathbf{Y}(T) - \mathbf{S}(T|1,n)\mathbf{b}\,(T-1|n)\right], \end{aligned}$$

or

$$\begin{aligned} &\mathbf{S}(T|1,n)\left\{[\mathbf{f}\,(T|n+1)]_{1,n} - \mathbf{f}\,(T|n) + \frac{K_{n+1}(T)}{\epsilon_b(T-1|n)}\,\mathbf{b}\,(T-1|n)\right\} \\ &+ [z^{-n-1}\mathbf{Y}(T)]\left[f_{n+1}(T|n+1) - \frac{K_{n+1}(T)}{\epsilon_b(T-1|n)}\right] = 0\ , \end{aligned} \tag{6.3.26}$$

where $[\mathbf{f}\,(T|n+1)]_{l,m}$ denotes the lth through mth components of $\mathbf{f}\,(T|n+1)$. Assuming the vectors $z^{-1}\mathbf{Y}(T), \ldots, z^{-n}\mathbf{Y}(T)$ are linearly independent implies that

$$f_{n+1}(T|n+1) = \frac{K_{n+1}(T)}{\epsilon_b(T-1|n)} \tag{6.3.27a}$$

and

$$[\mathbf{f}\,(T|n+1)]_{1,n} = \mathbf{f}\,(T|n) - f_{n+1}(T|n+1)\mathbf{b}\,(T-1|n)\ . \tag{6.3.27b}$$

Similarly, expanding (6.3.20b) in an analogous fashion gives the order update for the backward prediction vector,

$$b_1(T|n+1) = \frac{K_{n+1}(T)}{\epsilon_f(T|n)} \tag{6.3.28a}$$

$$[\mathbf{b}(T|n+1)]_{2,n} = \mathbf{b}(T-1|n) - b_1(T|n+1)\ \mathbf{f}(T|n)\ . \tag{6.3.28b}$$

Equations (6.3.27) and (6.3.28) are the LS analogues of the Levinson, or MMSE, recursions (4.3.6-8). It is particularly interesting to find from (6.3.20) that the prewindowed LS residuals $\mathbf{E}_f(T|n)$ and $\mathbf{E}_b(T|n)$ satisfy the lattice structure in figure 4-9.

Having completed the derivation of prewindowed LS order recursions, we proceed to the derivation of time updates. It is necessary to define the "causal" forward residual vector,

$$\mathbf{E}_f^o(T|n) \equiv \mathbf{P}^{\perp}_{M_{T-1}(T|1,n)}\mathbf{Y}(T) = \mathbf{Y}(T) - \mathbf{S}(T|1,n)\mathbf{f}(T-1|n)\ , \tag{6.3.29}$$

which is the forward residual vector at time T using the LS forward prediction vector computed at time $T-1$. The superscript "o" stands for "oblique", and will be explained shortly. A time update for the forward prediction vector $\mathbf{f}$ can now be obtained by replacing $M(T)$ by $M(T|1,n)$ in (6.2.34),

$$\begin{aligned} \mathbf{P}^{\perp}_{M(T|1,n)}\mathbf{Y}(T) &= \mathbf{P}^{\perp}_{M_{T-1}(T|1,n)}\mathbf{Y}(T) \\ &- \mathbf{P}_{M(T|1,n)}\ \mathbf{u}(T)\ \langle\mathbf{u}(T),\mathbf{P}^{\perp}_{M(T|1,n)}\mathbf{Y}(T)\rangle \sec^2\theta(T|1,n), \end{aligned} \tag{6.3.30}$$

or

$$\mathbf{E}_f(T|n) = \mathbf{E}_f^o(T|n) - [\mathbf{P}_{M(T|1,n)}\mathbf{u}(T)]\ e_f(T|n)\sec^2\theta(T|1,n)\ , \tag{6.3.31}$$

where $\theta(T|1,n)$ is the "angle" between the spaces spanned by the columns of $\mathbf{S}(T|1,n)$ and $\tilde{\mathbf{S}}(T|1,n)$. Equation (6.2.30) implies that

$$\begin{aligned} \sin^2\theta(T|1,n) &= \langle\mathbf{u}(T),\mathbf{P}_{M(T|1,n)}\mathbf{u}(T)\rangle \\ &= \mathbf{y}'(T-1|n)[\mathbf{S}'(T|1,n)\mathbf{S}(T|1,n)]^{-1}\mathbf{y}(T-1|n)\ , \end{aligned} \tag{6.3.32}$$

where $\mathbf{y}(T|n)$, defined in (3.0.2), is the vector of n most recent given data samples and

$$\sec^2\theta(T|1,n) = \frac{1}{1 - \sin^2\theta(T|1,n)}\ . \tag{6.3.33}$$

Notice that

$$\begin{aligned} [\mathbf{S}'(T|1,n)\mathbf{S}(T|1,n)]_{jk} &= [z^{-j}\mathbf{Y}(T)]'[z^{-k}\mathbf{Y}(T)]\ , \\ &= \sum_{m=0}^{T-j} y_m y_{m+(j-k)} \end{aligned} \tag{6.3.34}$$

where $1 \leqslant j \leqslant n$, $1 \leqslant k \leqslant n$. $\mathbf{S}'(T|1,n)\mathbf{S}(T|1,n)$ is therefore the sample covariance matrix, i.e.

$$\hat{\mathbf{\Phi}}(T-1|n) = \mathbf{S}'(T|1,n)\mathbf{S}(T|1,n)\ , \tag{6.3.35}$$

where $\hat{\mathbf{\Phi}}(T-1|n)$ with an arbitrary weighting sequence was defined by (3.4.13).

Using (6.3.6a), (6.3.29), and (6.3.35), equation (6.3.31) can be rewritten as

$$\begin{aligned} \mathbf{S}(T|1,n)\mathbf{f}\,(T|n) &= \mathbf{S}(T|1,n)\mathbf{f}\,(T-1|n) \\ &+ \mathbf{S}(T|1,n)[\hat{\mathbf{\Phi}}^{-1}(T-1|n)\mathbf{y}(T-1|n)]e_f(T|n)\sec^2\theta(T|1,n)\,. \end{aligned} \tag{6.3.36}$$

Premultiplying (6.3.36) by $\mathbf{S}'(T|1,n)$ and then by $\hat{\mathbf{\Phi}}^{-1}(T-1|n)$ gives the forward time update for the forward prediction vector,

$$\mathbf{f}\,(T|n) = \mathbf{f}\,(T-1|n) + \mathbf{g}\,(T-1|n)e_f(T|n)\sec^2\theta(T|1,n) \tag{6.3.37}$$

where

$$\mathbf{g}\,(T|n) \equiv \hat{\mathbf{\Phi}}^{-1}(T|n)\mathbf{y}(T|n) \tag{6.3.38}$$

and from (6.3.32),

$$\sin^2\theta(T|1,n) = \mathbf{g}'(T-1|n)\mathbf{y}(T-1|n)\,. \tag{6.3.39}$$

Looking at the top component of (6.3.31), or equivalently, taking the inner product of both sides with $\mathbf{u}(T)$ gives

$$\begin{aligned} e_f(T|n) &= e_f^o(T|n) - \langle\mathbf{u}(T),\mathbf{P}_{M(T)}\mathbf{u}(T)\rangle e_f(T|n)\sec^2\theta(T|1,n) \\ &= e_f^o(T|n) - e_f(T|n)[\sin^2\theta(T|1,n)][\sec^2\theta(T|1,n)]\,, \end{aligned} \tag{6.3.40}$$

or

$$\begin{aligned} e_f^o(T|n) &= \left[1 + \frac{\sin^2\theta(T|1,n)}{1-\sin^2\theta(T|1,n)}\right]e_f(T|n) \\ &= e_f(T|n)\sec^2\theta(T|1,n)\,, \end{aligned} \tag{6.3.41}$$

where $e_f^o(T|n)$ is the forward prediction error at time T given the prediction coefficients computed at time $T-1$, that is,

$$\begin{aligned} e_f^o(T|n) &= \langle\mathbf{u}(T),\mathbf{E}_f^o(T|n)\rangle \\ &= y(T) - \mathbf{f}'(T-1|n)\mathbf{y}(T-1|n)\,. \end{aligned} \tag{6.3.42}$$

Referring to (6.3.32), $\sin^2\theta(T|1,n)$ can equal one only if the unit vector $\mathbf{u}(T)$ is contained in the space $M(T|1,n)$, which would imply that $\mathbf{u}(T)$ can be expressed as a linear combination of $z^{-1}\mathbf{Y}(T), z^{-2}\mathbf{Y}(T), \ldots, z^{-n}\mathbf{Y}(T)$. In that case it can be shown that the LS residual $e_f(T|n) = 0$ (problem 6-2), and hence the right hand side of (6.3.41) is undefined. It is therefore assumed that the angle $\theta(T|1,n)$ is not "90 degrees". Because of the relation (6.3.41), and because the angle θ is assumed to be oblique, e_f^o has been referred to as the "oblique" forward residual. Substituting (6.3.41) into (6.3.37) yields

$$\mathbf{f}\,(T|n) = \mathbf{f}\,(T-1|n) + \mathbf{g}\,(T-1|n)e_f^o(T|n)\,, \tag{6.3.43}$$

which was derived algebraically in chapter 3. The vector $\mathbf{g}\,(T|n)$ can be interpreted as a Kalman gain which arises from the derivation of recursive LS algorithms using Kalman filtering theory [8].

To derive a time update for the backward prediction vector $\mathbf{b}$, we again use (6.2.34), replacing $M(T)$ by $M(T|0,n-1)$ and $\mathbf{Y}$ by $z^{-n}\mathbf{Y}$, to get

$$\mathbf{E}_b(T|n) = \mathbf{E}_b^o(T|n) + [\mathbf{P}_{M(T|0,n-1)}\mathbf{u}(T)]e_b(T|n)\sec^2\theta(T|0,n-1) , \tag{6.3.44}$$

where $\mathbf{E}_b^o(T|n)$ is the backward residual vector at time T using the backward prediction coefficients computed at time $T-1$, i.e.,

$$\mathbf{E}_b^o(T|n) = z^{-n}\mathbf{Y}(T) - \mathbf{S}(T|0,n-1)\mathbf{b}(T-1|n) . \tag{6.3.45}$$

By definition,

$$\begin{aligned}\mathbf{P}_{M(T|0,n-1)}\mathbf{u}(T) &= \mathbf{S}(T|0,n-1)\hat{\mathbf{\Phi}}^{-1}(T|n)\mathbf{y}(T|n) \\ &= \mathbf{S}(T|0,n-1)\mathbf{g}(T|n) .\end{aligned} \tag{6.3.46}$$

Examining the top component of (6.3.44) yields, in analogy with (6.3.41),

$$e_b^o(T|n) = e_b(T|n)\sec^2\theta(T|0,n-1) . \tag{6.3.47}$$

The angle $\theta(T|0,n-1)$ is assumed to be oblique for analogous reasons to those given following equation (6.3.41). In analogy with the forward prediction case, equations (6.3.44) through (6.3.47) can be manipulated to give the forward time update for the backward prediction vector,

$$\mathbf{b}(T|n) = \mathbf{b}(T-1|n) + \mathbf{g}(T|n)e_b^o(T|n) . \tag{6.3.48}$$

Time updates for the variables $K_n(T)$, $\epsilon_f(T|n)$, and $\epsilon_b(T|n)$ are obtained by using the forward time update formula for inner products (6.2.40). Replacing $\mathbf{v}(T)$ by $\mathbf{Y}(T)$, $\mathbf{Y}(T)$ by $z^{-n-1}\mathbf{Y}(T)$, and $M(T)$ by $M(T|1,n)$ gives the forward time update for the inner product $K_{n+1}(T)$,

$$\begin{aligned}K_{n+1}(T) &= <\mathbf{Y}(T), \mathbf{P}^{\perp}_{M(T|1,n)}[z^{-n-1}\mathbf{Y}(T)]> \\ &= wK_{n+1}(T-1) \\ &\quad +<\mathbf{u}(T),\mathbf{P}^{\perp}_{M(T|1,n)}\mathbf{Y}(T)><\mathbf{u}(T),\mathbf{P}^{\perp}_{M(T|1,n)}z^{-n-1}\mathbf{Y}(T)>\sec^2\theta(T|1,n) \\ &= wK_{n+1}(T-1) + e_f(T|n)e_b(T-1|n)\sec^2\theta(T|1,n) .\end{aligned} \tag{6.3.49}$$

Similarly, replacing $\mathbf{v}(T)$ by $\mathbf{Y}(T)$, and $M(T)$ by $M(T|1,n)$ gives the forward time update for the forward LS cost function,

$$\epsilon_f(T|n) = w\epsilon_f(T-1|n) + e_f^2(T|n)\sec^2\theta(T|1,n)$$

or

$$\epsilon_f(T|n) = w\epsilon_f(T-1|n) + e_f(T|n)e_f^o(T|n) , \tag{6.3.50}$$

and replacing both $\mathbf{v}(T)$ and $\mathbf{Y}(T)$ by $z^{-n}\mathbf{Y}(T)$, and $M(T)$ by $M(T|0,n-1)$ gives the forward time update for the backward LS cost function,

$$\epsilon_b(T|n) = w\epsilon_b(T-1|n) + e_b^2(T|n)\sec^2\theta(T|0,n-1) ,$$

or

$$\epsilon_b(T|n) = w\epsilon_b(T-1|n) + e_b(T|n)e_b^o(T|n) . \tag{6.3.51}$$

Equations (6.3.20), (6.3.21), (6.3.27), (6.3.28), (6.3.43), and (6.3.48-51) constitute a complete set of order and time updates for the variables **f**, **b**, e_f, e_b, ϵ_f, ϵ_b, and K. Order updates for the gains **g** and $\sin^2\theta(T)$ are now derived using (6.2.6). Substituting $M(T|1,n)$ for M, $\mathbf{u}(T)$ for $\mathbf{Y}(T)$, and the data vector $\mathbf{Y}(T)$ for $\mathbf{X}$ gives

$$\mathbf{P}_{M(T|0,n)}\mathbf{u}(T) = \mathbf{P}_{M(T|1,n)}\mathbf{u}(T) + \mathbf{P}_{\{\mathbf{E}_f(T|n)\}}\mathbf{u}(T) . \tag{6.3.52}$$

Similarly, replacing M by $M(T|0,n-1)$, $\mathbf{X}$ by $z^{-n}\mathbf{Y}(T)$, and $\mathbf{Y}(T)$ by $\mathbf{u}(T)$ yields

$$\mathbf{P}_{M(T|0,n)}\mathbf{u}(T) = \mathbf{P}_{M(T|0,n-1)}\mathbf{u}(T) + \mathbf{P}_{\{\mathbf{E}_b(T|n)\}}\mathbf{u}(T) . \tag{6.3.53}$$

Recursions for $\sin^2\theta(T)$ can be obtained by taking the inner products of both sides of (6.3.52) and (6.3.53) with $\mathbf{u}(T)$. Using (6.3.32) and (6.3.52) gives

$$\sin^2\theta(T|0,n) = \sin^2\theta(T|1,n) + \frac{e_f^2(T|n)}{\epsilon_f(T|n)} , \tag{6.3.54}$$

and using (6.3.32) and (6.3.53) gives

$$\sin^2\theta(T|0,n) = \sin^2\theta(T|0,n-1) + \frac{e_b^2(T|n)}{\epsilon_b(T|n)} . \tag{6.3.55}$$

For notational convenience, define the variable

$$\gamma(T|n) \equiv \sin^2\theta(T|0,n-1) = \mathbf{y}'(T|n)\hat{\mathbf{\Phi}}^{-1}(T|n)\mathbf{y}(T|n) , \tag{6.3.56}$$

so that (6.3.54) and (6.3.55) can be rewritten as

$$\gamma(T|n+1) = \gamma(T-1|n) + \frac{e_f^2(T|n)}{\epsilon_f(T|n)} \tag{6.3.57a}$$

and

$$\gamma(T|n+1) = \gamma(T|n) + \frac{e_b^2(T|n)}{\epsilon_b(T|n)} . \tag{6.3.57b}$$

An order update for **g** is obtained by rewriting (6.3.52) as,

$$\begin{aligned}\mathbf{S}(T|0,n)\mathbf{g}(T|n+1) &= \mathbf{S}(T|1,n)\mathbf{g}(T-1|n)\\ &\quad + \frac{e_f(T|n)}{\epsilon_f(T|n)}\,[\mathbf{Y}(T) - \mathbf{S}(T|1,n)\mathbf{f}(T|n)] ,\end{aligned}$$

or

$$\begin{aligned}&\mathbf{S}(T|1,n)\left\{[\mathbf{g}(T|n+1)]_{2,n+1} - \mathbf{g}(T-1|n) + \frac{e_f(T|n)}{\epsilon_f(T|n)}\,\mathbf{f}(T|n)\right\}\\ &\quad + \mathbf{Y}(T)\left\{[\mathbf{g}(T|n+1)]_1 - \frac{e_f(T|n)}{\epsilon_f(T|n)}\right\} = 0\end{aligned} \tag{6.3.58}$$

where $[\mathbf{g}(T|n+1)]_{l,m}$ is the vector formed by the lth through the mth components of $\mathbf{g}(T|n+1)$. Assuming

$$\mathbf{Y}(T),\ z^{-1}\mathbf{Y}(T),\ \ldots,\ z^{-n}\mathbf{Y}(T)$$

are linearly independent gives

$$[\mathbf{g}(T|n+1)]_1 = \frac{e_f(T|n)}{\epsilon_f(T|n)} \tag{6.3.59a}$$

$$[\mathbf{g}(T|n+1)]_{2,n+1} = \mathbf{g}(T-1|n) - [\mathbf{g}(T|n+1)]_1\mathbf{f}(T|n)\ . \tag{6.3.59b}$$

Similarly, using (6.3.53) gives

$$\begin{aligned} \mathbf{S}(T|0,n-1)&\left\{\mathbf{g}(T|n+1)_{1,n} - \mathbf{g}(T|n) + \frac{e_b(T|n)}{\epsilon_b(T|n)}\,\mathbf{b}(T|n)\right\} \\ &+ z^{-n}\mathbf{Y}(T)\left\{[\mathbf{g}(T|n+1)]_{n+1} - \frac{e_b(T|n)}{\epsilon_b(T|n)}\right\} = 0\ , \end{aligned} \tag{6.3.60}$$

and assuming $\mathbf{S}(T|0,n)$ is of full rank, it follows that

$$[\mathbf{g}(T|n+1)]_{n+1} = \frac{e_b(T|n)}{\epsilon_b(T|n)} \tag{6.3.61a}$$

$$\mathbf{g}(T|n) = [\mathbf{g}(T|n+1)]_{1,n} + [\mathbf{g}(T|n+1)]_{n+1}\mathbf{b}(T|n)\ . \tag{6.3.61b}$$

Because the backward prediction vector $\mathbf{b}(T|n)$ must be computed before $\mathbf{g}(T|n)$ can be computed from (6.3.61), and because the time update for $\mathbf{b}(T|n)$ given by (6.3.48) uses the value of $\mathbf{g}(T|n)$, it is necessary to combine (6.3.48) and (6.3.61) to obtain

$$[\mathbf{g}(T|n+1)]_{1,n} + [\mathbf{g}(T|n+1)]_{n+1}\,\mathbf{b}(T|n) = \frac{\mathbf{b}(T|n) - \mathbf{b}(T-1|n)}{e_b^o(T|n)}\ ,$$

or

$$\mathbf{b}(T|n) = \frac{\mathbf{b}(T-1|n) + e_b^o(T|n)[\mathbf{g}(T|n+1)]_{1,n}}{1 - e_b^o(T|n)[\mathbf{g}(T|n+1)]_{n+1}}\ . \tag{6.3.62}$$

This completes the derivation of prewindowed LS prediction recursions. Combining these recursions in an appropriate fashion gives both a LS fixed-order (transversal), or fast Kalman, algorithm and a LS order-recursive (lattice) algorithm. In particular, the fast Kalman algorithm is obtained by combining (6.3.42), (6.3.43) (6.3.50), (6.3.59), (6.3.61), and (6.3.62) as follows,

$$e_f^o(T) = y(T) - \mathbf{f}'(T-1)\mathbf{y}(T-1|N) \tag{6.3.63a}$$

$$\mathbf{f}(T) = \mathbf{f}(T-1) + e_f^o(T)\mathbf{g}(T-1|N) \tag{6.3.63b}$$

$$e_f(T) = y(T) - \mathbf{f}'(T)\mathbf{y}(T-1|N) \tag{6.3.63c}$$

$$\epsilon_f(T) = w\epsilon_f(T-1) + e_f(T)e_f^o(T) \tag{6.3.63d}$$

$$[\mathbf{g}(T|N+1)]_1 = \frac{e_f(T)}{\epsilon_f(T)} \tag{6.3.63e}$$

$$[\mathbf{g}(T|N+1)]_{2,N+1} = \mathbf{g}(T-1|N) - [\mathbf{g}(T|N+1)]_1\mathbf{f}(T) \tag{6.3.63f}$$

$$e_b^o(T) = y(T-N) - \mathbf{b}'(T-1)\mathbf{y}(T|N) \tag{6.3.63g}$$

$$\mathbf{b}(T) = \frac{\mathbf{b}(T-1) + e_b^o(T)[\mathbf{g}(T|N+1)]_{1,N}}{1 - e_b^o(T)[\mathbf{g}(T|N+1)]_{N+1}} \tag{6.3.63h}$$

$$\mathbf{g}(T|N) = [\mathbf{g}(T|N+1)]_{1,n} + [\mathbf{g}(T|N+1)]_{N+1}\mathbf{b}(T) \, . \tag{6.3.63i}$$

Where unspecified, the order of the variable is assumed to be N, the order of the filter. The fast Kalman algorithm essentially propagates the three vectors, or "transversal filters", **f**, **b**, and **g**. The LS lattice algorithm is obtained by combining (6.3.20) (top component only), (6.3.21), (6.3.49), and (6.3.57b),

$$K_{n+1}(T) = wK_{n+1}(T-1) + e_f(T|n)e_b(T-1|n)\frac{1}{1-\gamma(T-1|n)} \tag{6.3.64a}$$

$$k_{n+1}^f(T) = \frac{K_{n+1}(T)}{\epsilon_f(T|n)}, \qquad k_{n+1}^b(T) = \frac{K_{n+1}(T)}{\epsilon_b(T-1|n)} \tag{6.3.64b}$$

$$e_f(T|n+1) = e_f(T|n) - k_{n+1}^b(T)\, e_b(T-1|n) \tag{6.3.64c}$$

$$e_b(T|n+1) = e_b(T-1|n) - k_{n+1}^f(T)\, e_f(T|n) \tag{6.3.64d}$$

$$\epsilon_f(T|n+1) = \epsilon_f(T|n) - k_{n+1}^b(T)K_{n+1}(T) \tag{6.3.64e}$$

$$\epsilon_b(T|n+1) = \epsilon_b(T-1|n) - k_{n+1}^f(T)K_{n+1}(T) \tag{6.3.64f}$$

$$\gamma(T|n+1) = \gamma(T|n) + \frac{e_b^2(T|n)}{\epsilon_b(T|n)} \, . \tag{6.3.64g}$$

At each iteration $T>N$ these recursions are computed for $n=0$ to $n=N-1$. The interleaving of order and time updates for the LS lattice algorithm is illustrated in figure 6-7, which shows the sequence of updates for a particular variable as a path in two dimensions. For example, figure 6-7a corresponds to the computation of the residual $e_f(T|n)$. At each new iteration $e_f(T|0)$ is updated in time by the new received sample $y(T)$, and then the residuals $e_f(T|1)$, $e_f(T|2)$, and so forth, are computed via order updates. On the other hand, the coefficients $K_n(T)$, $n = 1, \ldots, N$, are updated in time only, which corresponds to the computation path shown in figure 6-7b.

Both algorithms (6.3.63) and (6.3.64) require on the order of N arithmetic operations per iteration. This represents a considerable savings over straightforward techniques for solving the prewindowed LS problem, such as inverting the sample covariance matrix at each iteration. If division is counted the same as a multiplication, the fixed-order algorithm (6.3.63) requires $8N + 5$ multiplications and $7N + 2$ additions per iteration, and the lattice algorithm (6.3.64)

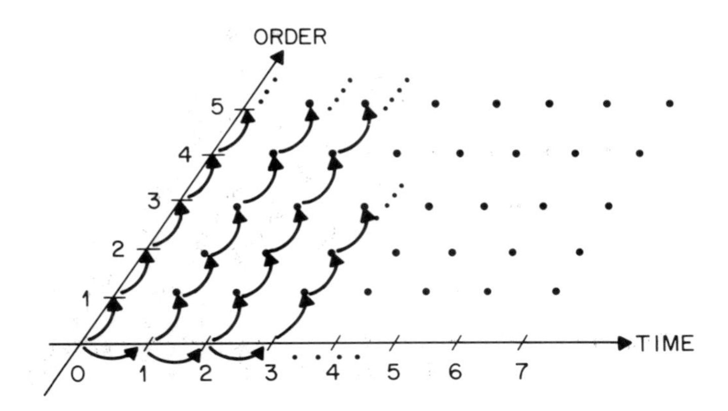

Figure 6-7a. Sequence of computations for time update followed by order updates.

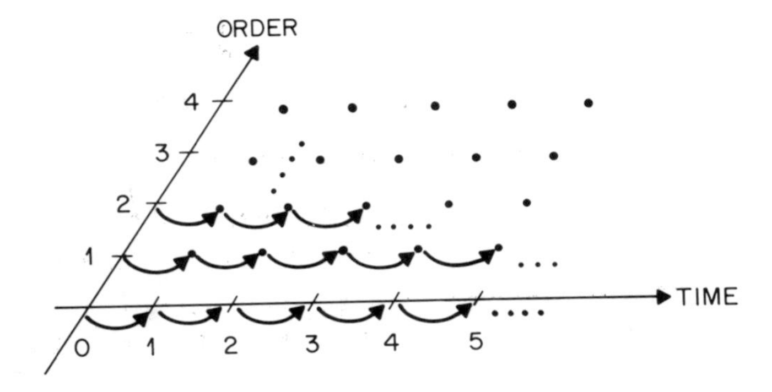

Figure 6-7b. Sequence of computations for time update only.

requires $11N$ multiplies and $6N$ additions per iteration. The order-recursive algorithm allows the order of the filter to be chosen dynamically; however, if the desired order of the adaptive filter is known in advance, the fixed order algorithm offers a modest savings in computational complexity.

Both algorithms (6.3.63) and (6.3.64) can be modified by using other recursions derived in this section. For example, the order updates (6.3.64e) and (6.3.64f) can be equivalently replaced by the time updates (6.3.50) and (6.3.51). The following modification to the transversal algorithm, which reduces the number of required multiplies per iteration, has been proposed [24]. Recursion (6.3.63c) can be replaced by the following equation, which is obtained from (6.3.41),

$$e_f(T|N) = e_f{}^o(T|N)\{1-\gamma(T-1|N)\} . \tag{6.3.65}$$

An additional update for γ is now required. Substituting (6.3.59a), (6.3.61a), and (6.3.47) into (6.3.57a) and (6.3.57b), and solving for $\gamma(T|n)$ gives

$$\gamma(T|N) = \frac{\gamma(T-1|N) + e_f(T|N)\,[\mathbf{g}(T|N+1)]_1 - e_b^o(T|N)[\mathbf{g}(T|N+1)]_{N+1}}{1 - e_b^o(T|N)\,[\mathbf{g}(T|N+1)]_{N+1}} . \tag{6.3.66}$$

More important than the modest reduction in multiplies per iteration offered by this modified algorithm is the fact that it is less sensitive to roundoff errors than the fast Kalman algorithm (6.3.63). More recently, additional, more extensive modifications have been proposed for the fast Kalman algorithm, resulting in new fixed-order algorithms with fewer multiplies per iteration and superior numerical properties [14, 13]. (See also problem 6-9.)

Initialization of the lattice algorithm can be accomplished as follows,

$$\gamma(-1|n) = e_b(-1|n) = K_{n+1}(-1) = 0 \quad \text{for } 0 \leqslant n \leqslant N . \tag{6.3.67a}$$

At each iteration T,

$$\gamma(T|0) = 0 , \tag{6.3.67b}$$

$$e_f(T|0) = e_b(T|0) = y(T) , \tag{6.3.67c}$$

and

$$\epsilon_b(T|0) = \epsilon_f(T|0) = w\epsilon_f(T-1|0) + y^2(T) . \tag{6.3.67d}$$

The order updates (6.3.64c) and (6.3.64d) cannot be computed unless $\epsilon_b(T-1|n) \neq 0$ and $\epsilon_f(T|n) \neq 0$. Each successive stage of the lattice must therefore be "turned on" at each successive time instant when $T<N$. In particular, the lattice recursions can only be computed up to order T. An *exact* LS solution is obtained at each iteration only by increasing the order of the filter by one at each successive iteration when $T < N$.

There is an alternative initialization technique, in which the cost functions $\epsilon_b(T|n)$ and $\epsilon_f(T|n)$ are initialized to some small constant $\delta>0$, and this ensures that the algorithm is stable. In this case, at each iteration recursions (6.3.64) are computed for $n=0$ to $n=N$. While this is not an exact LS solution, it generally does not cause significant deviation from the true LS solution.

In analogy with the lattice algorithm, the transversal algorithm can also be initialized in more than one way. One way is to set

$$\mathbf{f}(0) = \mathbf{b}(0) = \mathbf{g}(0) = \mathbf{0}, \text{ and } \epsilon_f(0) = \delta, \tag{6.3.68}$$

where $\mathbf{0}$ is the vector with all elements equal to zero and δ is a small constant which ensures stability. As with the lattice algorithm, this type of initialization destroys the exactness of the LS solution computed at each iteration. It can be shown, however, that the initialization (6.3.68) causes the minimization of a slightly modified LS cost function (see problem 6-5). In order to compute an *exact* LS solution at each iteration using the fast Kalman algorithm, the order of the filter must be increased by one at each successive iteration when $T < N$. Perhaps the most straightforward order-recursive initialization routine for (6.3.63) is to use the lattice algorithm (6.3.64), in conjunction with the order recursions for the prediction vectors (6.3.27) and (6.3.28). An order-recursive initialization routine for the fixed-order algorithm (6.3.63), (6.3.65), and (6.3.66) that requires much less computation than this straightforward approach has been proposed [13]. This initialization routine depends upon certain properties of the vectors **f**, **b**, and **g** when $T < N$. In particular, given the data samples

$$y(0), y(1), \ldots, y(T),$$

there exist T coefficients

$$f_1(T), f_2(T), \ldots, f_T(T)$$

such that

$$\begin{bmatrix} y(1) \\ y(2) \\ \cdot \\ \cdot \\ \cdot \\ \\ y(T) \end{bmatrix} = \begin{bmatrix} y(0) & 0 & 0 & \cdots 0 \\ y(1) & y(0) & 0 & \cdots 0 \\ \cdot & \cdot & y(0) & \cdot \\ \cdot & \cdot & \cdot & \cdot \\ \cdot & \cdot & \cdot & \cdot \\ & & & 0 \\ y(T-1) & y(T-2) & y(T-3) & \cdots y(0) \end{bmatrix} \begin{bmatrix} f_1(T) \\ f_2(T) \\ \cdot \\ \cdot \\ \cdot \\ \\ f_T(T) \end{bmatrix}, \tag{6.3.69}$$

and hence

$$e_f(j|T) = y(j) - \sum_{m=1}^{T} f_m(T)\, y(j-m) = 0, \quad 1 \leqslant j \leqslant T. \tag{6.3.70}$$

The coefficients $f_m(T)$, $m = 1, \ldots, T$, are therefore the LS prediction coefficients. Letting

$$e_f{}^o(j|j-1) = y(j) - \sum_{m=1}^{j-1} f_m(T)\, y(j-m), \tag{6.3.71}$$

it is easily verified that

$$f_j(T) = \frac{e_f{}^o(j|j-1)}{y(0)}, \qquad j = 1, \ldots, T. \tag{6.3.72}$$

The LS cost function at time $T < N$ is

$$\epsilon_f(T|T) = \sum_{j=0}^{T} e_f^2(j|T) = y^2(0), \tag{6.3.73}$$

since $e_f(0|T) = y(0)$. The coefficient $f_j(T)$ depends only upon the first $j+1$ data samples, and is *independent* of T, assuming $T > j$. The forward prediction vector $\mathbf{f}(T)$ can therefore be computed from (6.3.71) and (6.3.72) when $T < N$. Only one element, $f_T(T)$, needs to be computed at iteration T, since the remaining nonzero coefficients $f_1(T), \ldots, f_T(T-1)$ remain constant until $T = N+1$, in which case (6.3.63b) is used to compute $\mathbf{f}(N+1|N)$.

The backward prediction vector $\mathbf{b}(T)$ can be conveniently computed when $T < N$ by observing that

$$\epsilon_b(T-1|T) = \sum_{m=0}^{T-1} [y(m-T) - \mathbf{b}'(T-1|T)\mathbf{y}(m|T)]^2$$
$$= \sum_{m=0}^{T-1} [\mathbf{b}'(T-1|T)\mathbf{y}(m|T)]^2 \qquad (6.3.74)$$

is minimized by setting

$$\mathbf{b}(T-1|T) = \mathbf{0} \; . \qquad (6.3.75)$$

By definition,

$$e_b{}^o(T|T) = y(0) - \mathbf{b}'(T-1|T)\; \mathbf{y}(T|T) = y(0) \; , \qquad (6.3.76)$$

and hence from (6.3.48),

$$\mathbf{b}(T|T) = y(0)\; \mathbf{g}(T|T) \; . \qquad (6.3.77)$$

Combining (6.3.42), (6.3.43), (6.3.50), (6.3.57a), (6.3.59), (6.3.65), (6.3.71)-(6.3.73), and (6.3.77) gives the initialization routine for the fixed-order prewindowed LS algorithm,

$$\mathbf{f}(0|j) = \mathbf{b}(0|j) = \mathbf{g}(0|j) = \mathbf{0}, \quad j = 0, 1, \ldots, N \; . \qquad (6.3.78a)$$

$$\gamma(0|0) = 0 \qquad (6.3.78b)$$

For T=1, 2, . . . , N, compute

$$e_f{}^o(T|T-1) = y(T) - \mathbf{f}'(T-1|T-1)\mathbf{y}(T-1|T-1) \qquad (6.3.79a)$$

$$[\mathbf{f}(T|T)]_T = \frac{e_f{}^o(T|T-1)}{y(0)} \quad (\text{rest of } \mathbf{f}(T|T) = \mathbf{f}(T-1|T-1)\;) \qquad (6.3.79b)$$

$$\mathbf{f}(T|T-1) = \mathbf{f}(T-1|T-1) + e_f{}^o(T|T-1)\; \mathbf{g}(T-1|T-1) \qquad (6.3.79c)$$

$$e_f(T|T-1) = e_f{}^o(T|T-1)\; [1 - \gamma(T-1|T-1)] \qquad (6.3.79d)$$

$$\epsilon_f(T|T-1) = w\; y^2(0) + e_f(T|T-1)\; e_f{}^o(T|T-1) \qquad (6.3.79e)$$

$$[\mathbf{g}(T|T)]_1 = \frac{e_f(T|T-1)}{\epsilon_f(T|T-1)} \qquad (6.3.79f)$$

$$[\mathbf{g}(T|T)]_{2,T} = \mathbf{g}(T-1|T-1) - [\mathbf{g}(T|T)]_1\; \mathbf{f}(T|T-1) \qquad (6.3.79g)$$

$$\mathbf{b}(T|T) = y(0)\mathbf{g}(T|T) \qquad (\text{only when } T=N\;) \qquad (6.3.79h)$$

$$\gamma(T|T) = \gamma(T-1|T-1) + [\mathbf{g}(T|T)]_1\; e_f(T|T-1) \; . \qquad (6.3.79i)$$

Because only the first T elements of the vectors **f**, **b**, and **g** are computed at each iteration, this initialization routine requires approximately 20% of the computation required by the steady-state fixed-order algorithm. It has been found that the fixed-order steady state algorithm displays improved numerical properties when used with the order-recursive routine, (6.3.79), rather than with the initialization specified by (6.3.68) [13].

6.3.1. Prewindowed Whitening Filter Algorithm

In chapter 4 it was shown how the lattice structure can be conveniently used to generate the forward and backward prediction filter polynomials $F_n(z)$ and $B_n(z)$ given the (time invariant) PARCOR coefficients k_n, $n = 1,2,\ldots,N$. That discussion is now generalized by showing how the prediction filters which enter the prewindowed LS transversal algorithm can be generated from prewindowed LS lattice variables at time T. The vector of shift operators is first defined,

$$\mathbf{z}_n = [1 \;\; z^{-1} \;\; z^{-2} \;\; \cdots \;\; z^{-n+1}]' . \tag{6.3.80}$$

The LS forward and backward prediction, or whitening, filters can now be defined, respectively, as

$$\begin{aligned} F_{T|n}(z) &= 1 - \sum_{j=1}^{n} f_j(T) z^{-j} \\ &= 1 - \mathbf{f}'(T|n) \; [z^{-1} \; \mathbf{z}_n] \end{aligned} \tag{6.3.81a}$$

and

$$\begin{aligned} B_{T|n}(z) &= z^{-n} - \sum_{j=0}^{n-1} b_j(T|n) z^{-j} \\ &= z^{-n} - \mathbf{b}'(T|n)\mathbf{z}_n \;, \end{aligned} \tag{6.3.81b}$$

where the time index T indicates that the filters are time-varying.

An order recursion for the whitening filter $F_{T|n}(z)$ can be obtained by premultiplying both sides of (6.3.27b) by $\mathbf{z}_n'$, and adding (6.3.27a) to get

$$\begin{aligned} &[\mathbf{f}'(T|n+1)]_{1,n}(z^{-1}\mathbf{z}_n) + f_{n+1}(T|n+1)z^{-n-1} \\ &= \mathbf{f}'(T|n)(z^{-1}\mathbf{z}_n) - f_{n+1}(T|n+1)\left[\mathbf{b}'(T-1|n)(z^{-1}\mathbf{z}_n) - z^{-n-1}\right], \end{aligned} \tag{6.3.82}$$

or

$$F_{T|n+1}(z) = F_{T|n}(z) - k_{n+1}^{b}(T)[z^{-1}B_{T-1|n}(z)] \;, \tag{6.3.83}$$

where

$$k_{n+1}^{b}(T) = f_{n+1}(T|n+1) \;. \tag{6.3.84}$$

Similarly, using (6.3.28) it is easily verified that

$$B_{T|n+1}(z) = z^{-1}B_{T-1|n}(z) - k_{n+1}^f(T)F_{T|n}(z) , \quad \textbf{(6.3.85)}$$

where

$$k_{n+1}^f(T) = b_1(T|n) . \quad (6.3.86)$$

If our objective is to find an order recursive algorithm which computes $F_{T|n+1}(z)$ and $B_{T|n+1}(z)$ in terms of $F_{T|n}(z)$ and $B_{T|n}(z)$, given $k_{n+1}^f(T)$ and $k_{n+1}^b(T)$, then the order recursions (6.3.83) and (6.3.85) are not sufficient since $B_{T-1|n}(z)$ must also be computed. In order to compute $B_{T-1|n}(z)$, (6.3.48) is used as follows,

$$\mathbf{b}'(T-1|n)\mathbf{z}_n = \mathbf{b}'(T|n)\mathbf{z}_n - e_b^o(T|n)\mathbf{g}'(T|n)\mathbf{z}_n . \quad (6.3.87)$$

Subtracting both sides from z^{-n} and using (6.3.47) gives

$$B_{T-1|n}(z) = B_{T|n}(z) + e_b(T|n)G_{T|n}(z)\frac{1}{1-\gamma(T|n)} , \quad (6.3.88)$$

where

$$G_{T|n}(z) = \mathbf{g}'(T|n)\mathbf{z}_n . \quad (6.3.89)$$

Finally, an order recursion for $G_{T|n}(z)$ is obtained by using (6.3.61) to get

$$G_{T|n+1}(z) = G_{T|n}(z) + \frac{e_b(T|n)}{\epsilon_b(T|n)} B_{T|n}(z) . \quad (6.3.90)$$

The recursions (6.3.88), (6.3.83), (6.3.85), and (6.3.90) (in that order) can now be used to generate the polynomials $F_{T|n}(z)$ and $B_{T|n}(z)$ in an order-recursive fashion given the lattice "gains" $k_{n+1}^f(T)$, $k_{n+1}^b(T)$, $e_b(T|n)$, $\epsilon_b(T|n)$, and $\gamma(T|n)$, $0 \leqslant n \leqslant N-1$ [23]. The structure of the whitening filter algorithm is shown in figure 6-8. The complete filter is obtained by cascading N such sections. The data $\{y(T)\}$ can be fed into the upper (lattice)

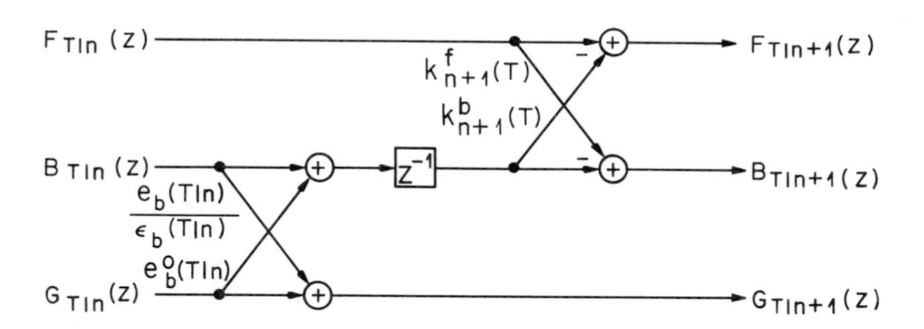

Figure 6-8. One stage of the order-recursive filter structure which generates the whitening filters $F_{T|n}(z)$ and $B_{T|n}(z)$.

section in order to compute the necessary lattice gains. The order recursions are initialized by setting

$$A_{T|0}(z) = B_{T|0}(z) = 1 \tag{6.3.91a}$$

and

$$G_{T|0}(z) = 0 . \tag{6.3.91b}$$

6.4. COVARIANCE LS ALGORITHMS

Covariance LS estimation is often used in speech processing applications where an accurate estimate of the short-term spectrum is desired given relatively few input samples. For these applications it is desirable to obtain an LS estimate based only upon available data, thereby eliminating any undesirable end effects caused by assuming the data is zero outside a given window, as is the case in prewindowed LS estimation. Recursions which apply to the covariance LS estimation problem are derived in this section. Combining these recursions in an appropriate fashion gives fixed-order (transversal) and order-recursive (lattice) algorithms that solve both the sliding window and growing memory covariance LS estimation problems described in chapter 3.

6.4.1. Covariance Least Squares Recursions

Recall that covariance LS estimation makes no assumptions about the input data outside the window of interest. Given the input samples $\{y(T_0), y(T_0+1), \ldots, y(T)\}$, the n coefficient prediction vectors $\mathbf{f}(T|n)$ and $\mathbf{b}(T|n)$ are desired which minimize the sums

$$\epsilon_f(T|n) = \sum_{j=T_0+n}^{T} [y(j) - \mathbf{f}'(T|n)\, \mathbf{y}(j-1|n)]^2 . \tag{6.4.1a}$$

and

$$\epsilon_b(T|n) = \sum_{j=T_0+n}^{T} [y(j-n) - \mathbf{b}'(T|n)\, \mathbf{y}(T|n)]^2 . \tag{6.4.1b}$$

The crucial difference between the case considered here and the prewindowed case is the lower limit of the summation. Changing this lower limit from $j=T_0$ (prewindowed case where $y(j)=0$, $j<T_0$) to $j=T_0+n$ increases the complexity of the resulting recursive LS algorithms.

Setting the derivatives of the cost functions $\epsilon_f(T|n)$ and $\epsilon_b(T|n)$ with respect to the prediction coefficients equal to zero results in the linear equations,

$$\hat{\mathbf{\Phi}}(T_0+n-1, T-1|n)\, \mathbf{f}(T|n) = \sum_{j=T_0+n}^{T} y(j)\mathbf{y}(j-1|n) \tag{6.4.2a}$$

and

$$\hat{\mathbf{\Phi}}(T_0+n, T|n)\, \mathbf{b}(T|n) = \sum_{j=T_0+n}^{T} y(j-n)\mathbf{y}(j|n) , \tag{6.4.2b}$$

where the sample covariance matrix is denoted as

$$\hat{\Phi}(T_0+n,T|n) \equiv \sum_{j=T_0+n}^{T} \mathbf{y}(j|n)\mathbf{y}'(j|n) . \tag{6.4.3}$$

The time updates in this section fall into two categories: forward and backward time updates. Given some LS parameter ξ (i.e., $\mathbf{f}$ or e_f), a forward time update for ξ expresses the value of ξ computed from the data samples $\{y(T_0), y(T_0+1), \ldots, y(T)\}$ in terms of ξ computed from the data samples $\{y(T_0), y(T_0+1), \ldots, y(T-1)\}$. A backward time update for ξ expresses the value of ξ computed from the data samples $\{y(T_0), y(T_0+1), \ldots, y(T)\}$ in terms of ξ computed from the data samples $\{y(T_0+1), y(T_0+2), \ldots, y(T)\}$. Associated with the variable ξ are an order index n and the time indices of the data used in the LS computation. If the data samples $\{y(T_0), y(T_0+1), \ldots, y(T)\}$ are used to compute ξ, then the indices T_0 and T must be specified. This is in contrast to the prewindowed case where it is necessary only to specify T since T_0 is always zero.

The recursions in this section are derived by first defining the necessary notation, and then using the projection and inner product updates in section 6.2 to systematically generate order and time updates for every state variable. The algorithms in section 6.4.2 are then constructed by properly organizing the recursions in this section. The data vector in this case is defined as

$$\mathbf{Y}(T_0+n,T) = [y(T)\ y(T-1) \cdots y(T_0+n)]' , \tag{6.4.4}$$

and the shifted data vector is

$$z^{-j}\mathbf{Y}(T_0+n,T) = [y(T-j)\ \ y(T-j-1)\ \ \cdots\ \ y(T_0+n-j)]' . \tag{6.4.5}$$

Equations (6.4.1) can now be rewritten as

$$\epsilon_f(T|n) = ||\ \mathbf{Y}(T_0+n,T) - \sum_{j=1}^{n} f_j(T|n)\ [z^{-j}\mathbf{Y}(T_0+n,T)]\ ||^2 \tag{6.4.6a}$$

and

$$\epsilon_b(T|n) = ||\ z^{-n}\mathbf{Y}(T_0+n,T) - \sum_{j=1}^{n} b_j(T|n)\ [z^{-j+1}\mathbf{Y}(T_0+n,T)]\ ||^2 . \tag{6.4.6b}$$

The matrix of shifted data vectors is denoted as

$$\mathbf{S}(T_0+n,T|l,n) = [z^{-l}\mathbf{Y}(T_0+n,T)\ \ z^{-l-1}\mathbf{Y}(T_0+n,T)\ \ \cdots\ \ z^{-n}\mathbf{Y}(T_0+n,T)] , \tag{6.4.7}$$

where $l<n$, and in particular,

$$\mathbf{S}(T_0+n,T|0,n) = \begin{bmatrix} y(T) & y(T-1) & \cdots & y(T-n+1) & y(T-n) \\ y(T-1) & y(T-2) & \cdots & y(T-n) & y(T-n-1) \\ \cdot & \cdot & & \cdot & \cdot \\ \cdot & \cdot & & \cdot & \cdot \\ \cdot & \cdot & & \cdot & \cdot \\ y(T_0+n) & y(T_0+n-1) & \cdots & y(T_0+1) & y(T_0) \end{bmatrix} . \qquad \textbf{(6.4.8)}$$

The space spanned by the columns of $\mathbf{S}(T_0+n,T|l,n)$, which is a subspace generated by past data values, is denoted as $M(T_0+n,T|l,n)$. For notational convenience the lower time index of $\mathbf{S}$ and M will be omitted and always assumed to be T_0+n. The covariance matrix defined by (6.4.3) can be written as

$$\hat{\mathbf{\Phi}}(T_0+n,T|n) = \mathbf{S}'(T|0,n-1)\mathbf{S}(T|0,n-1) . \qquad (6.4.9)$$

Throughout this chapter, the time indices of the generic parameter ξ will appear as function arguments. As before, the order of the variable will appear as a conditional function argument. As an example, $\xi(T_0,T|n)$ indicates that the data values $\{y(T_0), y(T_0+1), \ldots, y(T)\}$ are used to compute the nth order variable ξ. The following variables are needed to derive the covariance algorithms in the next section. Most are analogous to prewindowed variables defined in the last section.

Forward and backward prediction vectors (from (6.4.2)):

$$\mathbf{f}(T_0,T|n) = \hat{\mathbf{\Phi}}^{-1}(T_0+n-1,T-1|n)\ [\mathbf{S}'(T|1,n)\mathbf{Y}(T_0+n,T)] \qquad (6.4.10a)$$

$$\mathbf{b}(T_0,T|n) = \hat{\mathbf{\Phi}}^{-1}(T_0+n,T|n)\ \{\mathbf{S}'(T|0,n-1)[z^{-n}\mathbf{Y}(T_0+n,T)]\} . \qquad (6.4.10b)$$

Forward and backward prediction residual vectors:

$$\mathbf{E}_f(T_0,T|n) \equiv \mathbf{Y}(T_0+n,T) - \mathbf{S}(T|1,n)\ \mathbf{f}(T_0,T|n) \qquad (6.4.11a)$$

$$\mathbf{E}_b(T_0,T|n) \equiv z^{-n}\mathbf{Y}(T_0+n,T) - \mathbf{S}(T|0,n-1)\ \mathbf{b}(T_0,T|n) . \qquad (6.4.11b)$$

Forward and backward prediction residuals (scalars):

$$e_f(T_0,T|n) \equiv <\mathbf{u}(T),\mathbf{E}_f(T_0,T|n)> = y(T) - \mathbf{f}'(T_0,T|n)\ \mathbf{y}(T-1|n) \qquad (6.4.12a)$$

$$e_b(T_0,T|n) \equiv <\mathbf{u}(T),\mathbf{E}_b(T_0,T|n)> = y(T-n) - \mathbf{b}'(T_0,T|n)\ \mathbf{y}(T|n) . \qquad (6.4.12b)$$

Forward and backward LS cost functions:

$$\epsilon_f(T_0,T|n) \equiv ||\mathbf{E}_f(T_0,T|n)||^2, \quad \epsilon_b(T_0,T|n) \equiv ||\mathbf{E}_b(T_0,T|n)||^2 . \qquad (6.4.13)$$

Least Squares Partial Correlation coefficient:

$$K_n(T_0,T) \equiv <\mathbf{E}_f(T_0+1,T|n-1),\mathbf{E}_b(T_0,T-1|n-1)> . \qquad (6.4.14)$$

Auxiliary variables, or gains:

$$\mathbf{g}(T_0+1,T|n) \equiv \hat{\mathbf{\Phi}}^{-1}(T_0+n,T|n)\ \mathbf{y}(T|n) \qquad (6.4.15a)$$

$$\mathbf{h}(T_0{+}1,T|n) \equiv \hat{\mathbf{\Phi}}^{-1}(T_0{+}n,T|n)\ \mathbf{y}(T_0{+}n|n) \tag{6.4.15b}$$

$$\begin{aligned}\gamma(T_0{+}1,T|n) &\equiv \langle \mathbf{u}(T),\mathbf{P}_{M(T|0,n-1)}\mathbf{u}(T)\rangle \\ &= \mathbf{y}'(T|n)\ \hat{\mathbf{\Phi}}^{-1}(T_0{+}n,T|n)\ \mathbf{y}(T|n)\end{aligned} \tag{6.4.16a}$$

$$\begin{aligned}\gamma^*(T_0{+}1,T|n) &\equiv \langle \mathbf{u}(T_0),\mathbf{P}_{M(T|0,n-1)}\mathbf{u}(T_0)\rangle \\ &= \mathbf{y}'(T_0{+}n|n)\ \hat{\mathbf{\Phi}}^{-1}(T_0{+}n,T|n)\ \mathbf{y}(T_0{+}n|n)\end{aligned} \tag{6.4.16b}$$

$$\begin{aligned}\alpha(T_0{+}1,T|n) &\equiv \langle \mathbf{u}(T_0),P_{M(T|0,n-1)}\mathbf{u}(T)\rangle \\ &= \mathbf{y}'(T|n)\ \hat{\mathbf{\Phi}}^{-1}(T_0,T|n)\ \mathbf{y}(T_0{+}n|n)\ .\end{aligned} \tag{6.4.16c}$$

The LS state variables used to construct the algorithms which follow (and the algorithms in the last section) are defined by (6.4.10) through (6.4.16). The only variables appearing here which did not enter the prewindowed algorithms of the last section are $\mathbf{h}$, γ^*, and α. These variables are computed by forming a weighted linear combination of the elements of $\mathbf{y}(T_0{+}n|n)$.

The gains $\mathbf{g}$, $\mathbf{h}$, γ, γ^*, and α each involve the projection of either $\mathbf{u}(T)$ or $\mathbf{u}(T_0)$ onto some subspace. For example,

$$\mathbf{S}(T|0,n{-}1)\ \mathbf{g}(T_0{+}1,T|n) = \mathbf{P}_{M(T|0,n-1)}\mathbf{u}(T) \tag{6.4.17a}$$

and

$$\mathbf{S}(T|0,n{-}1)\ \mathbf{h}(T_0{+}1,T|n) = \mathbf{P}_{M(T|0,n-1)}\mathbf{u}(T_0)\ . \tag{6.4.17b}$$

From the definitions (6.4.15) and (6.4.16),

$$\gamma(T_0{+}1,T|n) = \mathbf{g}'(T_0{+}1,T|n)\ \mathbf{y}(T|n)\ , \tag{6.4.18a}$$

$$\gamma^*(T_0{+}1,T|n) = \mathbf{h}'(T_0{+}1,T|n)\ \mathbf{y}(T_0{+}n|n)\ , \tag{6.4.18b}$$

and

$$\begin{aligned}\alpha(T_0{+}1,T|n) &= \mathbf{g}'(T_0{+}1,T|n)\ \mathbf{y}(T_0{+}n|n) \\ &= \mathbf{h}'(T_0{+}1,T|n)\mathbf{y}(T|n)\ .\end{aligned} \tag{6.4.18c}$$

Using the notation in section 6.2, the "gains" γ and γ^* are respectively $\sin^2\theta(T)$ and $\sin^2\theta^*(T)$, where $\theta(T)$ and $\theta^*(T)$ are respectively the "angles" between $M(T|0,n{-}1)$ and $\tilde{M}(T|0,n{-}1)$ and between $M(T|0,n{-}1)$ and $M^*(T|0,n{-}1)$.

At each time interval our objective is to minimize the cost functions $\epsilon_f(T_0,T|n)$ and $\epsilon_b(T_0,T|n)$. From the discussion in section 6.1 it follows that the forward and backward residual vectors can be written as the orthogonal projections,

$$\mathbf{E}_f(T_0,T|n) = \mathbf{P}^{\perp}_{M(T|1,n)}\mathbf{Y}(T_0{+}n,T) \tag{6.4.19a}$$

and

$$\mathbf{E}_b(T_0,T|n) = \mathbf{P}^{\perp}_{M(T|0,n-1)}[z^{-n}\mathbf{Y}(T_0{+}n,T)]\ . \tag{6.4.19b}$$

The vectors,

$$\mathbf{E}_f^o(T_0,T|n) \equiv \mathbf{P}^{\perp}_{M_{T-1}(T|1,n)}\mathbf{Y}(T_0+n,T) = \mathbf{Y}(T_0+n,T) - \mathbf{S}(T|1,n)\,\mathbf{f}(T_0,T-1|n) \tag{6.4.20a}$$

and

$$\mathbf{E}_b^o(T_0,T|n) \equiv \mathbf{P}^{\perp}_{M_{T-1}(T|0,n-1)}[z^{-n}\mathbf{Y}(T_0+n,T)] = z^{-n}\mathbf{Y}(T_0+n,T) - \mathbf{S}(T|0,n-1)\,\mathbf{b}(T_0,T-1|n)\ , \tag{6.4.20b}$$

which are closely related to the prediction residual vectors and are analogous to the oblique residuals defined in the last section, are also needed. $\mathbf{E}_f^o$ and $\mathbf{E}_b^o$ are the forward and backward residual vectors that result from using prediction coefficients computed at the *preceding* time interval. The top components of $\mathbf{E}_f^o(T_0,T|n)$ and $\mathbf{E}_b^o(T_0,T|n)$ are, respectively,

$$e_f^o(T_0,T|n) \equiv \langle\mathbf{u}(T),\mathbf{E}_f^o(T_0,T|n)\rangle = y(T) - \mathbf{f}'(T_0,T-1|n)\,\mathbf{y}(T-1|n) \tag{6.4.21a}$$

and

$$e_b^o(T_0,T|n) \equiv \langle\mathbf{u}(T),\mathbf{E}_b^o(T_0,T|n)\rangle = y(T-n) - \mathbf{b}'(T_0,T-1|n)\mathbf{y}(T|n)\ . \tag{6.4.21b}$$

Unlike the prewindowed case, in the covariance case there are "backward time" counterparts to the residuals defined by (6.4.12), (6.4.20) and (6.4.21). The "backward" counterpart to the forward residual defined by (6.4.12a) is the nth order forward prediction residual computed at time T_0+n using the coefficient vector $\mathbf{f}(T_0,T|n)$, and is denoted as

$$e_f^*(T_0,T|n) \equiv \langle\mathbf{u}(T_0),\mathbf{E}_f(T_0,T|n)\rangle = y(T_0+n) - \mathbf{f}'(T_0,T|n)\mathbf{y}(T_0+n-1|n)\ . \tag{6.4.22}$$

The backward counterpart to the oblique residual vector defined by (6.4.20a) is the forward residual vector at time T that results from using the forward prediction vector calculated from the data samples $\{y(T_0+1),\ldots,y(T)\}$, and is denoted as

$$\mathbf{E}_f^{ob}(T_0,T|n) \equiv \mathbf{P}^{\perp}_{M_{T_0+1}(T|1,n)}\mathbf{Y}(T_0+n,T) = \mathbf{Y}(T_0+n,T) - \mathbf{S}(T|1,n)\mathbf{f}(T_0+1,T|n)\ . \tag{6.4.23}$$

The backward residuals $e_b^*(T_0,T|n)$ and $\mathbf{E}_b^{ob}(T_0,T|n)$ are similarly defined.

The time indices associated with a residual vector change in accordance with the projection space, i.e.,

$$\mathbf{P}^{\perp}_{M(T|1,n-1)}\mathbf{Y}(T_0+n,T) = \mathbf{E}_f(T_0+1,T|n-1) \tag{6.4.24a}$$

and

$$\mathbf{P}^{\perp}_{M(T|1,n-1)}[z^{-n}\mathbf{Y}(T_0+n,T)] = \mathbf{E}_b(T_0,T-1|n-1) . \qquad (6.4.24b)$$

The recursions needed to derive the algorithms in the next section can now be generated systematically. By appropriately defining the vectors and subspaces entering the projection order update (6.2.6), order updates are derived for all of the state variables. The forward and backward projection operator time updates (6.2.34) and (6.2.48) are then used to obtain forward and backward time updates for the fundamental vectors **f**, **b**, **g**, and **h**. Finally, the forward and backward time updates for inner products (6.2.40) and (6.2.49) are applied to $K_n(T_0,T)$, $\epsilon_f(T_0,T|n)$, and $\epsilon_b(T_0,T|n)$. Detailed derivations of the following recursions, in which the vectors and subspaces that must be substituted in the appropriate projection update are explicitly defined, are left as problems since in most cases the analogous substitutions were made in the prewindowed case. Consequently, only the results are stated with a few representative examples worked out in more detail.

Order Updates

The order updates are obtained by using the projection order update (6.2.6) (or equivalently (6.2.5)). Throughout this chapter $[\mathbf{f}(T_0,T|n)]_{l,m}$ denotes the lth through the mth component of $\mathbf{f}(T_0,T|n)$, $[\mathbf{f}(T_0,T|n)]_j$ is the jth component of $\mathbf{f}(T_0,T|n)$, and the same notation is used for the backward prediction vector $\mathbf{b}(T_0,T|n)$ and the gain vectors $\mathbf{g}(T_0,T|n)$ and $\mathbf{h}(T_0,T|n)$.

Residual vector order updates:

$$\mathbf{E}_f(T_0,T|n) = \mathbf{E}_f(T_0+1,T|n-1) - \frac{K_n(T_0,T)}{\epsilon_b(T_0,T-1|n-1)}\,\mathbf{E}_b(T_0,T-1|n-1) \qquad (6.4.25a)$$

$$\mathbf{E}_b(T_0,T|n) = \mathbf{E}_b(T_0,T-1|n-1) - \frac{K_n(T_0,T)}{\epsilon_f(T_0+1,T|n-1)}\,\mathbf{E}_f(T_0+1,T|n-1) . \qquad (6.4.25b)$$

Cost function order updates:

$$\epsilon_f(T_0,T|n) = \epsilon_f(T_0+1,T|n-1) - \frac{K_n^2(T_0,T)}{\epsilon_b(T_0,T-1|n-1)} \qquad (6.4.26a)$$

$$\epsilon_b(T_0,T|n) = \epsilon_b(T_0,T-1|n-1) - \frac{K_n^2(T_0,T)}{\epsilon_f(T_0+1,T|n-1)} . \qquad (6.4.26b)$$

Forward and backward prediction vector order updates:

$$[\mathbf{f}(T_0,T|n)]_n = \frac{K_n(T_0,T)}{\epsilon_b(T_0,T-1|n-1)} \qquad (6.4.27a)$$

$$\begin{aligned}[\mathbf{f}(T_0,T|n)]_{1,n-1} &= \mathbf{f}(T_0+1,T|n-1) \\ &\quad - [\mathbf{f}(T_0,T|n)]_n\,\mathbf{b}(T_0,T-1|n-1)\end{aligned} \qquad (6.4.27b)$$

$$[\mathbf{b}(T_0,T|n)]_1 = \frac{K_n(T_0,T)}{\epsilon_f(T_0+1,T|n-1)} \qquad (6.4.28a)$$

$$[\mathbf{b}(T_0,T|n)]_{2,n} = \mathbf{b}(T_0,T-1|n-1) - [\mathbf{b}(T_0,T|n)]_1 \, \mathbf{f}(T_0+1,T|n-1) \, . \tag{6.4.28b}$$

Order updates for vector gains:

$$[\mathbf{g}(T_0,T|n+1)]_{n+1} = \frac{e_b(T_0,T|n)}{\epsilon_b(T_0,T|n)} \tag{6.4.29a}$$

$$[\mathbf{g}(T_0,T|n+1)]_{1,n} = \mathbf{g}(T_0+1,T|n) - [\mathbf{g}(T_0,T|n+1)]_{n+1}\mathbf{b}(T_0,T|n) \tag{6.4.29b}$$

$$[\mathbf{g}(T_0,T|n+1)]_1 = \frac{e_f(T_0,T|n)}{\epsilon_f(T_0,T|n)} \tag{6.4.30a}$$

$$[\mathbf{g}(T_0,T|n+1)]_{2,n+1} = \mathbf{g}(T_0,T-1|n) - [\mathbf{g}(T_0,T|n+1)]_1 \, \mathbf{f}(T_0,T|n) \tag{6.4.30b}$$

$$[\mathbf{h}(T_0,T|n+1)]_{n+1} = \frac{e_b^*(T_0,T|n)}{\epsilon_b(T_0,T|n)} \tag{6.4.31a}$$

$$[\mathbf{h}(T_0,T|n+1)]_{1,n} = \mathbf{h}(T_0+1,T|n) - [\mathbf{h}(T_0,T|n+1)]_{n+1}\mathbf{b}(T_0,T|n) \tag{6.4.31b}$$

$$[\mathbf{h}(T_0,T|n+1)]_1 = \frac{e_f^*(T_0,T|n)}{\epsilon_f(T_0,T|n)} \tag{6.4.32a}$$

$$[\mathbf{h}(T_0,T|n+1)]_{2,n+1} = \mathbf{h}(T_0,T-1|n) - [\mathbf{h}(T_0,T|n+1)]_1 \, \mathbf{f}(T_0,T|n) \, . \tag{6.4.32b}$$

Order updates for scalar gains:

$$\gamma(T_0,T|n+1) = \gamma(T_0+1,T|n) + \frac{e_b^2(T_0,T|n)}{\epsilon_b(T_0,T|n)} \tag{6.4.33a}$$

$$\gamma(T_0,T|n+1) = \gamma(T_0,T-1|n) + \frac{e_f^2(T_0,T|n)}{\epsilon_f(T_0,T|n)} \tag{6.4.33b}$$

$$\gamma^*(T_0,T|n+1) = \gamma^*(T_0+1,T|n) + \frac{e_b^{*2}(T_0,T|n)}{\epsilon_b(T_0,T|n)} \tag{6.4.34a}$$

$$\gamma^*(T_0,T|n+1) = \gamma^*(T_0,T-1|n) + \frac{e_f^{*2}(T_0,T|n)}{\epsilon_f(T_0,T|n)} \tag{6.4.34b}$$

$$\alpha(T_0,T|n+1) = \alpha(T_0+1,T|n) + \frac{e_b(T_0,T|n) \, e_b^*(T_0,T|n)}{\epsilon_b(T_0,T|n)} \tag{6.4.35a}$$

$$\alpha(T_0,T|n+1) = \alpha(T_0,T-1|n) + \frac{e_f(T_0,T|n) \, e_f^*(T_0,T|n)}{\epsilon_f(T_0,T|n)} \, . \tag{6.4.35b}$$

As an example, (6.4.25a) is derived from (6.2.6), where M is replaced by $M(T|1,n-1)$, $\mathbf{X}$ is replaced by $z^{-n}\mathbf{Y}(T_0+n,T)$, and $\mathbf{Y}$ is replaced by $\mathbf{Y}(T_0+n,T)$, and by observing that since $\mathbf{E}_b(T_0,T-1|n-1)$ is orthogonal to $M(T|1,n-1)$,

$$<\mathbf{Y}(T_0+n,T),\mathbf{E}_b(T_0,T-1|n-1)> = <\mathbf{E}_f(T_0+1,T|n-1),\mathbf{E}_b(T_0,T-1|n-1)> \\ = K_n(T_0,T) \,. \tag{6.4.36}$$

Recursions (6.4.26) are obtained by taking norms of (6.4.25). The recursions (6.4.29-30) and (6.4.31-32) are obtained from (6.2.5), where $\mathbf{Y}$ is replaced by $\mathbf{u}(T)$ and $\mathbf{u}(T_0)$, respectively. Recursions (6.4.33-35) are obtained by making the same substitutions in (6.2.5) and then taking inner products with $\mathbf{u}(T)$ or $\mathbf{u}(T_0)$.

Forward Time Updates

The forward time updates are obtained from the (orthogonal) projection operator forward time update (6.2.34).

Residual vector forward time updates:

$$\mathbf{E}_f(T_0,T|n) = \mathbf{E}_f^o(T_0,T|n) - [\mathbf{P}_{M(T|1,n)}\mathbf{u}(T)] \frac{e_f(T_0,T|n)}{1-\gamma(T_0,T-1|n)} \tag{6.4.37a}$$

$$\mathbf{E}_b(T_0,T|n) = \mathbf{E}_b^o(T_0,T|n) - [\mathbf{P}_{M(T|0,n-1)}\mathbf{u}(T)] \frac{e_b(T|n)}{1-\gamma(T_0+1,T|n)} \,. \tag{6.4.37b}$$

Forward and backward prediction vector forward time updates:

$$\mathbf{f}(T_0,T|n) = \mathbf{f}(T_0,T-1|n) + \mathbf{g}(T_0,T-1|n)\frac{e_f(T_0,T|n)}{1-\gamma(T_0,T-1|n)} \tag{6.4.38a}$$

$$\mathbf{b}(T_0,T|n) = \mathbf{b}(T_0,T-1|n) + \mathbf{g}(T_0+1,T|n)\frac{e_b(T_0,T|n)}{1-\gamma(T_0+1,T|n)} \,. \tag{6.4.38b}$$

Forward time updates for gains:

$$\mathbf{h}(T_0,T|n) = \mathbf{h}(T_0,T-1|n) - \mathbf{g}(T_0,T|n)\frac{\alpha(T_0,T|n)}{1-\gamma(T_0,T|n)} \tag{6.4.39}$$

$$\gamma^*(T_0,T|n) = \gamma^*(T_0,T-1|n) - \frac{\alpha^{*2}(T_0,T|n)}{1-\gamma(T_0,T|n)} \tag{6.4.40}$$

$$\alpha(T_0,T|n) = \mathbf{y}'(T|n)\,\mathbf{h}(T_0,T-1|n)\,[1-\gamma(T_0,T|n)] \,. \tag{6.4.41}$$

Equation (6.4.39) is obtained from (6.2.33), where $M(T)$ is replaced by $M(T|0,n)$ and $\mathbf{Y}(T)$ is replaced by $\mathbf{u}(T_0)$. Equations (6.4.40) and (6.4.41) are obtained by making the same substitutions in (6.2.33) and then taking inner products with $\mathbf{u}(T_0)$ and $\mathbf{u}(T)$, or by premultiplying (6.4.39) by $\mathbf{y}'(T_0+n-1|n)$ and $\mathbf{y}'(T|n)$, respectively.

The following recursions are obtained by taking inner products of (6.4.37a) and (6.4.37b) with $\mathbf{u}(T)$ and $\mathbf{u}(T_0)$, respectively,

$$e_f^o(T_0,T|n) = \frac{e_f(T_0,T|n)}{1-\gamma(T_0,T-1|n)} \tag{6.4.42a}$$

$$e_b^o(T_0,T|n) = \frac{e_b(T_0,T|n)}{1-\gamma(T_0+1,T|n)} \tag{6.4.42b}$$

$$e_f^*(T_0,T|n) = e_f^*(T_0,T-1|n) - e_f(T_0,T|n)\frac{\alpha(T_0,T-1|n)}{1-\gamma(T_0,T-1|n)} \tag{6.4.43a}$$

$$e_b^*(T_0,T|n) = e_b^*(T_0,T-1|n) - e_b(T_0,T|n)\frac{\alpha(T_0+1,T|n)}{1-\gamma(T_0+1,T|n)}\,. \tag{6.4.43b}$$

Backward Time Updates

The following backward time updates are obtained from the projection operator backward time update (6.2.48).

Residual vector backward time updates:

$$\mathbf{E}_f(T_0,T|n) = \mathbf{E}_f^{ob}(T_0,T|n) - [\mathbf{P}_{M(T|1,n)}\mathbf{u}(T_0)]\frac{e_f^*(T_0,T|n)}{1-\gamma^*(T_0,T-1|n)} \tag{6.4.44a}$$

$$\mathbf{E}_b(T_0,T|n) = \mathbf{E}_b^{ob}(T_0,T|n) - [\mathbf{P}_{M(T|0,n-1)}\mathbf{u}(T_0)]\frac{e_b^*(T_0,T|n)}{1-\gamma^*(T_0+1,T|n)}\,. \tag{6.4.44b}$$

Forward and backward prediction vector backward time updates:

$$\mathbf{f}(T_0+1,T|n) = \mathbf{f}(T_0,T|n) - \mathbf{h}(T_0,T-1|n)\frac{e_f^*(T_0,T|n)}{1-\gamma^*(T_0,T-1|n)} \tag{6.4.45a}$$

$$\mathbf{b}(T_0+1,T|n) = \mathbf{b}(T_0,T|n) - \mathbf{h}(T_0+1,T|n)\frac{e_b^*(T_0,T|n)}{1-\gamma^*(T_0+1,T|n)}\,. \tag{6.4.45b}$$

Backward time updates for gains:

$$\mathbf{g}(T_0,T|n) = \mathbf{g}(T_0+1,T|n) - \mathbf{h}(T_0,T|n)\frac{\alpha(T_0,T|n)}{1-\gamma^*(T_0,T|n)} \tag{6.4.46}$$

$$\gamma(T_0,T|n) = \gamma(T_0+1,T|n) - \frac{\alpha^{*2}(T_0,T|n)}{1-\gamma^*(T_0,T|n)} \tag{6.4.47}$$

$$\alpha(T_0,T|n) = \mathbf{y}'(T_0+n-1|n)\mathbf{g}(T_0+1,T|n)[1-\gamma^*(T_0,T|n)]\,. \tag{6.4.48}$$

Equation (6.4.46) is obtained by replacing $\mathbf{Y}(T)$ by $\mathbf{u}(T)$ in (6.2.48). Equations (6.4.47) and (6.4.48) are obtained by premultiplying (6.4.46) by $\mathbf{y}'(T|n)$ and $\mathbf{y}'(T_0+n-1|n)$, respectively. The following recursions are obtained by taking inner products of (6.4.44a) and (6.4.44b) with $\mathbf{u}(T)$, respectively,

$$e_f(T_0+1,T|n) = e_f(T_0,T|n) + e_f^*(T_0,T|n)\frac{\alpha(T_0,T-1|n)}{1-\gamma^*(T_0,T-1|n)} \tag{6.4.49a}$$

$$e_b(T_0{+}1,T|n) = e_b(T_0,T|n) + e_b^*(T_0,T|n)\,\frac{\alpha(T_0{+}1,T|n)}{1-\gamma^*(T_0{+}1,T|n)}\,. \qquad \textbf{(6.4.49b)}$$

The recursions which result from taking inner products of (6.4.44a) and (6.4.44b) with $\mathbf{u}(T_0)$ will not be used and are omitted.

Inner Product Updates

The following recursions are obtained from the forward time update for inner products (6.2.40).

Least squares PARCOR coefficient forward time update:

$$K_n(T_0,T) = K_n(T_0,T-1) + e_f(T_0{+}1,T|n-1)\,e_b(T_0,T-1|n-1)\,\frac{1}{1-\gamma(T_0{+}1,T-1|n-1)}\,. \qquad (6.4.50)$$

Cost function forward time updates:

$$\epsilon_f(T_0,T|n) = \epsilon_f(T_0,T-1|n) + e_f^2(T_0,T|n)\,\frac{1}{1-\gamma(T_0,T-1|n)} \qquad (6.4.51a)$$

$$\epsilon_b(T_0,T|n) = \epsilon_b(T_0,T-1|n) + e_b^2(T_0,T|n)\,\frac{1}{1-\gamma(T_0{+}1,T|n)}\,. \qquad (6.4.51b)$$

The following recursions are obtained from the backward time update for inner products (6.2.49).

Least squares PARCOR coefficient backward time update:

$$K_n(T_0,T) = K_n(T_0{+}1,T) + e_f^*(T_0{+}1,T|n-1)\,e_b^*(T_0,T-1|n-1)\,\frac{1}{1-\gamma^*(T_0{+}1,T-1|n-1)}\,. \qquad (6.4.52)$$

Cost function backward time updates:

$$\epsilon_f(T_0,T|n) = \epsilon_f(T_0{+}1,T|n) + e_f^{*2}(T_0,T|n)\,\frac{1}{1-\gamma^*(T_0,T-1|n)} \qquad (6.4.53a)$$

$$\epsilon_b(T_0,T|n) = \epsilon_b(T_0,T|n) + e_b^{*2}(T_0,T|n)\,\frac{1}{1-\gamma^*(T_0{+}1,T|n)}\,. \qquad (6.4.53b)$$

Equations (6.4.50) and (6.4.52) are obtained by using (6.2.40) and (6.2.49) where $\mathbf{v}(T)$ is replaced by $\mathbf{Y}(T_0{+}n,T)$, $\mathbf{Y}(T)$ is replaced by $z^{-n}\mathbf{Y}(T_0{+}n,T)$, and $M(T)$ is replaced by $M(T|1,n-1)$, respectively.

The previous set of recursions (6.4.25-53) are complete in the sense that any of the existing computationally efficient LS algorithms can be derived by manipulating suitable subsets of these recursions.

6.4.2. Sliding Memory Covariance Algorithms

Given the infinite data sequence

$$\{ \cdots, y(-2), y(-1), y(0), y(1), y(2), \cdots \},$$

recall that sliding window covariance estimation computes LS estimates based upon data contained in a window of fixed length which "slides" across the sequence of data values one by one. Assuming a window of length M, at time T the LS estimation is based upon the data samples

$$y(T-M+1), y(T-M+2), \ldots, y(T),$$

whereas at time $T+1$, the estimation is based upon the data samples

$$y(T-M+2), y(T-M+3), \ldots, y(T+1),$$

and so forth. Past data samples outside the window are therefore totally "forgotten." This is in contrast to exponential weighting schemes which make use of *all* of the past data. The sliding window technique is therefore useful in situations where the statistics of the input data change abruptly and it is desirable to completely discard the effects of past data samples.

Combining (6.4.21a), (6.4.22), (6.4.29-34), (6.4.38), (6.4.42a), (6.4.45), (6.4.51a), and (6.4.53a) gives the following fixed-order sliding window LS algorithm for the prediction coefficients. Where unspecified, the order of the variable is assumed to be N, the order of the LS filter. Also, the starting time index is denoted as T_0. If the sliding window contains M data values, then $T_0 = T-M+1$. At each iteration $T > N$, the recursions are computed,

$$e_f^o(T_0,T) = y(T) - \mathbf{f}'(T_0,T-1)\,\mathbf{y}(T-1) \tag{6.4.54a}$$

$$\mathbf{f}(T_0,T) = \mathbf{f}(T_0,T-1) + \mathbf{g}(T_0,T-1)\,e_f^o(T_0,T) \tag{6.4.54b}$$

$$e_f(T_0,T) = e_f^o(T_0,T)\,[1 - \gamma(T_0,T-1)] \tag{6.4.54c}$$

$$e_f^*(T_0,T) = y(T_0+N) - \mathbf{f}'(T_0,T)\,\mathbf{y}(T_0+N-1) \tag{6.4.54d}$$

$$\epsilon_f(T_0,T) = \epsilon_f(T_0,T-1) + e_f^o(T_0,T)\,e_f(T_0,T) \tag{6.4.54e}$$

$$[\mathbf{g}(T_0,T|N+1)]_1 = \frac{e_f(T_0,T)}{\epsilon_f(T_0,T)} \tag{6.4.54f}$$

$$[\mathbf{g}(T_0,T|N+1)]_{2,N+1} = \mathbf{g}(T_0,T-1) - [\mathbf{g}(T_0,T|N+1)]_1\,\mathbf{f}(T_0,T) \tag{6.4.54g}$$

$$[\mathbf{h}(T_0,T|N+1)]_1 = \frac{e_f^*(T_0,T)}{\epsilon_f(T_0,T)} \tag{6.4.54h}$$

$$[\mathbf{h}(T_0,T|N+1)]_{2,N+1} = \mathbf{h}(T_0,T-1) - [\mathbf{h}(T_0,T|N+1)]_1\,\mathbf{f}(T_0,T) \tag{6.4.54i}$$

$$\mathbf{f}(T_0+1,T) = \mathbf{f}(T_0,T) - \mathbf{h}(T_0,T-1)\frac{e_f^*(T_0,T)}{1 - \gamma^*(T_0,T-1)} \quad \textbf{(6.4.54j)}$$

$$\epsilon_f(T_0+1,T) = \epsilon_f(T_0,T) - \frac{e_f^{*2}(T_0,T)}{1 - \gamma^*(T_0,T-1)} \quad (6.4.54k)$$

$$e_b^o(T_0,T) = y(T-N) - \mathbf{b}'(T_0,T-1)\,\mathbf{y}(T) \quad (6.4.54l)$$

$$\mathbf{b}(T_0,T) = \frac{\mathbf{b}(T_0,T-1) + e_b^o(T_0,T)\,[\mathbf{g}(T_0,T|N+1)]_{1,N}}{1 - e_b^o(T_0,T)\,[\mathbf{g}(T_0,T|N+1)]_{N+1}} \quad (6.4.54m)$$

$$e_b^*(T_0,T) = y(T_0), - \mathbf{b}'(T_0,T)\,\mathbf{y}(T_0+N) \quad (6.4.54n)$$

$$\mathbf{g}(T_0+1,T) = [\mathbf{g}(T_0,T|N+1)]_{1,N} + [\mathbf{g}(T_0,T|N+1)]_{N+1}\mathbf{b}(T_0,T) \quad (6.4.54o)$$

$$\mathbf{h}(T_0+1,T) = [\mathbf{h}(T_0,T|N+1)]_{1,N} + [\mathbf{b}(T_0,T|N+1)]_{N+1}\mathbf{b}(T_0,T) \quad (6.4.54p)$$

$$\gamma(T_0+1,T) = \quad (6.4.54q)$$

$$\frac{\gamma(T_0,T-1) + e_f(T_0,T)\,[\mathbf{g}(T_0,T|N+1)]_1 - e_b^o(T_0,T)\,[\mathbf{g}(T_0,T|N+1)]_{N+1}}{1 - e_b^o(T_0,T)\,[\mathbf{g}(T_0,T|N+1)]_{N+1}}$$

$$\gamma^*(T_0+1,T) = \quad (6.4.54r)$$

$$\gamma^*(T_0,T-1) + e_f^*(T_0,T)\,[\mathbf{h}(T_0,T|N+1)]_1 - e_b^*(T_0,T)\,[\mathbf{h}(T_0,T|N+1)]_N$$

$$\mathbf{b}(T_0+1,T) = \mathbf{b}(T_0,T) - \mathbf{h}(T_0+1,T)\frac{e_b^*(T_0,T)}{1 - \gamma^*(T_0+1,T)}\,. \quad (6.4.54s)$$

The recursions (6.4.54m) and (6.4.54q) were not listed in the previous section, but are obtained by solving (6.4.38b) and (6.4.29b) simultaneously for $\mathbf{b}(T_0,T)$, and by substituting (6.4.29a), (6.4.30a), and (6.4.42b) into (6.4.33), and solving for $\gamma(T_0+1,T|n)$, respectively.

Combining (6.4.25) (top and bottom components), (6.4.26), (6.4.33a), (6.4.34a), (6.4.50), and (6.4.52) gives the order-recursive (lattice) sliding window LS algorithm,

$$K_n(T_0+N-n,T) = \quad (6.4.55a)$$

$$K_n(T_0+N-n,T-1) + e_f(T|n-1)e_b(T-1|n-1)\,\frac{1}{1 - \gamma(T_0+N-n+1,T-1|n-1)}$$

$$k_n^f(T) = \frac{K_n(T_0+N-n,T)}{\epsilon_f(T|n-1)}, \quad k_n^b(T) = \frac{K_n(T_0+N-n,T)}{\epsilon_b(T-1|n-1)} \quad (6.4.55b)$$

$$e_f(T|n) = e_f(T|n-1) - k_n^b(T)e_b(T-1|n-1) \tag{6.4.55c}$$

$$e_b(T|n) = e_b(T-1|n-1) - k_n^f(T)e_f(T|n-1) \tag{6.4.55d}$$

$$\epsilon_f(T|n) = \epsilon_f(T|n-1) - k_n^f(T)K_n(T_0+N-n,T) \tag{6.4.55e}$$

$$\epsilon_b(T|n) = \epsilon_b(T-1|n-1) - k_n^f(T)K_n(T_0+N-n,T) \tag{6.4.55f}$$

$$\gamma(T_0+N-n,T|n+1) = \gamma(T_0+N-n+1,T|n) + \frac{e_b^2(T|n)}{\epsilon_b(T|n)} \tag{6.4.55g}$$

$$e_f^*(T|n) = e_f^*(T|n-1) - k_n^b(T)e_b^*(T-1|n-1) \tag{6.4.55h}$$

$$e_b^*(T|n) = e_b^*(T-1|n-1) - k_n^f(T)e_f^*(T|n-1) \tag{6.4.55i}$$

$$\gamma^*(T_0+N-n,T|n+1) = \gamma^*(T_0+N-n+1,T|n) + \frac{e_b^{*2}(T|n)}{\epsilon_b(T|n)} \tag{6.4.55j}$$

$$K_n(T_0+N-n+1,T) = K_n(T_0+N-n,T) \tag{6.4.55k}$$

$$- e_f^*(T|n-1)e_b^*(T-1|n-1)\,\frac{1}{1-\gamma^*(T_0+N-n+1,T-1|n-1)} .$$

The lower time index has been omitted in most cases since it is uniquely determined by the upper time index, the length of the window (M), and the order of the variable. Letting N denote the order of the filter, suppose that $e_f(T|N)$ is computed from the data $y(T_0), \ldots, y(T)$ (i.e., $M = T-T_0+1$). Referring to equation (6.4.8), where n is replaced by N, $e_f(T|N)$ is the top component of the orthogonal projection of the first column onto the space spanned by the remaining N columns. In contrast, the residual $e_f(T|n)$, where $n < N$, is the top component of the orthogonal projection of the first column onto the space spanned by the adjacent n columns, and hence has lower time index T_0+N-n. The column vectors $[\mathbf{Y}(T), z^{-1}\mathbf{Y}(T), z^{-2}\mathbf{Y}(T), \ldots, z^{-n}\mathbf{Y}(T)]$ used to generate each variable in the lattice algorithm (6.4.55) have the same dimension, *independent* of the parameters T_0, T, and the order of the variable n. The number of samples in the data window used to compute a variable of order n at time T is therefore different from the number of samples in the data window used to compute the same variables of order $m(\neq n)$ at time T. Referring again to (6.4.8), the nth order forward residual is

$$\begin{aligned} e_f(T|n) &= \langle \mathbf{u}(T), \mathbf{P}^{\perp}_{M(T|1,n)}\mathbf{Y}(T) \rangle \\ &= e_f(T_0+N-n,T|n) \ , \ 1 \leqslant n \leqslant N \ , \end{aligned} \tag{6.4.56a}$$

where $T_0 = T-M+1$. Similarly, the nth order backward residual is

$$\begin{aligned} e_b(T|n) &= \langle \mathbf{u}(T), \mathbf{P}^{\perp}_{M(T|0,n-1)}z^{-n}\mathbf{Y}(T) \rangle \\ &= e_b(T_0+N-n,T|n) \ , \ 1 \leqslant n \leqslant N \ . \end{aligned} \tag{6.4.56b}$$

The lower time index for $\epsilon_f(T|n)$ and $\epsilon_b(T|n)$ must also be T_0+N-n. Incrementing T automatically increments T_0, so that the length of the data window used to compute each variable remains constant, assuming the order is fixed. The lower time indices for $K_n(T)$, $\gamma(T|n)$, and $\gamma^*(T|n)$ can be similarly obtained from the definitions (6.4.14) and (6.4.16); however, these have been specified in the recursions (6.4.55a-55k).

The set of recursions, (6.4.55), are computed at each iteration $T > N$ for $n=0$ to $n=N$. It is straightforward to verify that evaluation of all variables at time $T-1$ (for $0 \leqslant n \leqslant N$) enables the recursive computation of the same variables at the Tth iteration. The flow diagram for the algorithm appears in figure 6-9. It consists of two prewindowed type lattice filters with identical time varying coefficients.

Both algorithms (6.4.54) and (6.4.55) require that all data samples in the sliding window $\{y(T-M+1), \ldots, y(T)\}$ be stored. If division is counted as multiplication, then the fixed-order algorithm requires $13N+13$ multiplies and $12N+7$ additions at each iteration. The sliding window lattice predictor requires $16N$ multiplies and $10N$ additions per iteration.

Because sliding window algorithms have finite memory, initialization for these algorithms is basically the same as for the prewindowed case, i.e., the data $y(T)$ can be assumed to be zero for $T < 0$. After M iterations, where M is the window length, these data samples are discarded. The algorithm (6.4.55) can therefore be initialized by setting the gains γ and γ^*, and the elements of the vectors **f**, **b**, **g**, and **h** equal to zero, and letting

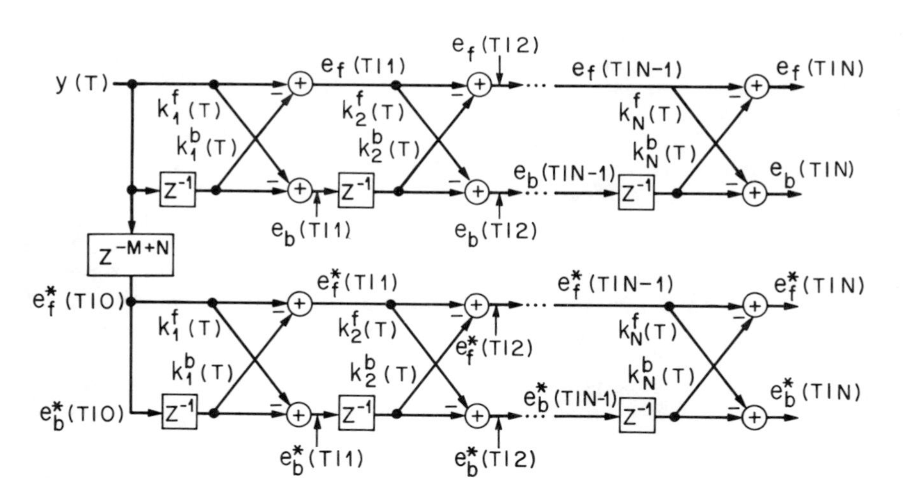

Figure 6-9. Flow diagram for the sliding window covariance lattice algorithm.

$$\epsilon_f(0,-1) = \delta\ , \tag{6.4.57}$$

where δ is a small constant chosen to ensure that the algorithm remains stable. The lattice algorithm is initialized by setting

$$K_{n+1}(-1) = e_b(-1|n) = e_b^*(-1|n) = \gamma(-1|n) = \gamma^*(-1|n) = 0, \quad 0 \leqslant n \leqslant N\ , \tag{6.4.58a}$$

and at each iteration T,

$$e_f(T|0) = e_b(T|0) = y(T)\ , \tag{6.4.58b}$$

$$e_f^*(T|0) = e_b^*(T|0) = y(T_0+N)\ , \tag{6.4.58c}$$

$$\gamma(T_0+N+1,T|0) = \gamma^*(T_0+N+1,T|0) = 0\ , \tag{6.4.58d}$$

and

$$\epsilon_f(T|0) = \epsilon_b(T|0) = \epsilon_f(T-1|0) + y^2(T) - y^2(T_0+N-1)\ . \tag{6.4.58e}$$

It is easily verified that when $T<M-N-1$, the algorithms (6.4.54) and (6.4.55) become the same as, respectively, prewindowed LS transversal and lattice algorithms, which were presented in section 6.3.

6.4.3. Growing Memory Covariance Algorithms

The algorithms presented in this section recursively solve the growing memory covariance LS estimation problem. Recall from chapter 3 that in this case the prediction vector $\mathbf{f}(T|N)$ which minimizes the sum

$$\epsilon_f(0,T|N) = \sum_{j=N}^{T} [y(j) - \mathbf{f}'(T|N)\,\mathbf{y}(j-1|N)]^2 \tag{6.4.59}$$

is desired, where N is the prediction order. The lower index of the time window (T_0) is therefore a constant equal to zero.

A fixed-order growing memory covariance algorithm can be obtained by combining (6.4.21), (6.4.29b), (6.4.30), (6.4.38), (6.4.39), (6.4.41), (6.4.46), (6.4.48), and (6.4.51a). For notational convenience the variables are defined,

$$\beta(T_0,T|n) \equiv \frac{\alpha(T_0,T|n)}{1 - \gamma(T_0,T|n)} \tag{6.4.60a}$$

and

$$\beta^*(T_0,T|n) \equiv \frac{\alpha(T_0,T|n)}{1 - \gamma^*(T_0,T|n)}\ . \tag{6.4.60b}$$

Where unspecified, the lower time index and the order of the variables are equal to zero and N, respectively. At each iteration $T > N$ the recursions

$$e_f^0(T) = y(T) - \mathbf{f}'(T-1)\,\mathbf{y}(T-1) \tag{6.4.61a}$$

$$\mathbf{f}(T) = \mathbf{f}(T-1) + \mathbf{g}(T-1)\, e_f^o(T) \tag{6.4.61b}$$

$$e_f(T) = y(T) - \mathbf{f}'(T)\, \mathbf{y}(T-1) \tag{6.4.61c}$$

$$\epsilon_f(T) = \epsilon_f(T-1) + e_f(T)\, e_f^o(T) \tag{6.4.61d}$$

$$[\mathbf{g}(T|N+1)]_1 = \frac{e_f(T)}{\epsilon_f(T)} \tag{6.4.61e}$$

$$[\mathbf{g}(T|N+1)]_{2,N+1} = \mathbf{g}(T-1) - [\mathbf{g}(T|N+1)]_1\, \mathbf{f}(T) \tag{6.4.61f}$$

$$e_b^o(T) = y(T-N) - \mathbf{b}'(T-1)\, \mathbf{y}(T) \tag{6.4.61g}$$

$$\mathbf{b}(T) = \frac{\mathbf{b}(T-1) + e_b^o(T)\, [\mathbf{g}(T|N+1)]_{1,N}}{1 - e_b^o(T)\, [\mathbf{g}(T|N+1)]_{N+1}} \tag{6.4.61h}$$

$$\mathbf{g}(1,T) = [\mathbf{g}(T|N+1)]_{1,N} + [\mathbf{g}(T|N+1)]_{N+1}\mathbf{b}(T) \tag{6.4.61i}$$

$$\beta(T) = \mathbf{y}'(T)\, \mathbf{h}(T-1) \tag{6.4.61j}$$

$$\beta^*(T) = \mathbf{y}'(N-1)\, \mathbf{g}(1,T) \tag{6.4.61k}$$

$$\mathbf{g}(T) = \frac{\mathbf{g}(1,T) - \beta^*(T)\, \mathbf{h}(T-1)}{1 - \beta(T)\, \beta^*(T)} \tag{6.4.61l}$$

$$\mathbf{h}(T) = \mathbf{h}(T-1) - \beta(T)\, \mathbf{g}(T) \tag{6.4.61m}$$

are computed.

A growing memory covariance lattice algorithm is obtained from (6.4.25) (top component only), (6.4.26), (6.4.33), (6.4.34b), (6.4.35b), (6.4.43a), (6.4.49a), (6.4.50), and (6.4.53a). Where unspecified, the lower time index is assumed to be zero. At each iteration $T > N$,

$$K_n(T) = K_n(T-1) + e_f(1,T|n-1)\, e_b(T-1|n-1) \frac{1}{1-\gamma(1,T-1|n-1)} \tag{6.4.62a}$$

$$k_n^f(T) = \frac{K_n(T)}{\epsilon_f(1,T|n-1)} \quad , \quad k_n^b(T) = \frac{K_n(T)}{\epsilon_b(T-1|n-1)} \tag{6.4.62b}$$

$$e_f(T|n) = e_f(1,T|n-1) - k_n^b(T)\, e_b(T-1|n-1) \tag{6.4.62c}$$

$$e_b(T|n) = e_b(T-1|n-1) - k_n^f(T)\, e_f(1,T|n-1) \tag{6.4.62d}$$

$$\epsilon_f(T|n) = \epsilon_f(1,T|n-1) - k_n^b(T)\, K_n(T) \tag{6.4.62e}$$

$$\epsilon_b(T|n) = \epsilon_b(T-1|n-1) - k_n^b(T)\, K_n(T) \tag{6.4.62f}$$

$$e_f^*(T|n) = e_f^*(T-1|n) - e_f(T|n)\,\frac{\alpha(T-1|n)}{1-\gamma(T-1|n)} \tag{6.4.62g}$$

$$e_f(1,T|n) = e_f(T|n) + e_f^*(T|n)\,\frac{\alpha(T-1|n)}{1-\gamma^*(T-1|n)} \tag{6.4.62h}$$

$$\epsilon_f(1,T|n) = \epsilon_f(T|n) - e_f^{*2}(T|n)\,\frac{1}{1-\gamma^*(T-1|n)} \tag{6.4.62i}$$

$$\gamma(T|n+1) = \gamma(T-1|n) + \frac{e_f^2(T|n)}{\epsilon_f(T|n)} \tag{6.4.62j}$$

$$\gamma^*(T|n+1) = \gamma^*(T-1|n) + \frac{e_f^{*2}(T|n)}{\epsilon_f(T|n)} \tag{6.4.62k}$$

$$\alpha(T|n+1) = \alpha(T-1|n) + \frac{e_f^*(T|n)e_f(T|n)}{\epsilon_f(T|n)} \tag{6.4.62l}$$

$$\gamma(1,T|n) = \gamma(T|n+1) - \frac{e_b^2(T|n)}{\epsilon_b(T|n)} \tag{6.4.62m}$$

are computed for $n = 1, 2, \ldots, N$. The flow diagram for this algorithm is shown in figure 6-10. Counting division as multiplication, the fixed-order predictor requires $13N+6$ multiplies and $11N+1$ additions per iteration, and requires that the first N samples be stored. The growing memory lattice

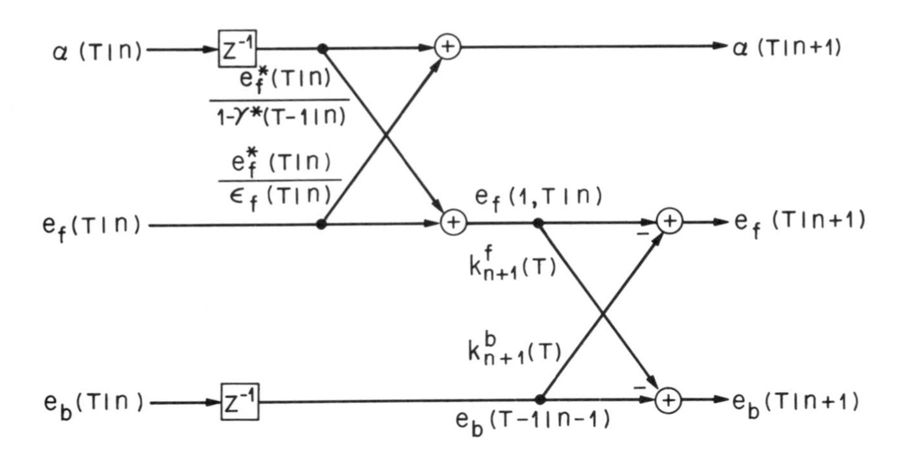

Figure 6-10. Flow diagram for one stage of the growing memory covariance lattice predictor.

predictor requires $21N$ multiplies and $12N$ additions per iteration. The modification proposed for the fixed-order prewindowed algorithm in section 6.3 can also be applied to the fixed-order growing memory covariance algorithm. In particular, equation (6.4.42a) can be used instead of (6.4.61c) to compute $e_f(T)$, and the recursion (6.4.54q) can be added to compute $\gamma(1,T)$. (Premultiplying (6.4.61l) by $\mathbf{y}'(T)$ enables the computation of $\gamma(T)$ from $\gamma(1,T)$). This modification results in a modest reduction of multiplies required at each iteration if the order of the filter is large. Further modifications to the fixed-order algorithm are discussed in problem 6-15.

Derivation of initial conditions for the algorithms (6.4.61) and (6.4.62), which guarantees an exact LS solution at all iterations $T > 0$, is more complicated than for the cases discussed so far. This is because for $T{=}n$, the matrix $\hat{\mathbf{\Phi}}(n,T|n)$ is singular, and hence all variables that depend on the inverse covariance matrix are undefined. A convenient solution to this startup problem has been presented [23]. If the covariance matrix $\hat{\mathbf{\Phi}}$ is singular, then the LS solution for **f** is not uniquely defined. Nevertheless, by always insisting on a *minimum norm* solution for **f**, which is the LS solution for **f** which simultaneously minimizes $||\mathbf{f}||$, the LS problem is well-defined even for $T = n$. (This problem and its solution is closely related to the problem of a singular autocorrelation matrix discussed in section 3.2.2.4.) Similarly, by using a "generalized inverse" of a singular or nonsingular matrix, rather than the standard inverse [25, 26], the LS projection operator **P**, defined by (6.1.13), can be defined even when the matrix $\mathbf{S}'\mathbf{S}$ is singular. It is always assumed that the generalized inverse of a nonsingular matrix is the same as the standard inverse. If this generalized inverse is defined appropriately, then the resulting solution for **f** (and **b**) can be guaranteed to be a minimum norm solution. It was shown in [23] that by appropriately redefining the projection operator with a generalized inverse, the projection updates in this chapter hold even when the covariance matrix is singular, and that the minimum norm solutions for **f** and **b** are thereby obtained. This implies that all of the recursions in (6.4.62) can be used from $T{=}0$ with the initial conditions,

$$K_n(-1) = 0, \; 1 \leqslant n \leqslant N \tag{6.4.63a}$$

$$\gamma(-1|0) = \gamma^*(-1|0) = \alpha(-1|0) = 0 \, . \tag{6.4.63b}$$

At each iteration T,

$$e_f^*(0|n) = \begin{cases} y(0) & \text{for } n = 0 \\ 0 & \text{for } n > 0 \end{cases} , \tag{6.4.63c}$$

$$e_f(T|0) = e_b(T|0) = y(T) \, , \tag{6.4.63e}$$

and

$$\epsilon_f(T|0) = \epsilon_b(T|0) = \epsilon_f(T-1|0) + y^2(T) \, , \tag{6.4.63f}$$

and when $T < N$, the recursions (6.4.62) are computed for $n{=}0$ to $n{=}T$.

The fixed-order algorithm can be applied only if $T>N$. Otherwise, because $y(T)$ is unknown when $T<0$, the variables of order N are undefined and cannot be used to compute the same variables at the successive time interval. Initialization of this algorithm can be performed, however, by using the order-recursive algorithm (6.4.62) along with order updates for $\mathbf{f}$ and $\mathbf{b}$, to increase the order of the filter by one at each successive time iteration $T<N$. An exact order-recursive initialization routine for the fixed-order order growing memory covariance algorithm is left as a problem (problem 6-13). Order-recursive computation of $\mathbf{f}$ and $\mathbf{b}$ requires on the order of N^2 arithmetic operations per iteration, rather than order N operations per iteration as required by the algorithms (6.4.61) and (6.4.62). Not all N components of the vectors $\mathbf{f}$, $\mathbf{b}$, and $\mathbf{h}$ need to be updated at each iteration for $T<N$, however. If data is first received at time $T=0$, the order-recursive initialization procedure can be performed for $n=0$ up to $n=T$. At time $T=N$ all of the variables which are needed by the fixed-order algorithm can easily be computed.

Another way to initialize (6.4.61) is to use the lattice algorithm for $T=0$ to $T=N$, whereupon the fixed-order variables $\mathbf{f}$, $\mathbf{b}$, $\mathbf{g}$, and $\mathbf{h}$ can be computed from lattice variables via the whitening filter updates in the next section. An approximate initialization routine, which is analogous to the initialization for the fixed-order sliding window algorithm discussed earlier, is to let $\epsilon_f(0) = \delta$, where δ is chosen large enough so that the algorithm does not diverge, and to initialize the remaining variables using the variable definitions. The details are left as a problem (problem 6-12).

6.4.4. Covariance Whitening Filter Algorithms

Order-recursive filter structures are now derived which generate the forward and backward prediction filters for the sliding window and growing memory covariance LS cases. This section parallels section 6.3.1 in which an order-recursive whitening filter algorithm was derived for the prewindowed LS case. The covariance forward and backward whitening filters are defined respectively as

$$F_{T_0,T|n}(z) = 1 - \mathbf{f}'(T_0,T|n)\ [z^{-1}\mathbf{z}_n] \tag{6.4.64a}$$

and

$$B_{T_0,T|n}(z) = z^{-n} - \mathbf{b}'(T_0,T|n)\ \mathbf{z}_n\ , \tag{6.4.64b}$$

where $\mathbf{z}_n$ is defined by (6.3.80). In addition, the "gain" filters are defined,

$$G_{T_0,T|n}(z) = \mathbf{g}'(T_0,T|n)\ \mathbf{z}_n \tag{6.4.65a}$$

and

$$H_{T_0,T|n}(z) = \mathbf{h}'(T_0,T|n)\ \mathbf{z}_n\ . \tag{6.4.65b}$$

Order updates, and forward and backward time updates for the filters $F_{T_0,T|n}(z)$, $B_{T_0,T|n}(z)$, $G_{T_0,T|n}(z)$, and $H_{T_0,T|n}(z)$ can be derived in an analogous fashion as the prewindowed updates derived in section 6.3.1. The order

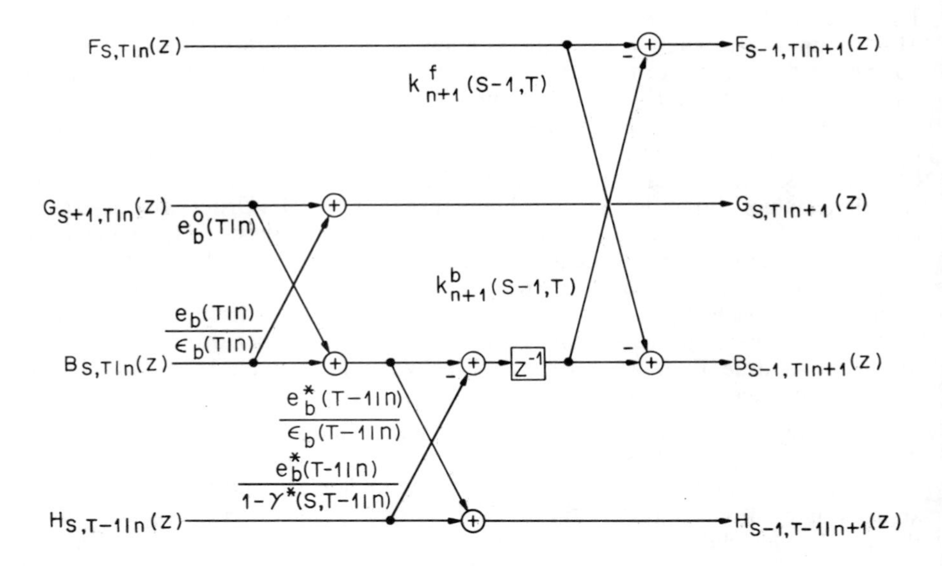

Figure 6-11. Flow diagram for one stage of the sliding window covariance whitening filter algorithm.

updates are obtained from (6.4.27-32),

$$F_{T_0,T|n+1}(z) = F_{T_0+1,T|n}(z) - k^b_{n+1}(T_0,T)\ [z^{-1}B_{T_0,T-1|n}(z)] \tag{6.4.66a}$$

$$B_{T_0,T|n+1}(z) = z^{-1}B_{T_0,T-1|n}(z) - k^f_{n+1}(T_0,T)F_{T_0+1,T|n}(z) \tag{6.4.66b}$$

$$G_{T_0,T|n+1}(z) = G_{T_0+1,T|n}(z) + \frac{e_b(T_0,T|n)}{\epsilon_b(T_0,T|n)}\, B_{T_0,T|n}(z) \tag{6.4.67a}$$

$$G_{T_0,T|n+1}(z) = z^{-1}G_{T_0,T-1|n}(z) + \frac{e_f(T_0,T|n)}{\epsilon_f(T_0,T|n)}\, F_{T_0,T|n}(z) \tag{6.4.67b}$$

$$H_{T_0,T|n+1}(z) = H_{T_0+1,T|n}(z) + \frac{e^*_b(T_0,T|n)}{\epsilon_b(T_0,T|n)}\, B_{T_0,T|n}(z) \tag{6.4.68a}$$

$$H_{T_0,T|n+1}(z) = z^{-1}H_{T_0,T-1|n}(z) + \frac{e^*_f(T_0,T|n)}{\epsilon_f(T_0,T|n)}\, F_{T_0,T|n}(z)\ . \tag{6.4.68b}$$

The forward time updates are obtained from (6.4.38-39),

$$F_{T_0,T-1|n}(z) = F_{T_0,T|n}(z) + z^{-1}G_{T_0,T-1|n}(z)\,\frac{e_f(T_0,T|n)}{1-\gamma(T_0,T-1|n)} \qquad \textbf{(6.4.69)}$$

$$B_{T_0,T-1|n}(z) = B_{T_0,T|n}(z) + G_{T_0+1,T|n}(z)\,\frac{e_b(T_0,T|n)}{1-\gamma(T_0+1,T|n)} \qquad (6.4.70)$$

$$H_{T_0,T-1|n}(z) = H_{T_0,T|n}(z) + G_{T_0,T|n}(z)\,\frac{\alpha(T_0,T|n)}{1-\gamma(T_0,T|n)}\,. \qquad (6.4.71)$$

The backward time updates are obtained from (6.4.45-46),

$$F_{T_0+1,T|n}(z) = F_{T_0,T|n}(z) + z^{-1}H_{T_0,T-1|n}(z)\,\frac{e_f^*(T_0,T|n)}{1-\gamma^*(T_0,T-1|n)} \qquad (6.4.72)$$

$$B_{T_0+1,T|n}(z) = B_{T_0,T|n}(z) + H_{T_0+1,T|n}(z)\,\frac{e_b^*(T_0,T|n)}{1-\gamma^*(T_0+1,T|n)} \qquad (6.4.73)$$

$$G_{T_0+1,T|n}(z) = G_{T_0,T|n}(z) + H_{T_0,T|n}(z)\,\frac{\alpha(T_0,T|n)}{1-\gamma^*(T_0,T|n)}\,. \qquad (6.4.74)$$

Combining (6.4.70), (6.4.73), (6.4.66a), (6.4.66b), (6.4.67a), and (6.4.68a) (in that order) gives the order-recursive whitening filter updates for the sliding window covariance case. Given the lattice variables $e_b(T_0,T|n)$, $e_b^*(T_0,T|n)$, $\gamma(T_0+1,T|n)$, $\gamma^*(T_0+1,T|n)$, $\epsilon_f(T_0,T|n)$, $\epsilon_b(T_0,T|n)$, $k_n^b(T_0,T)$, and $k_n^f(T_0,T)$ for $1 \leqslant n \leqslant N$, this algorithm can be used to compute the fixed-order state variables $\mathbf{g}(T_0,T|N)$, $\mathbf{h}(T_0,T|N)$, $\mathbf{f}(T_0,T|N)$, and $\mathbf{b}(T_0,T|N)$. The lower time index of the sliding window for each variable is determined in exactly the same fashion as for the variables entering the sliding window lattice algorithm (6.4.55). In particular, suppose the width of the sliding window is M. At time T, the filters to be computed are $F_{S,T|n}(z)$ and $B_{S,T|n}(z)$ where $S = (T-M+1) + (N-n)$. The window of data values used to compute a filter of order n therefore has width $T-S+1 = M - (N-n)$. Replacing T_0 in (6.4.70) and (6.4.67a) by S and T_0 in (6.4.73), (6.4.66a), (6.4.66b), and (6.4.68a) by $S-1$ makes the algorithm consistent. Initialization is accomplished as follows,

$$F_{T|0}(z) = B_{T|0}(z) = 1\,, \qquad (6.4.75a)$$

$$G_{S+1,T|0}(z) = H_{S,T|0}(z) = 0\,. \qquad (6.4.75b)$$

The resulting filter structure is shown in figure 6-11.

An order-recursive whitening filter algorithm for the growing memory covariance case is obtained by combining (6.4.74), (6.4.71), (6.4.72), (6.4.70), (6.4.66a), (6.4.66b), (6.4.67a), and (6.4.68b), where T_0 is replaced by zero. Initialization is again accomplished via (6.4.75). The resulting filter structure is shown in figure 6-12.

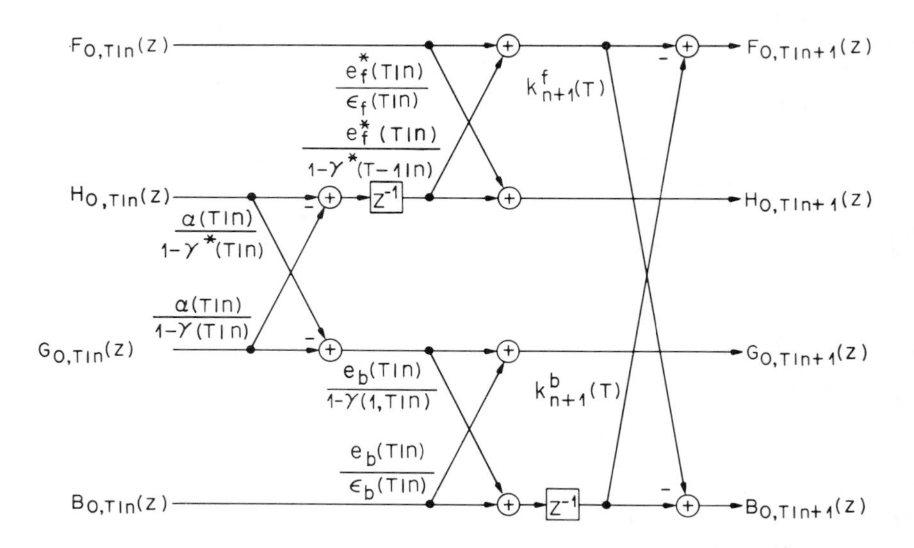

Figure 6-12. Flow diagram for one stage of the growing memory covariance whitening filter algorithm.

6.5. LS JOINT PROCESS ESTIMATION

The algorithms presented so far solve the LS prediction problem wherein the sums in (6.4.1) are minimized. In applications such as channel equalization and echo and noise cancellation, samples from two sequences, $\{d(j)\}$ and $\{y(j)\}$, are given and our objective is to estimate the desired sequence, $\{d(j)\}$, from the sequence $\{y(j)\}$. The desired data vector

$$\mathbf{D}(T_0+n,T) \equiv [d(T) \;\; d(T-1) \;\; \cdots \;\; d(T_0+n)]' \tag{6.5.1}$$

is therefore defined along with the vector of estimation errors,

$$\begin{aligned}\mathbf{E}_c(T_0+1,T|n) &\equiv \mathbf{D}(T_0+n,T) - \sum_{j=0}^{n-1} c_{j+1}(T|n)\;[z^{-j}\mathbf{Y}(T_0+n,T)] \\ &= \mathbf{D}(T_0+n,T) - \mathbf{S}(T|0,n-1)\;\mathbf{c}(T_0+1,T|n)\,,\end{aligned} \tag{6.5.2}$$

where $\mathbf{c}(T_0+1,T|n)$ is the n-dimensional vector of regression coefficients at time T used to estimate $\mathbf{D}(T_0+n,T)$, and the lower time index of $\mathbf{E}_c$ and $\mathbf{c}$ denotes the time index of the starting sample from the "y" sequence (i.e., $y(T_0+1)$), which is used in the LS computation. (This subscript can be ignored in the prewindowed case.) Our objective is to choose $\mathbf{c}(T_0+1,T|n)$ so as to minimize the sum

$$\epsilon_c(T_0+1,T|n) \equiv ||\mathbf{E}_c(T_0+1,T|n)||^2 .$$
$$= \sum_{j=T_0+n}^{T} [d(j) - \mathbf{c}'(T_0+1,T|n)\, \mathbf{y}(j|n)]^2 . \tag{6.5.3}$$

Setting the gradient of ϵ_c with respect to $\mathbf{c}$ equal to zero gives

$$\hat{\mathbf{\Phi}}(T_0+n,T|n)\, \mathbf{c}(T_0+1,T|n) = \sum_{j=T_0+n}^{T} d(T)\, \mathbf{y}(j|n) , \tag{6.5.4}$$

where the covariance matrix $\hat{\mathbf{\Phi}}$ is defined by (6.4.3). The discussion in section 6.1 implies that the Hilbert space residual vector can be written as the orthogonal projection,

$$\mathbf{E}_c(T_0+1,T|n) = \mathbf{P}^{\perp}_{M(T|0,n-1)}\mathbf{D}(T_0+n,T) . \tag{6.5.5}$$

The projection updates in section 6.2 are now used to derive order and time updates for $\mathbf{E}_c(T_0,T|n)$ and $\mathbf{c}(T_0,T|n)$ assuming first prewindowed, and then covariance LS estimation is to be performed. Combining the recursions in this section with the prediction algorithms of sections 6.3 and 6.4 results in recursive algorithms that solve the LS joint process estimation problem.

6.5.1. Prewindowed Joint Process Estimation

In this case the lower time subscript can be omitted, since both $d(T)$ and $y(T)$ are assumed to be zero when $T < 0$. The shift operator z^{-j} is defined by (6.3.2). An order update for $\mathbf{E}_c(T|n)$ is obtained by using (6.2.6), where $\mathbf{Y}$ is replaced by $\mathbf{D}(T)$, M is replaced by $M(T|0,n-1)$, and $\mathbf{D}$ is replaced by $z^{-n}\mathbf{Y}(T)$,

$$\mathbf{E}_c(T|n+1) = \mathbf{E}_c(T|n) - \frac{K^c_{n+1}(T)}{\epsilon_b(T|n)}\, \mathbf{E}_b(T|n) \tag{6.5.6}$$

where

$$K^c_{n+1}(T) = \langle \mathbf{D}(T),\mathbf{E}_b(T|n) \rangle$$
$$= \langle \mathbf{E}_c(T|n),\mathbf{E}_b(T|n) \rangle , \tag{6.5.7}$$

and is analogous to $K_{n+1}(T)$ given by (6.3.19). Taking norms of both sides of (6.5.6) gives

$$\epsilon_c(T|n+1) = \epsilon_c(T|n) - \frac{K^{c^2}_{n+1}(T)}{\epsilon_b(T|n)} . \tag{6.5.8}$$

Expanding (6.5.6) in terms of the data vectors (i.e., (6.5.2) and (6.3.6b)) gives the order update for $\mathbf{c}(T|n)$,

$$[\mathbf{c}(T|n+1)]_{n+1} = \frac{K^c_{n+1}(T)}{\epsilon_b(T|n)} \tag{6.5.9a}$$

$$[\mathbf{c}(T|n+1)]_{1,n} = \mathbf{c}(T|n) - [\mathbf{c}(T|n+1)]_{n+1}\mathbf{b}(T|n) . \tag{6.5.9b}$$

Time updates for $\mathbf{c}(T|n)$, $K_{n+1}^c(T)$, and $\epsilon_c(T|n)$ are easily obtained from (6.2.33) and (6.2.40). The results are

$$\mathbf{c}(T|n) = \mathbf{c}(T-1|n) + e_c^o(T|n)\mathbf{g}(T|n) , \tag{6.5.10}$$

$$K_{n+1}^c(T) = wK_{n+1}^c(T-1) + e_c(T|n)e_b(T|n)\frac{1}{1-\gamma(T|n)} , \tag{6.5.11}$$

and

$$\epsilon_c(T|n) = w\epsilon_c(T-1|n) + e_c^2(T|n)\frac{1}{1-\gamma(T|n)} , \tag{6.5.12}$$

where the current joint process residual is

$$\begin{aligned} e_c(T|n) &= <\mathbf{u}(T),\mathbf{E}_c(T|n)> \\ &= d(T) - \mathbf{c}'(T|n)\mathbf{y}(T|n) \end{aligned} \tag{6.5.13}$$

and the oblique joint process residual is

$$e_c^o(T|n) = d(T) - \mathbf{c}'(T-1|n)\mathbf{y}(T|n) . \tag{6.5.14}$$

Using (6.2.34), it is easy to show that

$$e_c^o(T|n) = e_c(T|n)\frac{1}{1-\gamma(T|n)} , \tag{6.5.15}$$

which is analogous to (6.3.41) and (6.3.47).

Combining (6.5.14), (6.5.10), and (6.5.13) with the fast Kalman prediction algorithm (6.3.63) gives a fast Kalman joint process estimation algorithm. Alternatively, (6.5.15) can be used to solve for $e_c(T|N)$, instead of (6.5.13), when the modifications to the fast Kalman algorithm discussed in section 6.3 are used. Adding (6.5.11) and the top component of (6.5.6) to the LS lattice algorithm (6.3.64) gives a prewindowed LS lattice joint process estimation algorithm. The additional recursions required in the (modified) fixed-order case bring the computational complexity to a total of $8N+9$ multiplies and $8N+4$ additions per iteration. The computational complexity of the lattice joint process estimator is $16N$ multiplies and $8N$ additions per iteration.

Initial conditions for the additional recursions entering the lattice algorithm are

$$K_n^c(-1) = 0, \quad 1 \leqslant n \leqslant N , \tag{6.5.16}$$

and at each iteration

$$e_c(T|0) = d(T) . \tag{6.5.17}$$

The flow diagram for the prewindowed LS lattice joint process estimator is identical to that shown in figure 4-12b where the MMSE lattice PARCOR coefficients are replaced by the analogous time varying LS coefficients and each coefficient k_n^c is replaced by

$$\frac{K_n^c(T)}{\epsilon_b(T|n-1)} , \quad 1 \leqslant n \leqslant N .$$

The fast Kalman joint process estimator can be initialized by using (6.3.68), in addition to setting the elements of $\mathbf{c}(0|N)$ equal to zero. As pointed out in section 6.3, however, an *exact* LS estimate for $\mathbf{c}(T|N)$ cannot be obtained with this initialization procedure. Rather, the LS filter must be constructed in an order-recursive fashion when $T < N$. The order-recursive initialization routine (6.3.79) accomplishes this task for the LS prediction case. In order to modify this routine so as to apply to the joint process estimation case, the discussion preceding (6.3.72) can be repeated with e_f replaced by e_c and $\mathbf{f}$ replaced by $\mathbf{c}$. It follows that for $T < N$ there exists a vector $\mathbf{c}(T|T+1)$ such that each element of the residual vector

$$\mathbf{E}_c(T|T+1) = \mathbf{D}(T) - \mathbf{S}(T|0,T)\,\mathbf{c}(T|T+1) \qquad (6.5.18)$$

is equal to zero. It is easily verified that

$$[\mathbf{c}(T|T+1)] = \frac{e_c^o(T|T+1)}{y(0)}\,, \qquad (6.5.19)$$

where e_c^o is defined by (6.5.14), and

$$[\mathbf{c}(T|T+1)]_j = [\mathbf{c}(T-1|T)]_j\,, \quad 1 \leqslant j \leqslant T\,. \qquad (6.5.20)$$

Appending (6.5.14), (6.5.19), and (6.5.20) to the order-recursive initialization routine (6.3.79) enables the order-recursive computation of $\mathbf{c}(T|N)$ when $T<N$. When $T=N$, (6.5.10) can be used to compute $\mathbf{c}(N|N)$ from $\mathbf{c}(N-1|N)$.

6.5.2. Covariance Joint Process Estimation

The vector of estimation errors is now defined by (6.5.2) where the starting time subscript T_0 must be included. The following notation, which is analogous to the notation in the last section, is also needed.

Cross-correlation coefficient:

$$\begin{aligned} K_{n+1}^c(T_0,T) &\equiv \langle \mathbf{D}(T_0+n,T),\ \mathbf{E}_b(T_0,T|n)\rangle \\ &= \langle \mathbf{E}_c(T_0+1,T|n),\ \mathbf{E}_b(T_0,T|n)\rangle\,. \end{aligned} \qquad (6.5.21)$$

Current residual (scalar):

$$\begin{aligned} e_c(T_0,T|n) &= \langle \mathbf{u}(T),\ \mathbf{E}_c(T_0,T|n)\rangle\,. \\ &= d(T) - \mathbf{c}'(T_0,T|n)\,\mathbf{y}(T|n)\,. \end{aligned} \qquad (6.5.22)$$

Past residual (scalar):

$$\begin{aligned} e_c^*(T_0+1,T|n) &= \langle \mathbf{u}(T_0), \mathbf{E}_c(T_0+1,T|n)\rangle \\ &= d(T_0+n) - \mathbf{c}'(T_0+1,T|n)\,\mathbf{y}(T_0+n|n)\,. \end{aligned} \qquad (6.5.23)$$

Oblique residual:

$$e_c^o(T_0,T|n) \equiv d(T) - \mathbf{c}'(T_0,T-1|n)\,\mathbf{y}(T|n)\,. \qquad (6.5.24)$$

The order recursions are obtained from (6.2.6),

$$\mathbf{E}_c(T_0,T|n+1) = \mathbf{E}_c(T_0+1,T|n) - \frac{K_{n+1}^c(T_0,T)}{\epsilon_b(T_0,T|n)} \mathbf{E}_b(T_0,T|n) \tag{6.5.25}$$

$$\epsilon_c(T_0,T|n+1) = \epsilon_c(T_0+1,T|n) - \frac{K_{n+1}^{c2}(T_0,T)}{\epsilon_b(T_0,T|n)} \tag{6.5.26}$$

$$[\mathbf{c}(T_0,T|n+1)]_{n+1} = \frac{K_{n+1}^c(T_0,T)}{\epsilon_b(T_0,T|n)} \tag{6.5.27a}$$

$$[\mathbf{c}(T_0,T|n+1)]_{1,n} = \mathbf{c}(T_0+1,T|n) - [\mathbf{c}(T_0,T|n+1)]_{n+1}\mathbf{b}(T_0,T|n). \tag{6.5.27b}$$

Derivation of the forward time updates involves a straightforward application of (6.2.33) and (6.2.40), where $\mathbf{Y}(T)$ is replaced by $\mathbf{D}(T_0,T)$ and $M(T)$ is replaced by $M(T|0,n-1)$,

$$\mathbf{c}(T_0,T|n) = \mathbf{c}(T_0,T-1|n) + e_c^o(T_0,T|n)\mathbf{g}(T_0,T|n) \tag{6.5.28}$$

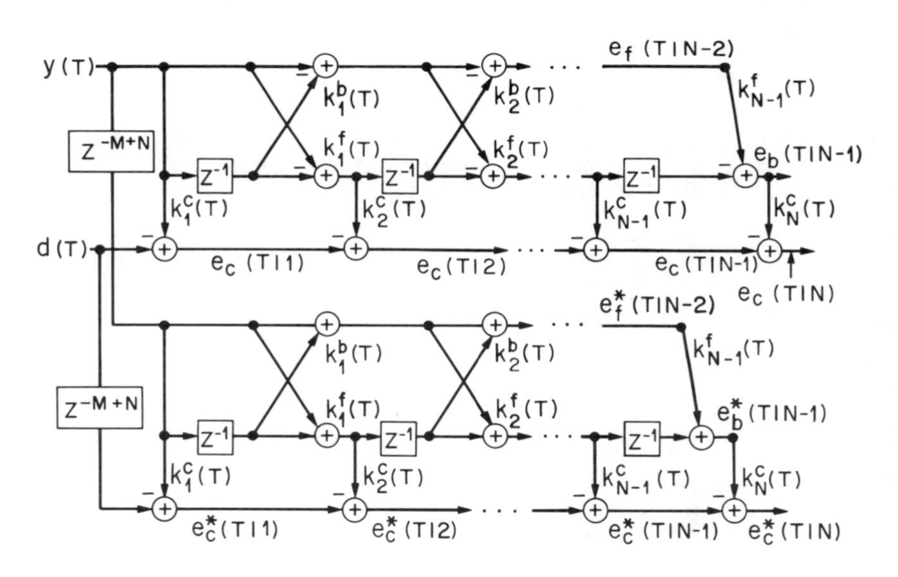

Figure 6-13. Flow diagram for the sliding window lattice joint process estimator.

$$K_{n+1}^c(T_0,T) = K_{n+1}^c(T_0,T-1) + e_c(T_0+1,T|n)e_b(T_0,T|n)\frac{1}{1-\gamma(T_0+1,T|n)} \tag{6.5.29}$$

$$\epsilon_c(T_0,T|n) = \epsilon_c(T_0,T-1|n) + e_c^2(T_0,T|n)\frac{1}{1-\gamma(T_0,T|n)} \tag{6.5.30}$$

$$e_c^*(T_0,T|n) = e_c^*(T_0,T-1|n) - e_c(T_0,T|n)\frac{\alpha(T_0,T|n)}{1-\gamma(T_0,T|n)} \tag{6.5.31}$$

$$e_c^o(T_0,T|n) = \frac{e_c(T_0,T|n)}{1-\gamma(T_0,T|n)} . \tag{6.5.32}$$

The backward time updates are similarly obtained from (6.2.48) and (6.2.49),

$$\mathbf{c}(T_0,T|n) = \mathbf{c}(T_0+1,T|n) + \mathbf{h}(T_0,T|n)\frac{e_c^*(T_0,T|n)}{1-\gamma^*(T_0,T|n)} \tag{6.5.33}$$

$$e_c(T_0+1,T|n) = e_c(T_0,T|n) + e_c^*(T_0,T|n)\frac{\alpha(T_0,T|n)}{1-\gamma^*(T_0,T|n)} \tag{6.5.34}$$

$$K_{n+1}^c(T_0+1,T) = K_{n+1}^c(T_0,T) - e_c^*(T_0+1,T|n)e_b^*(T_0,T|n)\frac{1}{1-\gamma^*(T_0+1,T|n)} \tag{6.5.35}$$

$$\epsilon_c(T_0+1,T|n) = \epsilon_c(T_0,T|n) - e_c^{*2}(T_0,T|n)\frac{1}{1-\gamma^*(T_0,T|n)} . \tag{6.5.36}$$

Combining (6.5.24), (6.5.28), (6.5.23), and (6.5.33) (in that order) with the fixed-order sliding window algorithm (6.4.54) gives the corresponding sliding window joint process estimation algorithm. Adding these additional recursions results in a total computational complexity of $17N+14$ multiplies and $9N+14$ additions per iteration. Combining (6.5.25) (top and bottom components), (6.5.29), and (6.5.35) with the sliding window lattice prediction algorithm (6.4.55) gives the corresponding sliding window lattice joint process estimation algorithm. The lower time subscripts in the latter case are changed according to the rules specified earlier for the sliding window lattice predictor. In particular, referring again to (6.4.8),

$$\begin{aligned} e_c(T|n) &= <\mathbf{u}(T),\mathbf{P}^{\perp}_{M(T|0,n-1)}\mathbf{D}(T)> \\ &= e_c(T_0+N-n,T|n) , \end{aligned} \tag{6.5.37}$$

where $T_0 = T-M+1$. The computational complexity of this algorithm is $23N$ multiplies and $14N$ additions per iteration. The flow diagram for the sliding window lattice joint process estimator is shown in figure 6-13.

Initialization of these recursions is accomplished in an analogous fashion to the prediction recursions. In particular, the data $y(T)$ and $d(T)$ is assumed to be zero for $T<0$, and

$$\mathbf{c}(-1|n) = \mathbf{0} \,, \tag{6.5.38}$$

$$K_n^c(-1) = 0 \,, \quad 1 \leqslant n \leqslant N \,, \tag{6.5.39}$$

and at each iteration,

$$e_c(T_0,T|0) = d(T), \quad e_c^*(T_0+1,T|0) = d(T_0+N) \,, \tag{6.5.40}$$

where N is the filter order.

The fixed-order growing memory algorithm (6.4.61) is extended to the joint process estimation case by adding the recursions (6.5.24) and (6.5.28). The lattice prediction algorithm (6.4.62) is extended to the joint process estimation case by adding the recursions (6.5.25) (top component only), (6.5.29), and (6.5.31). In each case the variable T_0=0. Adding (6.5.24) and (6.5.28) to the fixed-order algorithm results in a total computational complexity of $15N+6$ multiplies and $13N+1$ additions per iteration. The growing memory covariance lattice joint process estimator requires $27N$ multiplies and $15N$ additions per iterations. Initialization of the additional recursions for the lattice algorithm is accomplished as follows,

$$K_n^c(-1) = 0, \quad 1 \leqslant n \leqslant N \,, \tag{6.5.41a}$$

$$\mathbf{c}(-1|n) = \mathbf{0}, \quad 0 \leqslant n \leqslant N \,, \tag{6.5.41b}$$

and

$$e_c^*(0|n) = \begin{cases} d(0) & \text{for } n=0 \\ 0 & \text{for } n>0 \end{cases} . \tag{6.5.41c}$$

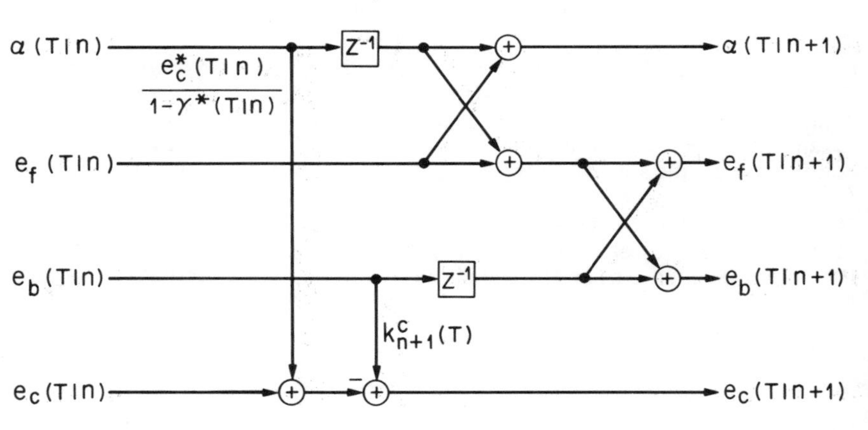

Figure 6-14. Flow diagram for the growing memory covariance lattice joint process estimat

At each iteration,

$$e_c(T|0) = e_c(1,T|0) = d(T) . \tag{6.5.41d}$$

As in the prediction case, the fixed-order algorithm can again be initialized by using an order-recursive algorithm for $T < N$ (problem 6-13), or by an approximate initialization procedure (problem 6-12). The flow diagram for the growing memory lattice joint process estimator is shown in figure 6-14.

6.5.3. Joint Process Whitening Filter Algorithms

The whitening filter algorithms in section 6.3.1 and 6.4.4 can be extended to the joint process estimation case by defining the filter

$$C_{T_0,T|n}(z) = \mathbf{c}'(T_0,T|n)\ \mathbf{z}_n . \tag{6.5.42}$$

Equations (6.5.27), (6.5.28), and (6.5.33) can be used to obtain, respectively,

$$C_{T_0,T|n+1}(z) = C_{T_0+1,T|n}(z) + k_{n+1}^c(T_0,T)\ B_{T_0,T|n}(z) , \tag{6.5.43}$$

$$C_{T_0,T|n}(z) = C_{T_0,T-1|n}(z) + e_c^o(T_0,T|n)G_{T_0|n}(T|n) , \tag{6.5.44}$$

and

$$C_{T_0+1,T|n}(z) = C_{T_0,T|n}(z) - H_{T_0,T|n}(z)\ \frac{e_c^*(T_0,T|n)}{1-\gamma^*(T_0,T|n)} , \tag{6.5.45}$$

where

$$k_{n+1}^c(T_0,T) = [\mathbf{c}(T_0,T|n+1)]_{n+1} = \frac{K_{n+1}^c(T_0,T)}{\epsilon_b(T_0,T|n)} . \tag{6.5.46}$$

The recursions (6.5.43) and (6.5.44) also result from the prewindowed recursions (6.5.9) and (6.5.10) if the lower time subscript T_0 is ignored. Combining (6.5.43) with the prewindowed predictor recursions specified in section 6.3.1 (figure 6-8) results in the filter structure shown in figure 6-15, from which the prewindowed polynomial $C_{T|n}(z)$ can be computed in an order-recursive fashion. Combining (6.5.43), the sliding window predictor recursions specified in section 6.4.4, (6.5.43) and (6.5.45) (where $T_0 = 0$), and the growing memory predictor recursions in section 6.4.4, results in analogous modifications of the predictor flow diagrams in figures 6-11 and 6-12, respectively. The resulting joint process whitening filter algorithms compute $C_{S,T|n}(z)$ and $C_{0,T|n}(z)$, respectively, in an order-recursive fashion. The recursions (6.5.43-45) are initialized by setting

$$C_{T_0,T|0}(z) = 0 . \tag{6.5.47}$$

6.6. NORMALIZED LS ALGORITHMS

Although the algorithms derived in the last three sections solve the associated LS problem in a computationally efficient manner, one can be easily overwhelmed by the number of recursions required in each case. In this section it is shown that by appropriately transforming, or normalizing, the residuals

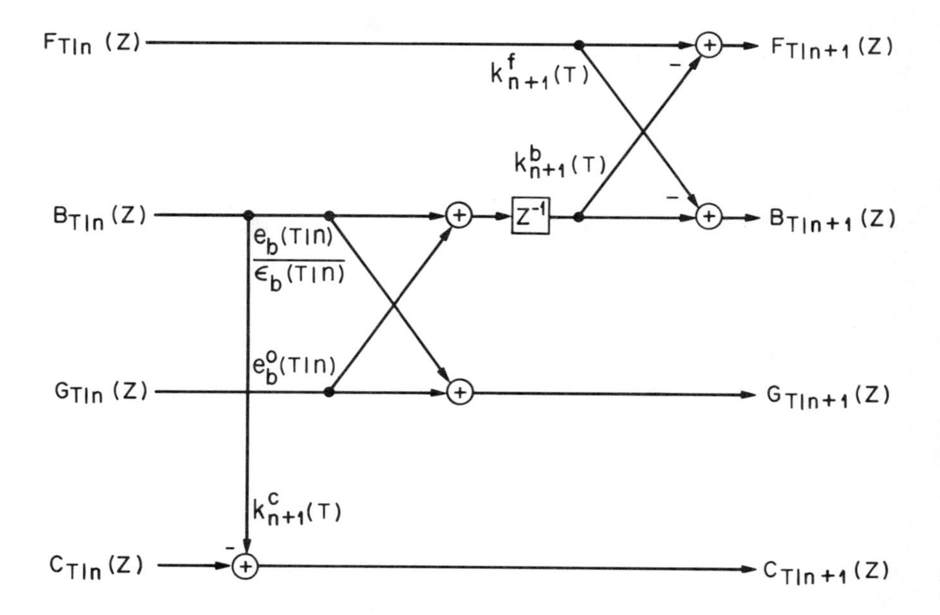

Figure 6-15. Filter structure that computes the prewindowed joint process filter $C_{T|n}(z)$.

in the order-recursive algorithms presented in the last three sections, new LS algorithms result which involve fewer recursions and are much less sensitive to finite precision errors than the previous "unnormalized" algorithms. (The normalized prewindowed lattice variables are in fact always less than unity in magnitude, which is a significant advantage for fixed precision implementations.) The variables which appear in these "normalized" algorithms are related in a nonlinear fashion to the variables entering the corresponding unnormalized LS algorithms presented thus far. However, the normalized algorithms are simpler in appearance than the corresponding unnormalized algorithms.

The normalized algorithms in this section stem from a normalized *fixed-coefficient* lattice filter [27] with reduced sensitivity to round-off errors when compared with its unnormalized counterpart derived in chapter 4. A normalization of the prewindowed LS lattice algorithm reduced the six "unnormalized" recursions in (6.3.64) to only three normalized recursions [11]. Subsequently, normalized versions of the lattice covariance algorithms in the last section [23] and normalized fixed-order LS algorithms have been derived [13, 28]. In contrast with the normalization of order-recursive algorithms, however, normalization of the fixed-order algorithms does not reduce the number of recursions which need to be computed at each iteration. Nevertheless, the normalized fixed-order algorithms have been reported to have excellent numerical properties.

The presentation of normalized LS algorithms in this section is essentially chronological. A normalized fixed-coefficient MMSE lattice filter is first derived. This is followed by a derivation of the normalized prewindowed LS lattice algorithm. Finally, normalized covariance algorithms are presented.

6.6.1. Mathematical Techniques

Before proceeding to the derivation of normalized algorithms, the available mathematical techniques which can be used are first outlined. As with the unnormalized LS recursions, an algebraic, geometric, or projection operator approach can be used to derive the associated normalized recursions. The algebraic approach is perhaps the most straightforward and involves merely the substitution of normalized variables into unnormalized LS recursions already derived. Algebraic manipulation subsequently yields the normalized version of the corresponding unnormalized algorithm. In contrast to algebraic derivations of the unnormalized LS algorithms, no matrix manipulations are required in the scalar data case.

A geometric approach to deriving normalized recursions examines the geometric relations between the Hilbert space vectors $\mathbf{E}_f$, $\mathbf{E}_b$, and $\mathbf{E}_c$ defined in previous sections. This approach was used to derive the normalized MMSE lattice algorithm. A more sophisticated version of this approach has been presented in [22].

Before proceeding, we briefly describe how the generalized projection operator approach outlined in section 6.2.4 can be modified so as to generate normalized rather than unnormalized recursions [4]. Given an arbitrary subspace M, and arbitrary vectors $\mathbf{v}$ and $\mathbf{w}$, a normalized version of the expression $\mathbf{v}'\mathbf{P}_M^\perp\mathbf{w}$ is defined to be

$$\rho_M(\mathbf{v},\mathbf{w}) \equiv \frac{\mathbf{v}'\mathbf{P}_M^\perp\mathbf{w}}{\left\{\left[\mathbf{v}'\mathbf{P}_M^\perp\mathbf{v}\right]\left[\mathbf{w}'\mathbf{P}_M^\perp\mathbf{w}\right]\right\}^{1/2}} . \tag{6.6.1}$$

Replacing the (unnormalized) expressions in (6.2.52) by the corresponding normalized quantities $\rho_{M+\mathbf{X}}(\mathbf{v},\mathbf{Y})$, $\rho_M(\mathbf{X},\mathbf{Y})$, $\rho_M(\mathbf{v},\mathbf{Y})$, and $\rho_M(\mathbf{v},\mathbf{X})$, and using the fact that $\rho_M(\mathbf{v},\mathbf{w}) = \rho_M(\mathbf{w},\mathbf{v})$ gives the generalized order update (problem 6-16),

$$\rho_{M\oplus\{\mathbf{X}\}}(\mathbf{v},\mathbf{Y}) = \frac{\rho_M(\mathbf{v},\mathbf{Y}) - \rho_M(\mathbf{v},\mathbf{X})\rho_M(\mathbf{X},\mathbf{Y})}{\left\{\left[1 - \rho_M^2(\mathbf{v},\mathbf{X})\right]\left[1 - \rho_M^2(\mathbf{X},\mathbf{Y})\right]\right\}^{1/2}} . \tag{6.6.2}$$

As an example, the prewindowed forward residual is

$$e_f(T|n) = \mathbf{u}'(T)\ \mathbf{P}_{M(T|1,n)}^\perp\mathbf{Y}(T) , \tag{6.6.3}$$

where $\mathbf{u}(T)$, $\mathbf{Y}(T)$, and $M(T|1,n)$ have been defined in section 6.3. The normalized forward residual is obtained by using (6.6.1), where $M = M(T|1,n)$, $\mathbf{v} = \mathbf{u}(T)$ and $\mathbf{w} = \mathbf{Y}(T)$, and is given by

$$\bar{e}_f(T|n) = \rho_{M(T|1,n)}[\mathbf{u}(T),\mathbf{Y}(T)]$$

$$= \frac{\mathbf{u}'(T)\mathbf{P}^{\perp}_{M(T|1,n)}\mathbf{Y}(T)}{\left\{\left[\mathbf{u}'(T)\mathbf{P}^{\perp}_{M(T|1,n)}\mathbf{u}(T)\right]\left[\mathbf{Y}'(T)\mathbf{P}^{\perp}_{M(T|1,n)}\mathbf{Y}(T)\right]\right\}^{1/2}} \tag{6.6.4}$$

$$= \frac{e_f(T|n)}{\{[1-\gamma(T-1|n)]\,\epsilon_f(T|n)\}^{1/2}} .$$

Using (6.6.2) gives an order update for $\bar{e}_f(T|n)$ in terms of other normalized LS variables. As with the unnormalized update (6.2.52), the normalized recursion (6.6.2) can be used to derive all of the normalized algorithms in this chapter by making the appropriate substitutions. The interested reader is referred to [4] for details (see also problem 6-16).

6.6.2. Normalized Lattice Structure

Before deriving normalized LS lattice algorithms, it is instructive to first consider a normalized version of a fixed-coefficient lattice filter. Normalization of the MMSE fixed coefficient lattice recursions (4.2.9) and (4.2.10) is therefore first performed. Continuing with the geometric approach used in previous sections, the MMSE order recursions derived in chapter 4 are repeated,

$$\mathbf{E}_f(T|n+1) = \mathbf{E}_f(T|n) - k^b_{n+1}\,[z^{-1}\mathbf{E}_b(T|n)] \tag{6.6.5a}$$

$$\mathbf{E}_b(T|n+1) = z^{-1}\mathbf{E}_b(T|n) - k^f_{n+1}\,\mathbf{E}_f(T|n) , \tag{6.6.5b}$$

where $\mathbf{E}_f$ and $\mathbf{E}_b$ are viewed as elements in the appropriately defined (MMSE) Hilbert space. The PARCOR coefficients are

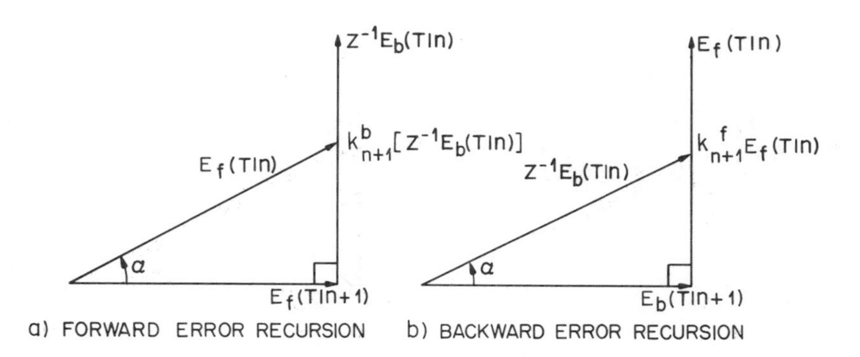

Figure 6-16. Geometric illustration of MMSE lattice recursions.

$$k_{n+1}^{b} = \frac{<\mathbf{E}_f(T|n), z^{-1}\mathbf{E}_b(T|n)>}{||z^{-1}\mathbf{E}_b(T|n)||^2} \tag{6.6.6a}$$

and

$$k_{n+1}^{f} = \frac{<\mathbf{E}_f(T|n), z^{-1}\mathbf{E}_b(T|n)>}{||\mathbf{E}_f(T|n)||^2} . \tag{6.6.6b}$$

From the discussion in chapter 4,

$$<\mathbf{E}_f(T|n+1), z^{-1}\mathbf{E}_b(T|n)> = 0 \tag{6.6.7a}$$

and

$$<\mathbf{E}_b(T|n+1), \mathbf{E}_f(T|n)> = 0 , \tag{6.6.7b}$$

and hence the order recursions (6.6.5) can be illustrated geometrically as in figure 6-16. Using (6.6.5-7), the angle α shown in figure 6-16 is given by

$$\begin{aligned} \sin\alpha &= \frac{||k_{n+1}^{b}\ [z^{-1}\mathbf{E}_b(T|n)]||}{||\mathbf{E}_f(T|n)||} \\ &= \frac{<\mathbf{E}_f(T|n), z^{-1}\mathbf{E}_b(T|n)>}{||z^{-1}\mathbf{E}_b(T|n)||\ ||\mathbf{E}_f(T|n)||} , \end{aligned} \tag{6.6.8}$$

and is the same in figures 6-16a and 6-16b. Defining the norms,

$$\epsilon_f(T|n) = ||\mathbf{E}_f(T|n)||^2 \tag{6.6.9a}$$

and

$$\epsilon_b(T|n) = ||\mathbf{E}_b(T|n)||^2 , \tag{6.6.9b}$$

the normalized residuals are defined,

$$\tilde{\mathbf{E}}_f(T|n) = \frac{\mathbf{E}_f(T|n)}{\sqrt{\epsilon_f(T|n)}} \tag{6.6.10a}$$

and

$$\tilde{\mathbf{E}}_b(T|n) = \frac{\mathbf{E}_b(T|n)}{\sqrt{\epsilon_b(T|n)}} . \tag{6.6.10b}$$

This substitution is illustrated geometrically for the forward error recursion (6.6.5a) in figure 6-17, where the vectors $\tilde{\mathbf{E}}_f(T|n)$ and $\tilde{\mathbf{E}}_f(T|n+1)$ have length unity. A normalized version of the order recursions (6.6.5) can therefore be represented as a rotation. The segment $\overline{CD}$ has been drawn parallel to $\overline{BE}$ and $\overline{CE}$ has been drawn parallel to $z^{-1}\mathbf{E}_b(T|n)$, so that $\overline{BD}$ is perpendicular to $\overline{AC}$, $\overline{CE}$ is perpendicular to $\overline{AD}$, and the angles are as indicated. Referring to figure 6-17, it follows that

$$\tilde{\mathbf{E}}_f(T|n+1) = \overline{AC} - \overline{EC} . \tag{6.6.11}$$

Now $\overline{AC}$ is parallel to $\tilde{\mathbf{E}}_f(T|n)$ and has length

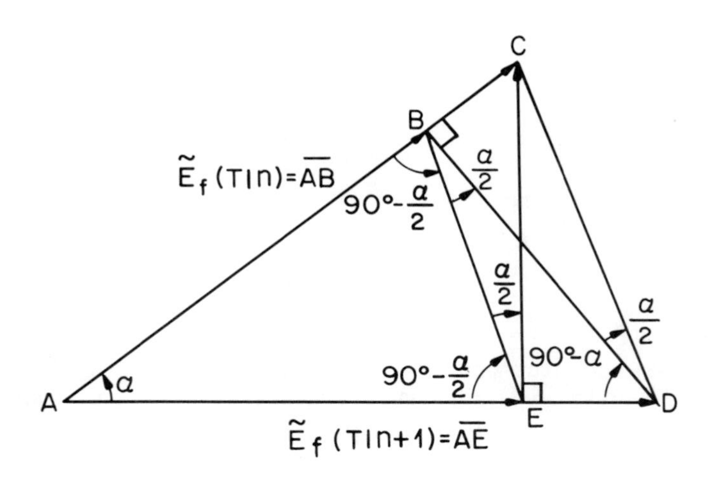

Figure 6-17. Geometric illustration of the forward error order recursion (6.6.5a).

$$AC = \frac{||\tilde{\mathbf{E}}_f(T|n+1)||}{\cos\ \alpha} = \frac{1}{\cos\ \alpha}, \qquad \textbf{(6.6.12)}$$

and $\overline{CE}$ is parallel to $z^{-1}\tilde{\mathbf{E}}_b(T|n)$ and has length

$$CE = ||\tilde{\mathbf{E}}_f(T|n+1)||\ \tan\ \alpha = \tan\ \alpha. \qquad (6.6.13)$$

Substituting (6.6.12) and (6.6.13) into (6.6.11) gives

$$\tilde{\mathbf{E}}_f(T|n+1) = \frac{\tilde{\mathbf{E}}_f(T|n) - \sin\ \alpha\ [z^{-1}\tilde{\mathbf{E}}_b(T|n)]}{\cos\ \alpha}. \qquad (6.6.14a)$$

Similarly, it is straightforward to show that

$$\tilde{\mathbf{E}}_b(T|n+1) = \frac{z^{-1}\tilde{\mathbf{E}}_b(T|n) - \sin\ \alpha\ \tilde{\mathbf{E}}_f(T|n)}{\cos\ \alpha}. \qquad (6.6.14b)$$

Letting

$$\bar{k}_{n+1} \equiv \sin\ \alpha, \qquad (6.6.15)$$

the corresponding MMSE time domain recursions are

$$\tilde{e}_f(T|n+1) = \frac{\tilde{e}_f(T|n) - \bar{k}_{n+1}\ \tilde{e}_b(T-1|n)}{\sqrt{1-\bar{k}_{n+1}^2}} \qquad (6.6.16a)$$

and

$$\tilde{e}_b(T|n+1) = \frac{\tilde{e}_b(T-1|n) - \bar{k}_{n+1}\tilde{e}_f(T|n)}{\sqrt{1 - \bar{k}_{n+1}^2}}, \tag{6.6.16b}$$

where

$$\tilde{e}_f(T|n) \equiv \frac{e_f(T|n)}{\sqrt{\epsilon_f(T|n)}} \tag{6.6.17a}$$

and

$$\tilde{e}_b(T|n) \equiv \frac{e_b(T|n)}{\sqrt{\epsilon_b(T|n)}}. \tag{6.6.17b}$$

In this case the MMSE prediction residuals $e_f(T|n)$ and $e_b(T|n)$ are computed from *fixed* coefficient filters. The normalized lattice structure corresponding to recursion (6.6.17) is shown in figure 6-18. The input to the lattice at each iteration T is

$$\tilde{e}_f(T|0) = \tilde{e}_b(T|0) = \frac{y(T)}{\{E[y^2(T)]\}^{1/2}}. \tag{6.6.18}$$

6.6.3. Normalized Prewindowed LS Algorithm

The derivation in the last section can be immediately generalized to the prewindowed LS case by simply changing the Hilbert space from the sample space of a scalar random process to the Euclidean space introduced in sections 6.1 and 6.3. The recursions (6.6.16) therefore also apply to the prewindowed LS case where the normalized residuals are defined by (6.6.17), the variables e_f, e_b, ϵ_f, ϵ_b, are the same as those defined in section 6.3, and the coefficients

$$\bar{k}_{n+1}(T) = \frac{<\mathbf{E}_f(T|n), z^{-1}\mathbf{E}_b(T|n)>}{\sqrt{\epsilon_f(T|n)\epsilon_b(T-1|n)}} \tag{6.6.19}$$

are time varying.

Although the normalized recursions in (6.6.16) are in an appealing form, it is readily discovered that normalization (6.6.10) does not reduce the number of recursions from the unnormalized lattice algorithm (6.3.64) which are needed to compute the normalized LS residuals $\tilde{e}_f(T|N)$ and $\tilde{e}_b(T|N)$. Rather, for the prewindowed LS case the following substitutions are made,

$$\bar{\mathbf{E}}_f(T|n) = \frac{\mathbf{E}_f(T|n)}{\sqrt{[1 - \gamma(T-1|n)]\epsilon_f(T|n)}} \tag{6.6.20a}$$

and

$$\bar{\mathbf{E}}_b(T|n) = \frac{\mathbf{E}_b(T|n)}{\sqrt{[1 - \gamma(T|n)]\epsilon_b(T|n)}}. \tag{6.6.20b}$$

Substituting normalized residuals for the unnormalized residuals in the order update (6.3.20a) gives

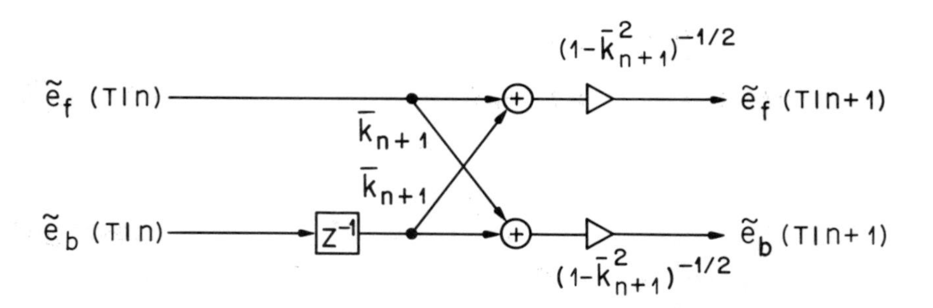

Figure 6-18. Flow diagram for normalized lattice predictor.

$$\bar{\mathbf{E}}_f(T|n+1) = \left\{\bar{\mathbf{E}}_f(T|n) - \bar{k}_{n+1}(T)[z^{-1}\bar{\mathbf{E}}_b(T|n)]\right\}\left\{\frac{[1-\gamma(T-1|n)]}{[1-\gamma(T-1|n+1)]}\frac{\epsilon_f(T|n)}{\epsilon_f(T|n+1)}\right\}^{1/2}. \tag{6.6.21}$$

From (6.3.21a) and (6.6.19) it follows that

$$\frac{\epsilon_f(T|n+1)}{\epsilon_f(T|n)} = 1 - \bar{k}^2_{n+1}(T), \tag{6.6.22}$$

and from (6.3.57b),

$$\frac{1-\gamma(T-1|n+1)}{1-\gamma(T-1|n)} = 1 - \bar{e}_b^2(T-1|n), \tag{6.6.23}$$

where

$$\bar{e}_b(T|n) \equiv \langle \mathbf{u}(T), \bar{\mathbf{E}}_b(T|n)\rangle. \tag{6.6.24}$$

Substituting (6.6.22) and (6.6.23) into (6.6.21) gives an order update for the normalized forward residual,

$$\bar{\mathbf{E}}_f(T|n+1) = \frac{\bar{\mathbf{E}}_f(T|n) - \bar{k}_{n+1}(T)[z^{-1}\bar{\mathbf{E}}_b(T|n)]}{\sqrt{[1-\bar{k}^2_{n+1}(T)][1-\bar{e}_b^2(T-1|n)]}}. \tag{6.6.25a}$$

Similarly, it can be shown from (6.3.20b) that

$$\bar{\mathbf{E}}_b(T|n+1) = \frac{z^{-1}\bar{\mathbf{E}}_b(T|n) - \bar{k}_{n+1}(T)\bar{\mathbf{E}}_f(T|n)}{\sqrt{[1-\bar{k}^2_{n+1}(T)][1-\bar{e}_f^2(T|n)]}}. \tag{6.6.25b}$$

To obtain a recursion for $\bar{k}_{n+1}(T)$, (6.3.64a) is rewritten as

$$\bar{k}_{n+1}(T) = w\,\bar{k}_{n+1}(T) + \bar{e}_f(T|n)\bar{e}_b(T-1|n)\,[\epsilon_f(T|n)\epsilon_b(T-1|n)]^{1/2}, \tag{6.6.26}$$

or

$$\bar{k}_{n+1}(T) = w\bar{k}_{n+1}(T-1)\left[\frac{\epsilon_f(T-1|n)\epsilon_b(T-2|n)}{\epsilon_f(T|n)\epsilon_b(T-1|n)}\right]^{1/2} + \bar{e}_f(T|n)\bar{e}_b(T-1|n). \tag{6.6.27}$$

From (6.3.50) and (6.3.51), it follows that

$$\frac{\epsilon_f(T-1|n)}{\epsilon_f(T|n)} = \frac{1}{w}\ [1 - \bar{e}_f^2(T|n)] \tag{6.6.28}$$

and

$$\frac{\epsilon_b(T-1|n)}{\epsilon_b(T|n)} = \frac{1}{w}\ [1 - \bar{e}_b^2(T|n)]\ , \tag{6.6.29}$$

where

$$\bar{e}_f(T|n) = <\mathbf{u}(T),\ \bar{\mathbf{E}}_f(T|n)> . \tag{6.6.30}$$

Substituting (6.6.28) and (6.6.29) into (6.6.27) gives

$$\bar{k}_{n+1}(T) = \bar{k}_{n+1}(T-1)\left\{\left[1 - \bar{e}_f^2(T|n)\right]\left[1 - \bar{e}_b^2(T-1|n)\right]\right\}^{1/2} + \bar{e}_f(T|n)\bar{e}_b(T-1|n)\ . \tag{6.6.31}$$

The top components of the order recursions (6.6.25) combined with (6.6.31) constitute the normalized prewindowed LS lattice predictor. The "residuals" $\bar{e}_f$ and $\bar{e}_b$ are related in a nonlinear fashion to the unnormalized LS residuals e_f and e_b; however, the coefficients $\bar{k}_n(T)$, $1 \leqslant n \leqslant N$, given by (6.6.19) can be used directly in applications such as vocal tract modeling or linear system identification. The exponential weighting constant w in (6.6.26-29) does not appear in the normalized recursions (6.6.25) and (6.6.31). It appears only in the initialization procedure,

$$e_b(-1|n) = \bar{k}_{n+1}(-1) = 0, \quad 0 \leqslant n \leqslant N-1\ . \tag{6.6.32a}$$

At each iteration,

$$\epsilon(T) = w\ \epsilon(T-1) + y^2(T), \quad \epsilon(-1) = 0\ , \tag{6.6.32b}$$

and

$$\bar{e}_f(T|0) = \bar{e}_b(T|0) = \frac{y(T)}{\sqrt{\epsilon(T)}}\ . \tag{6.6.32c}$$

Although the normalized lattice predictor just derived contains fewer recursions than the unnormalized lattice predictor, it requires significantly more computation. In particular, counting division as multiplication, the three normalized lattice recursions require $12N$ multiplies, $6N$ additions, and $3N$ square roots per iteration. In contrast, the unnormalized prewindowed LS lattice predictor requires $11N$ multiplies and $6N$ additions per iteration.

A normalized fixed-order prewindowed LS algorithm is discussed in reference [13] and in problem 6-19. As with the normalized lattice algorithms, normalized fixed-order algorithms also require somewhat more computation than the associated unnormalized algorithms; however, the normalized algorithms are less sensitive to round-off errors.

A summary of the computational complexity of the recursive LS algorithms considered in this chapter is given in table 6.1.

6.6.3.1. Normalized Joint Process Estimator

The prewindowed LS joint process estimation algorithm derived in section 6.5.1 can be similarly normalized. The normalized joint process residual vector is defined as

$$\overline{\mathbf{E}}_c(T|n) \equiv \frac{\mathbf{E}_c(T|n)}{\sqrt{[1-\gamma(T|n)]\epsilon_c(T|n)}} . \tag{6.6.33}$$

From (6.5.6) it follows that

$$\overline{\mathbf{E}}_c(T|n+1) = [\overline{\mathbf{E}}_c(T|n) - \overline{k}^c_{n+1}(T)\overline{\mathbf{E}}_b(T|n)]\left\{\frac{[1-\gamma(T|n)]}{[1-\gamma(T|n+1)]}\,\frac{\epsilon_c(T|n)}{\epsilon_c(T|n+1)}\right\}^{1/2} , \tag{6.6.34}$$

where the normalized coefficient,

$$\overline{k}^c_{n+1}(T) = \frac{<\mathbf{E}_c(T|n), \mathbf{E}_b(T|n)>}{\sqrt{\epsilon_c(T|n)\epsilon_b(T|n)}} , \tag{6.6.35}$$

and must be less than unity in magnitude. Using (6.5.8) and (6.6.35),

$$\frac{\epsilon_c(T|n+1)}{\epsilon_c(T|n)} = 1 - [\overline{k}^c_{n+1}(T)]^2 , \tag{6.6.36}$$

so that combining (6.6.23), (6.6.34), and (6.6.36) gives

$$\overline{\mathbf{E}}_c(T|n+1) = \frac{\overline{\mathbf{E}}_c(T|n) - \overline{k}^c_{n+1}(T)\,\overline{\mathbf{E}}_b(T|n)}{\sqrt{[1-\overline{e}_b^2(T|n)][1-\overline{k}_n^{c\,2}(T)]}} . \tag{6.6.37}$$

From the time update (6.5.11) we also have that

$$\overline{k}^c_{n+1}(T) = w\overline{k}^c_{n+1}(T-1)\left[\frac{\epsilon_c(T-1|n)\epsilon_b(T-1|n)}{\epsilon_c(T|n)\epsilon_b(T|n)}\right]^{1/2} + \overline{e}_c(T|n)\overline{e}_b(T|n), \tag{6.6.38}$$

where the normalized joint process residual at time T is

$$\overline{e}_c(T|n) = <\mathbf{u}(T), \overline{\mathbf{E}}_c(T|n)> . \tag{6.6.39}$$

Using (6.6.29) and (6.5.12) it follows that

	Prediction			Joint Process Estimation		
	Mults./ Divides	Adds.	Square Roots	Mults./ Divides	Adds.	Square Roots
SG Transversal (3.0.3), (3.2.3)	2N+1	2N		2N+1	2N	
SG Lattice (4.2.9-10),(4.5.3)	5N	4N		8N	6N	
Modified Fast Kalman[2] (P6-9)	5N+13	5N+3		7N+14	7N+3	
Fast Kalman (6.3.63)	8N+5	7N+2		10N+5	9N+2	
PLS[1] Lattice (6.3.64)	11N	6N		16N	8N	
GMC[1] Transversal[3] (6.4.61)	13N+6	11N+1		15N+6	13N+1	
SWC[1] Transversal (6.4.54)	13N+13	12N+7		17N+14	9N+14	
SWC[1] Lattice (6.4.55)	16N	10N		23N	14N	
GMC[1] Lattice (6.4.62)	21N	12N		27N	15N	
Normalized PLS[1] Transversal (P6-19)	9N+17	5N+2	2	11N+19	7N+2	2
Normalized PLS[1] Lattice (6.6.25,31)	12N	6N	3N	21N	11N	5N
Normalized GMC[1] Transversal (P6-22)	12N+16	7N+2	3	14N+18	9N+2	3
Normalized SWC[1] Lattice (6.6.57)	23N	11N	5N	40N	20N	9N
Normalized GMC[1] Lattice (6.6.60)	24N	12N	6N	42N	22N	10N

[1]PLS= Prewindowed LS, GMC= Growing Memory Covariance, SWC= Sliding Window Covariance.

[2]Modified fast Kalman algorithms requiring on the order of 5N arithmetic operations per iteration have also been presented in [29] and [47].

[3]The GMC Transversal predictor in problem 6-15 requires 8N+16 multiplies/divides per iteration. GMC and SWC transversal algorithms requiring less computation than shown here have been presented in [14] and [28].

Table 6.1. Number of arithmetic operations per iteration required by some adaptive algorithms (N is filter order). An effort has been made to count redundant computations only once.

$$\bar{k}^c_{n+1}(T) = \bar{k}^c_{n+1}(T-1)\left\{\left[1-\bar{e}_c^2(T|n)\right]\left[1-\bar{e}_b^2(T|n)\right]\right\}^{1/2}$$
$$+\ \bar{e}_c(T|n)\bar{e}_b(T|n)\,. \qquad \textbf{(6.6.40)}$$

Combining the predictor recursions (6.6.25) and (6.6.31) with (6.6.40) and the top component of (6.6.37) completes the normalized prewindowed LS lattice joint process estimator. At each iteration $T > N$, all three recursions are computed for $n{=}0$ to $n{=}N$. Initialization of (6.6.37) and (6.6.38) is accomplished as follows,

$$\bar{k}^c_n(-1) = 0, \quad 1 \leqslant n \leqslant N\,. \qquad (6.6.41a)$$

When $d(T)$ is observed,

$$\epsilon_c(T) = w\,\epsilon_c(T-1) + d^2(T), \quad \epsilon_c(-1) = 0\,, \qquad (6.6.41b)$$

and

$$\bar{e}_c(T|0) = \frac{d(T)}{\sqrt{\epsilon_c(T)}}\,. \qquad (6.6.41c)$$

When $T < N$, the normalized lattice recursions are computed from $n{=}0$ to $n{=}T$.

6.6.3.2. Normalized SG Lattice Algorithm

It is now shown that an analogous normalization procedure can be applied to a SG lattice algorithm resulting in a slightly simpler version of the normalized prewindowed LS lattice algorithm. The normalized MMSE order recursions (6.6.16) assume known and stationary input statistics. In the case of unknown or nonstationary input statistics, the inner products in (6.6.6) and (6.6.9) that appear in (6.6.16), i.e., $E[e_f(T|n)e_b(T-1|n)]$, $E[e_f^2(T|n)]$, and $E[e_b^2(T|n)]$, can be respectively estimated by the time averages,

$$K_{n+1}(T) = wK_{n+1}(T-1) + e_f(T|n)e_b(T-1|n) \qquad (6.6.42a)$$

$$F(T|n) = wF(T-1|n) + e_f^2(T|n) \qquad (6.6.42b)$$

$$B(T|n) = wB(T-1|n) + e_b^2(T|n) \qquad (6.6.42c)$$

$$k^f_{n+1}(T) = \frac{K_{n+1}(T)}{F(T|n)} \quad \text{and} \quad k^b_{n+1}(T) = \frac{K_{n+1}(T)}{B(T-1|n)}\,. \qquad (6.6.42d)$$

The estimates $k^f_{n+1}(T)$ and $k^b_{n+1}(T)$ are very similar to the SG estimates (4.5.11) and (4.5.12), where the adaptation constant β is properly normalized. The only difference is that (4.5.11) and (4.5.12) indicate that the coefficients are updated *after* the order recursions (4.2.9) and (4.2.10), whereas the coefficient updates (6.6.42) occur *before* the residual order recursions, which is analogous to the LS lattice time updates. Of course, it is also possible to compute the updates (6.6.42) after the residual order updates. The normalized gradient lattice algorithm which follows can be modified accordingly. Further discussion of the order in which gradient lattice recursions are computed is given in section 7.8. The difference between recursions (6.6.42) and the analogous prewindowed LS recursions is the absence of the weighting term $\gamma(T|n)$. In the MMSE case these gains are assumed to be zero.

Defining the normalized residuals

$$\tilde{e}_f(T|n) \equiv \frac{e_f(T|n)}{[F(T|n)]^{1/2}} \quad \text{and} \quad \tilde{e}_b(T|n) \equiv \frac{e_b(T|n)}{[B(T|n)]^{1/2}}, \tag{6.6.43}$$

it follows that

$$\bar{k}_{n+1}(T) = w\bar{k}_{n+1}(T-1)\left[\frac{F(T-1|n)B_n(T-2)}{F(T|n)B(T-1|n)}\right]^{1/2} + \tilde{e}_f(T|n)\tilde{e}_b(T-1|n) \tag{6.6.44}$$

where

$$\bar{k}_{n+1}(T) = \frac{K_{n+1}(T)}{[F(T|n)B(T-1|n)]^{1/2}}. \tag{6.6.45}$$

It is easily verified that (6.6.44) can be rewritten as

$$\bar{k}_{n+1}(T) = \bar{k}_{n+1}(T-1)\left\{[1 - \tilde{e}_f^2(T|n)][1 - \tilde{e}_b^2(T-1|n)]\right\}^{1/2} + \tilde{e}_f(T|n)\tilde{e}_b(T-1|n), \tag{6.6.46}$$

which is identical to the LS version (6.6.31). The order recursions (6.6.16) and the time recursion (6.6.46) therefore constitute a normalized SG lattice algorithm. The initialization procedure is the same as that for the normalized LS algorithm.

A normalized SG lattice joint process estimator is derived in an analogous fashion as the corresponding LS algorithm where the gain $\gamma(T|n)$ is assumed to be zero. A normalized order recursion is easily obtained from the MMSE recursion (4.4.10),

$$\tilde{e}_c(T|n+1) = \frac{\tilde{e}_c(T|n) - \bar{k}_{n+1}^c(T)\tilde{e}_b(T|n)}{[1 - \bar{k}_{n+1}^{c\,2}(T)]^{1/2}}, \tag{6.6.47a}$$

where

$$\tilde{e}_c(T|n) \equiv \frac{e_c(T|n)}{[C(T|n)]^{1/2}}, \qquad \bar{k}_{n+1}^c(T) \equiv \frac{K_{n+1}^c(T)}{[C(T|n)B(T|n)]^{1/2}}, \tag{6.6.47b}$$

and $K_{n+1}^c(T)$ and $C(T|n)$ are obtained recursively via the time averages,

$$K_{n+1}^c(T) = wK_{n+1}^c(T-1) + e_c(T|n)e_b(T|n) \tag{6.6.48}$$

and

$$C(T|n) = wC(T-1|n) + e_c^2(T|n) . \tag{6.6.49}$$

Normalization of (6.6.48) is accomplished in an analogous fashion as in the LS case giving the time recursion

$$\bar{k}_{n+1}^c = wk_{n+1}^c(T-1)\left\{[1 - \tilde{e}_c^2(T|n)][1 - \tilde{e}_b^2(T|n)]\right\}^{1/2} + \tilde{e}_c(T|n)\tilde{e}_b(T|n) , \tag{6.6.50}$$

which is identical to the LS version. Initialization is again the same as in the LS case.

6.6.3.3. Normalized Prewindowed Whitening Filter Algorithms

In order to normalize the whitening filter algorithm derived in section 6.3.1, the filters $F_{T|n}(z)$, $B_{T|n}(z)$, and $G_{T|n}(z)$ are normalized as follows,

$$\bar{F}_{T|n}(z) \equiv \frac{F_{T|n}(z)}{\sqrt{\epsilon_f(T|n)}} \tag{6.6.51a}$$

$$\bar{B}_{T|n}(z) \equiv \frac{B_{T|n}(z)}{\sqrt{\epsilon_b(T|n)}} \tag{6.6.51b}$$

$$\bar{G}_{T|n}(z) \equiv \frac{G_{T|n}(z)}{\sqrt{1 - \gamma(T|n)}} . \tag{6.6.51c}$$

The filters $F_{T|n}(z)$ and $B_{T|n}(z)$, are normalized by the norms of the respective output vectors when excited by the input sequence $y(0), y(1), \ldots, y(T)$. The resulting forward and backward prediction errors obtained using $\bar{F}_{T|n}(z)$ and $\bar{B}_{T|n}(z)$ are the *variance* normalized prediction errors $\tilde{e}_f(T|n)$ and $\tilde{e}_b(T|n)$ given by (6.6.17). It is easy to verify that the following recursions result from substituting (6.6.51) into the recursions (6.3.88), (6.3.83), (6.3.85), and (6.3.90), respectively,

$$\bar{B}_{T-1|n}(z) = \frac{\bar{B}_{T|n}(z) + \bar{e}_b(T|n)\bar{G}_{T|n}(z)}{\sqrt{1 - \bar{e}_b^2(T|n)}} \tag{6.6.52a}$$

$$\bar{F}_{T|n+1}(z) = \frac{\bar{F}_{T|n}(z) - \bar{k}_{n+1}(T)[z^{-1}\bar{B}_{T-1|n}(z)]}{\sqrt{1 - \bar{k}_{n+1}^2(T)}} \tag{6.6.52b}$$

$$\bar{B}_{T|n+1}(z) = \frac{z^{-1}\bar{B}_{T-1|n}(z) - \bar{k}_{n+1}(T)\bar{F}_{T|n}(z)}{\sqrt{1 - \bar{k}_{n+1}^2(T)}} \tag{6.6.52c}$$

$$\bar{G}_{T|n+1}(z) = \frac{\bar{G}_{T|n}(z) + \bar{e}_b(T|n)\bar{B}_{T|n}(z)}{\sqrt{1 - \bar{e}_b^2(T|n)}} . \tag{6.6.52d}$$

For joint process estimation, the normalized filter $\bar{C}_{T|n}(z)$ is defined as

$$\bar{C}_{T|n}(z) \equiv \frac{C_{T|n}(z)}{\sqrt{\epsilon_c(T|n)}}, \tag{6.6.53}$$

and substituted into (6.5.43) to give

$$\bar{C}_{T|n+1}(z) = \frac{\bar{C}_{T|n}(z) + \bar{k}^c_{n+1}(T)[z^{-1}\bar{B}_{T|n}(z)]}{\sqrt{[1 - \bar{e}_b^2(T|n)][1 - \bar{k}^{c\,2}_{n+1}(T)]}}. \tag{6.6.54}$$

The recursions (6.6.52-54)) are initialized as follows,

$$\bar{F}_{T|0}(z) = \bar{B}_{T|0}(z) = \left[\sum_{j=0}^{T} y^2(j)\right]^{-1/2} \tag{6.6.55a}$$

$$\bar{C}_{T|0}(z) = \left[\sum_{j=0}^{T} d^2(j)\right]^{-1/2} \tag{6.6.55b}$$

$$\bar{G}_{T|0}(z) = 0. \tag{6.6.55c}$$

The algorithm recursions (6.6.52) and (6.6.54) can be used to compute the normalized filters $\bar{F}$, $\bar{B}$, $\bar{G}$, and $\bar{C}$ given normalized lattice variables at time T.

6.6.4. Normalized Covariance Algorithms

Normalization of the order-recursive sliding window and growing memory covariance algorithms and the associated whitening filter algorithms is accomplished in an analogous fashion to the prewindowed algorithms. To normalize the sliding window algorithm (6.4.55) the normalized variables are defined,

$$\bar{e}_f(T_0,T|n) \equiv \frac{e_f(T_0,T|n)}{\sqrt{\epsilon_f(T_0,T|n)[1 - \gamma(T_0,T-1|n)]}} \tag{6.6.56a}$$

$$\bar{e}_b(T_0,T|n) \equiv \frac{e_b(T_0,T|n)}{\sqrt{\epsilon_b(T_0,T|n)[1 - \gamma(T_0+1,T|n)]}} \tag{6.6.56b}$$

$$\bar{e}_f^*(T_0,T|n) \equiv \frac{e_f^*(T_0,T|n)}{\sqrt{\epsilon_f(T_0,T|n)[1 - \gamma^*(T_0,T-1|n)]}} \tag{6.6.56c}$$

$$\bar{e}_b^*(T_0,T|n) \equiv \frac{e_b^*(T_0,T|n)}{\sqrt{\epsilon_b(T_0,T|n)[1 - \gamma^*(T_0+1,T|n)]}} \tag{6.6.56d}$$

$$\bar{k}_n(T_0,T) \equiv \frac{K_n(T_0,T)}{\sqrt{\epsilon_f(T_0+1,T|n-1)\epsilon_b(T_0,T-1|n-1)}}. \tag{6.6.56e}$$

Replacement of the unnormalized variables in (6.4.55) by their normalized counterparts results in the algorithm.

$$\bar{k}_{n+1}(T_0,T) = \tag{6.6.57a}$$

$$\bar{k}_{n+1}(T_0,T-1)\left\{\left[1-\bar{e}_f^2(T_0+1,T|n)\right]\left[1-\bar{e}_b^2(T_0,T-1|n)\right]\right\}^{1/2}$$

$$+\ \bar{e}_f(T_0+1,T|n)\bar{e}_b(T_0,T-1|n)$$

$$\bar{k}_{n+1}(T_0+1,T) = \frac{\bar{k}_{n+1}(T_0,T) - \bar{e}_f^{\bullet}(T_0+1,T|n)\bar{e}_b^{\bullet}(T_0,T-1|n)}{\sqrt{[1-\bar{e}_f^{\bullet 2}(T_0+1,T|n)][1-\bar{e}_b^{\bullet 2}(T_0,T-1|n)]}} \tag{6.6.57b}$$

$$\bar{e}_f(T_0,T|n+1) = \frac{\bar{e}_f(T_0+1,T|n) - \bar{k}_{n+1}(T_0,T)\bar{e}_b(T_0,T-1|n)}{\sqrt{[1-\bar{k}_{n+1}^2(T_0,T)][1-\bar{e}_b^2(T_0,T-1|n)]}} \tag{6.6.57c}$$

$$\bar{e}_b(T_0,T|n+1) = \frac{\bar{e}_b(T_0,T-1|n) - \bar{k}_{n+1}(T_0,T)\bar{e}_f(T_0+1,T|n)}{\sqrt{[1-\bar{k}_{n+1}^2(T_0,T)][1-\bar{e}_f^2(T_0+1,T|n)]}} \tag{6.6.57d}$$

$$\bar{e}_f^{\bullet}(T_0,T|n+1) = \frac{\bar{e}_f^{\bullet}(T_0+1,T|n) - \bar{k}_{n+1}(T_0,T)\bar{e}_b^{\bullet}(T_0,T-1|n)}{\sqrt{[1-\bar{k}_{n+1}^2(T_0,T)][1-\bar{e}_b^{\bullet 2}(T_0,T-1|n)]}} \tag{6.6.57e}$$

$$\bar{e}_b^{\bullet}(T_0,T|n+1) = \frac{\bar{e}_b^{\bullet}(T_0,T-1|n) - \bar{k}_{n+1}(T_0,T)\bar{e}_f^{\bullet}(T_0+1,T|n)}{\sqrt{[1-\bar{k}_{n+1}^2(T_0,T)][1-\bar{e}_f^{\bullet 2}(T_0+1,T|n)]}}\ . \tag{6.6.57f}$$

This algorithm can be viewed as two prewindowed lattice algorithms, one of which updates the current residuals $\bar{e}_f(T)$ and $\bar{e}_b(T)$, and the other which updates the past residuals $\bar{e}_f^{\bullet}(T)$ and $\bar{e}_b^{\bullet}(T)$. The derivation is similar to the derivation of the normalized prewindowed LS algorithm and is therefore left as a problem.

To normalize the growing memory algorithm (6.4.62), the normalized variable

$$\bar{\alpha}(T_0,T|n) \equiv \frac{\alpha(T_0,T|n)}{\sqrt{[1-\gamma(T_0,T|n)][1-\gamma^{\bullet}(T_0,T|n)]}}\ , \tag{6.6.58}$$

is needed, in addition to

$$\frac{1-\gamma(T_0+1,T|n)}{1-\gamma(T_0,T|n)} = \frac{1-\gamma^{\bullet}(T_0,T-1|n)}{1-\gamma^{\bullet}(T_0,T|n)} = 1-\bar{\alpha}^2(T_0,T|n)\ , \tag{6.6.59}$$

which is obtained from (6.4.40) and (6.4.47). Setting $T_0 = 0$ in (6.6.56), (6.6.58), and (6.6.59), and substituting into (6.4.62) results in the algorithm,

$$\bar{k}_n(T) = \bar{k}_n(T-1)\left\{\left[1-\bar{e}_f^2(1,T|n-1)\right]\left[1-\bar{e}_b^2(T-1|n-1)\right]\right\}^{1/2}$$

$$+\ \bar{e}_f(1,T|n-1)\bar{e}_b(T-1|n-1) \tag{6.6.60a}$$

$$\bar{e}_f(T|n) = \frac{\bar{e}_f(1,T|n-1) - \bar{k}_n(T)\bar{e}_b(T-1|n-1)}{\sqrt{[1-\bar{e}_b^2(T-1|n-1)][1-\bar{k}_n^2(T)]}} \tag{6.6.60b}$$

$$\bar{e}_b(T|n) = \frac{\bar{e}_b(T-1|n-1) - \bar{k}_n(T)\bar{e}_f(1,T|n-1)}{\sqrt{[1 - \bar{e}_f^2(1,T|n-1)][1 - \bar{k}_n^2(T)]}} \tag{6.6.60c}$$

$$\bar{e}_f^*(T|n) = \left\{\left[1 - \bar{e}_f^2(T|n)\right]\left[1 - \bar{\alpha}^2(T-1|n)\right]\right\}^{1/2} \bar{e}_f^*(T-1|n) - \bar{\alpha}(T-1|n)\bar{e}_f(T|n) \tag{6.6.60d}$$

$$\bar{e}_f(1,T|n) = \frac{\bar{e}_f(T|n) + \bar{\alpha}(T-1|n)\bar{e}_f^*(T|n)}{\sqrt{[1 - \bar{e}_f^{*2}(T|n)][1 - \bar{\alpha}^2(T-1|n)]}} \tag{6.6.60e}$$

$$\bar{\alpha}(T|n+1) = \frac{\bar{\alpha}(T-1|n) + \bar{e}_f^*(T|n)\bar{e}_f(T|n)}{\sqrt{[1 - \bar{e}_f^{*2}(T|n)][1 - \bar{e}_f^2(T|n)]}} . \tag{6.6.60f}$$

The starting time index is zero where unspecified. Derivation of the normalized growing memory lattice algorithm is left as a problem.

Normalization of the covariance joint process estimation recursions in section 6.5 is accomplished by using the substitutions,

$$\bar{e}_c(T_0,T|n) = \frac{e_c(T_0,T|n)}{\sqrt{[1 - \gamma(T_0,T|n)]\epsilon_c(T_0,T|n)}} \tag{6.6.61a}$$

$$\bar{e}_c^*(T_0,T|n) = \frac{e_c^*(T_0,T|n)}{\sqrt{[1 - \gamma^*(T_0,T|n)]\epsilon_c(T_0,T|n)}} \tag{6.6.61b}$$

$$\bar{k}_n^c(T_0,T) = \frac{K_n^c(T_0,T)}{\sqrt{[\epsilon_c(T_0+1,T|n-1)\epsilon_b(T_0,T-1|n-1)]}} . \tag{6.6.61c}$$

Substituting (6.6.61a-61c) in (6.5.25), (6.5.29), (6.5.31), (6.5.34), and (6.5.35) gives, respectively,

$$\bar{e}_c(T_0,T|n+1) = \frac{\bar{e}_c(T_0+1,T|n) - \bar{k}_{n+1}^c(T_0,T)\bar{e}_b(T_0,T|n)}{\sqrt{[1 - \bar{e}_b^2(T_0,T|n)][1 - \bar{k}_{n+1}^{c2}(T_0,T)]}} \tag{6.6.62}$$

$$\bar{e}_c^*(T_0,T|n+1) = \frac{\bar{e}_c^*(T_0+1,T|n) - \bar{k}_{n+1}^c(T_0,T)\bar{e}_b^*(T_0,T|n)}{\sqrt{[1 - \bar{e}_b^{*2}(T_0,T|n)][1 - \bar{k}_{n+1}^{c2}(T_0,T)]}} \tag{6.6.63}$$

$$\bar{k}_{n+1}^c(T_0,T) = \left\{\left[1 - \bar{e}_c^2(T_0+1,T|n)\right]\left[1 - \bar{e}_b^2(T_0,T|n)\right]\right\}^{1/2} \bar{k}_{n+1}^c(T_0,T-1) + \bar{e}_c(T_0+1,T|n)\bar{e}_b(T_0,T|n) \tag{6.6.64}$$

$$\bar{e}_c^*(T_0,T|n) = \left\{\left[1 - \bar{e}_c^2(T_0,T|n)\right]\left[1 - \bar{\alpha}^2(T_0,T|n)\right]\right\}^{1/2} \bar{e}_c^*(T_0,T-1|n) - \bar{\alpha}(T_0,T|n)\bar{e}_c(T_0,T|n) \tag{6.6.65}$$

$$\bar{e}_c(T_0+1,T|n) = \frac{\bar{e}_c(T_0,T|n) + \bar{e}_c^*(T_0,T|n)\bar{\alpha}(T_0,T|n)}{\sqrt{[1 - \bar{e}_c^{*2}(T_0,T|n)][1 - \bar{\alpha}^2(T_0,T|n)]}} \tag{6.6.66}$$

$$\bar{k}_{n+1}^c(T_0+1,T) = \frac{\bar{k}_{n+1}^c(T_0,T) - \bar{e}_c^*(T_0+1,T|n)\bar{e}_b^*(T_0,T|n)}{\sqrt{[1 - \bar{e}_c^{*2}(T_0+1,T|n)][1 - \bar{e}_b^{*2}(T_0,T|n)]}}\,. \tag{6.6.67}$$

Combining (6.6.62), (6.6.63), (6.6.64), and (6.6.67) with (6.6.57) gives a normalized sliding window joint process estimation algorithm. Combining (6.6.62), (6.6.64), (6.6.65), and (6.6.66) with (6.6.60) gives a normalized growing memory joint process estimation algorithm.

Initialization of the normalized recursions in this section is accomplished by combining the initialization of the analogous unnormalized recursions with the substitutions (6.6.56), (6.6.58), and (6.6.61). For the sliding memory algorithms,

$$\bar{k}_n(0,-1) = \bar{k}_n^c(0,-1) = 0, \quad 0 \leqslant n \leqslant N \tag{6.6.68a}$$

$$\epsilon_f(T|0) = w\,\epsilon_f(T-1|0) + y^2(T) - y^2(T-M) \tag{6.6.68b}$$

$$\epsilon_c(T|0) = w\,\epsilon_c(T-1|0) + d^2(T) - d^2(T-M) \tag{6.6.68c}$$

$$\bar{e}_f(T|0) = \bar{e}_b(T|0) = \frac{y(T)}{[\epsilon_f(T|0)]^{1/2}} \tag{6.6.68d}$$

$$\bar{e}_f^*(T|0) = \bar{e}_b^*(T|0) = \frac{y(T-M+1)}{[\epsilon_f(T|0)]^{1/2}} \tag{6.6.68e}$$

$$\bar{e}_c(T|0) = \frac{d(T)}{[\epsilon_c(T|0)]^{1/2}} \tag{6.6.68f}$$

$$\bar{e}_c^*(T|0) = \frac{d(T-M+1)}{[\epsilon_c(T|0)]^{1/2}}\,, \tag{6.6.68g}$$

where $d(T) = 0$ and $y(T) = 0$ for $T < 0$ and M is the width of the sliding window (i.e., $M = T - T_0 + 1$). For the growing memory algorithms,

$$\bar{k}_n(-1) = \bar{k}_n^c(-1) = 0, \quad 0 \leqslant n \leqslant N \tag{6.6.69a}$$

$$\bar{\alpha}(T|0) = 0 \tag{6.6.69b}$$

$$\bar{e}_c^{\bullet}(0|n) = \bar{e}_f^{\bullet}(0|n) = \begin{cases} 1 \text{ if } n = 0 \\ 0 \text{ if } n > 0 \end{cases} . \tag{6.6.69c}$$

$$\epsilon_f(T|0) = w\,\epsilon_f(T-1|0) + y^2(T) \tag{6.6.69d}$$

$$\epsilon_c(T|0) = w\,\epsilon_c(T-1|0) + d^2(T) \tag{6.6.69e}$$

$$\bar{e}_f(T|0) = \bar{e}_b(T|0) = \frac{y(T)}{[\epsilon_f(T|0)]^{1/2}} \tag{6.6.69f}$$

$$\bar{e}_c(T|0) = \frac{d(T)}{[\epsilon_c(T|0)]^{1/2}} . \tag{6.6.69g}$$

The whitening filter recursions for the covariance case can be normalized by using the substitutions (6.6.51) and (6.6.53), where the lower time index T_0 is added. The resulting sliding and growing memory algorithms for the (variance) normalized filters $\bar{F}_{T_0,T|n}(z)$, $\bar{B}_{T_0,T|n}(z)$, $\bar{G}_{T_0,T|n}(z)$, $\bar{H}_{T_0,T|n}(z)$, and $\bar{C}_{T_0,T|n}(z)$ are easily obtained from the unnormalized recursions (6.4.66-74) and (6.5.43-45), and are therefore left as a problem.

In analogy with the prewindowed lattice algorithms, the normalized covariance lattice algorithms contain fewer recursions than the associated unnormalized algorithms; however, they require more computation (see table 6.1). Normalized fixed-order covariance algorithms have have also been proposed [28]. These also involve somewhat more computation than the associated unnormalized algorithms; however, they offer a substantial improvement in numerical properties. The interested reader is referred to reference [28] and to problem 6-22 for further discussion.

6.7. APPLICATION: PITCH DETECTION

The recursive LS algorithms can be considered as one-for-one replacements for the gradient or block LS algorithms in many applications. The major considerations in whether to use the LS algorithms are speed of convergence, computational complexity, and numerical precision and stability. The speed of convergence will be discussed in detail in chapter 7.

Occasionally an opportunity arises to do more than simply substitute an LS algorithm for a gradient algorithm. An example of this is in pitch detection of speech, where an algorithm has been proposed which uses to advantage variables developed internally to the recursive LS algorithm [30]. We will review this proposed algorithm in this section.

Pitch detection refers to the need in the analysis of voiced sounds to estimate the period of the quasi-periodic waveform which excites the vocal tract. The frequency of this fundamental is called the *pitch* of the speech. This pitch estimate can be combined with the prediction coefficients estimated from a segment of the sampled speech waveform to synthesize the voiced sound. Since the algorithm about to be described uses parameters entering a recursive LS algorithm to estimate the pitch, when used in conjunction with a recursive LS predictor this pitch detection technique requires little additional computational overhead.

The pitch detection technique assumes a specific statistical model of the input speech in order to decide whether or not a particular sample is a pitch pulse. This is in contrast to more conventional techniques such as autocorrelation or cepstral methods [31], which make no statistical assumptions about the speech signal. Perhaps the simplest way to statistically model the periodic plus noise-like behavior of speech is to assume that it can be described approximately as a Gaussian noise source with a superimposed jump process feeding an all-pole time-varying linear filter. Using this model, the decision as to whether or not the current sample is a pitch pulse, or "jump," is equivalent to deciding whether or not the current sample deviates from a Gaussian distribution.

Assume that the samples $\{y(T-n+1), y(T-n), \ldots, y(T)\}$ are given, and we wish to decide whether or not $y(T)$ is a pitch pulse. Assuming the data is Gaussian, the joint probability density is

$$P[y(T-n+1), y(T-n), \ldots, y(T)] = \frac{1}{(2\pi|\mathbf{\Phi}_n|)^{1/2}} \exp\left\{-\frac{1}{2}\mathbf{y}'(T|n)\ \mathbf{\Phi}_n^{-1}\ \mathbf{y}(T|n)\right\}, \tag{6.7.1}$$

where the autocorrelation matrix is defined as

$$\mathbf{\Phi}_n \equiv E[\mathbf{y}(T|n)\ \mathbf{y}'(T|n)]\ . \tag{6.7.2}$$

The "log-likelihood variable" is now defined,

$$\begin{aligned}\nu(T|n) &\equiv -2\ \ln\left\{P[y(T-n+1), y(T-n), \ldots, y(T)]\right\} - \ln\ (2\pi) \\ &= \ln|\mathbf{\Phi}_n| + \mathbf{y}'(T|n)\ \mathbf{\Phi}_n^{-1}\mathbf{y}(T|n)\ ,\end{aligned} \tag{6.7.3}$$

and the "log-likelihood ratio" is defined as

$$\nu(T|n) - \nu(T-1|n) = \ln\left[\frac{P[y(T-n+1), \ldots, y(T)]}{P[y(T-n), \ldots, y(T-1)]}\right]. \tag{6.7.4}$$

In the case being considered, the statistics of the input process are unknown; however, the autocorrelation matrix $\mathbf{\Phi}_n$ can be estimated (within a multiplicative constant) by the sample covariance matrix,

$$\hat{\mathbf{\Phi}}(T-M+1,T|n) = \sum_{T=T-M+1}^{T} \mathbf{y}(T|n)\mathbf{y}'(T|n)\ , \tag{6.7.5}$$

where a window of M data samples is assumed. In order to obtain an estimate for the determinant of $\mathbf{\Phi}_n$, the vector of backward errors is first defined,

$$\begin{aligned}\mathbf{e}_b(T) &= [e_b(T|0)\ \ e_b(T|1) \cdots e_b(T|n-1)]' \\ &= [\mathbf{L}\ \mathbf{y}(T|n)]\ ,\end{aligned} \tag{6.7.6}$$

where $\mathbf{L}$ is a lower triangular matrix with ones along the diagonal, so that the determinant of $\mathbf{L}$, $|\mathbf{L}| = 1$.

The determinant of $\mathbf{\Phi}_n$ is therefore

$$\begin{aligned}|\mathbf{\Phi}_n| &= |E[\mathbf{y}(T|n)\mathbf{y}'(T|n)]| \\ &= |\mathbf{L}|\ |E[\mathbf{y}(T|n)\mathbf{y}'(T|n)]|\ |\mathbf{L}'| \\ &= |E[\mathbf{L}\ \mathbf{y}(T|n)\mathbf{y}'(T|n)\ \mathbf{L}']| \\ &= |E[\mathbf{e}_b(T)\mathbf{e}_b{}'(T)]| \\ &= \prod_{j=1}^{n} E[e_b^2(T|j-1)]\ ,\end{aligned} \tag{6.7.7}$$

where the last step follows from the fact that $E[\mathbf{e}_b(T)\mathbf{e}_b{}'(T)]$ is a diagonal matrix. Assuming the data of interest is in a window of length M, $E[e_b^2(T|j)]$ can be estimated (within a multiplicative constant) by

$$\epsilon_b(T|j) = \sum_{l=T-M+1}^{T} w^{T-l} e_b^2(l|j)\ . \tag{6.7.8}$$

The log-likelihood variable $\nu(T|n)$ defined in (6.7.3) can therefore be estimated from LS variables (within an additive constant) as follows,

$$\hat{\nu}(T|n) = \sum_{j=0}^{n-1} \ln\ [\epsilon_b(T|j)] + \gamma(T|n)\ . \tag{6.7.9}$$

The constants of proportionalities relating $\mathbf{\Phi}_n$, $E[e_b^2(T|j)]$, and the respective estimates $\hat{\mathbf{\Phi}}(T+M+1,T|n)$ and $\epsilon_b(T|j)$ can be ignored since only the estimated log-likelihood *ratio* $\hat{\nu}(T|n) - \hat{\nu}(T-1|n)$ is of interest. If the magnitude of the estimated log-likelihood ratio is relatively large, then the current sample $y(T)$ is unlikely (i.e., a statistical outlier) assuming that the distribution of each sample $y(T)$ is Gaussian, and that the first and second moments have been estimated from the given samples

$$y(T-M+1), y(T-M+2), \ldots, y(T-1)\ .$$

It is therefore reasonable to expect that the estimated log-likelihood ratio is a good statistic for separating a Gaussian component from "jump" components.

The basic pitch detection algorithm [32] is:

1. Compute $\hat{\nu}(T|n)$.
2. If $\hat{\nu}(T|n) - \hat{\nu}(T-1|n)$ falls above a predetermined threshold ν_0, store the input data sample $y(T)$.
3. If $y(T)$ falls above an exponentially decaying window initialized at the value of the last pitch pulse, record a new pitch pulse.

The last step was originally proposed in [33] and presumes that whenever a pitch pulse is detected, a variable $W(T)$ is set equal to the current data value $y(T)$. At each successive iteration before the next pitch pulse is detected, $W(j) = vW(j-1)$, where v is a predetermined constant slightly less than one.

In this way if a pulse is detected at time T, then it is less likely that a pulse will be incorrectly detected at time $T+m$, where m is small relative to the true pitch period. Extra steps can be included to safeguard against obvious errors the algorithm may make. The parameters of this pitch detection scheme (i.e., ν, ν_0, and the LS exponential weight w) must in general be selected experimentally, and the optimal settings vary with different types of speech inputs (i.e., fricatives, vowels, and different speakers). The computation of $\hat{\nu}(T|n)$ can be performed via any of the LS algorithms derived in this chapter, depending upon which windowing scheme is desired. The performance of this pitch detection scheme in a speech modeling and synthesis system has been reported in [32, 30].

6.8. FURTHER READING

As mentioned in the introduction to this chapter, there are many variations and extensions of the algorithms presented here. Perhaps the most important is the extension of recursive LS algorithms to the ARMA, or pole-zero, case [20, 34, 4, 35, 36]. Other interesting variations on the recursive LS algorithms presented in this chapter include adaptive LS FIR algorithms that satisfy a linear phase constraint [37, 38], lattice algorithms for instrumental variable recursions [19, 39], LS algorithms with an arbitrary prediction delay [40, 41, 42], and a computationally efficient multi-channel lattice algorithm [43]. The modeling of nonstationary processes using lattice filters is a relatively new research area [44]. A comprehensive survey of adaptive lattice algorithms for both AR and ARMA models using the generalized projection operator approach discussed in section 6.2 is available [4]. The projection operator approach is also discussed elsewhere [23, 13, 12]. These references treat the more general multi-channel case, where the input data at each iteration is assumed to be a vector rather than a scalar. A geometric approach, similar to the one used in this chapter, has been used [21] to derive the prewindowed LS lattice algorithm. An excellent comprehensive treatment of Hilbert space has been given in the book by Naylor and Sell [45].

APPENDIX 6-A

PROOF OF (6.2.21)

By definition,

$$\begin{aligned}\mathbf{P}_{M_{T-1}(T)}\mathbf{Y}(T) &= \mathbf{P}^{\perp}_{U(T)}\mathbf{P}_{M_{T-1}(T)}\mathbf{Y}(T) + \mathbf{P}_{U(T)}\mathbf{P}_{M_{T-1}(T)}\mathbf{Y}(T) \\ &= \mathbf{P}_{\tilde{M}(T)}\mathbf{Y}(T) + \mathbf{P}_{U(T)}\mathbf{P}_{M_{T-1}(T)}\mathbf{Y}(T)\,,\end{aligned} \tag{A.1}$$

where $\tilde{M}$ is the subspace spanned by the column vectors of $\tilde{\mathbf{S}}(T)$. Projecting both sides of (A.1) onto $M(T)$ gives

$$\mathbf{P}_{M(T)}\mathbf{P}_{M_{T-1}(T)}\mathbf{Y}(T) = \mathbf{P}_{M(T)}\mathbf{P}_{\tilde{M}(T)}\mathbf{Y}(T) + \mathbf{P}_{M(T)}\mathbf{P}_{U(T)}\mathbf{P}_{M_{T-1}(T)}\mathbf{Y}(T)\,. \tag{A.2}$$

Now $\mathbf{P}_{M_{T-1}(T)}\mathbf{Y}(T)$ lies in $M(T)$, and hence

$$\mathbf{P}_{M(T)}\mathbf{P}_{M_{T-1}(T)}\mathbf{Y}(T) = \mathbf{P}_{M_{T-1}(T)}\mathbf{Y}(T) . \tag{A.3}$$

Also,

$$\begin{aligned}\mathbf{P}_{M(T)}\mathbf{P}_{\tilde{M}(T)}\mathbf{Y}(T) &= \mathbf{S}(T)[\mathbf{S}'(T)\mathbf{S}(T)]^{-1}[\mathbf{S}'(T)\tilde{\mathbf{S}}(T)][\tilde{\mathbf{S}}'(T)\tilde{\mathbf{S}}(T)]^{-1}\tilde{\mathbf{S}}'(T)\mathbf{Y}(T) \\ &= \mathbf{S}(T)[\mathbf{S}'(T)\mathbf{S}(T)]^{-1}\tilde{\mathbf{S}}'(T)\mathbf{Y}(T) \\ &= \mathbf{P}_{M(T)}[\mathbf{Y}(T) - \mathbf{P}_{U(T)}\mathbf{Y}(T)] .\end{aligned} \tag{A.4}$$

Combining (A.1-4) gives

$$\begin{aligned}\mathbf{P}_{M(T)}\mathbf{Y}(T) &= \mathbf{P}_{M_{T-1}(T)}\mathbf{Y}(T) + \mathbf{P}_{M(T)}\mathbf{P}_{U(T)}\mathbf{P}^{\perp}_{M_{T-1}(T)}\mathbf{Y}(T) \\ &= \mathbf{P}_{M_{T-1}(T)}\mathbf{Y}(T) + [\mathbf{P}_{M(T)}\mathbf{u}(T)] <\mathbf{u}(T),\mathbf{P}^{\perp}_{M_{T-1}(T)}\mathbf{Y}(T)> .\end{aligned} \tag{A.5}$$

Subtracting both sides of (A.5) from $\mathbf{Y}(T)$, and then taking inner products of both sides with $\mathbf{u}(T)$ gives

$$<\mathbf{u}(T),\mathbf{P}^{\perp}_{M(T)}\mathbf{Y}(T)> = <\mathbf{u}(T),\mathbf{P}^{\perp}_{M_{T-1}(T)}\mathbf{Y}(T)> [1 - <\mathbf{u}(T),\mathbf{P}_{M(T)}\mathbf{u}(T)>] . \tag{A.6}$$

Combining (A.5) and (A.6), and using the definition (6.2.30) gives (6.2.31).

PROBLEMS

6-1 a. Prove that any vector in the space M' spanned by the vectors $\mathbf{X}_1 \cdots \mathbf{X}_n$ and $\mathbf{Y}$ can be written as the sum of a vector in M, which is spanned by $\mathbf{X}_1 \cdots \mathbf{X}_n$, and a constant times $\mathbf{P}^{\perp}_M\mathbf{Y}$.

b. Suppose the vector $\mathbf{Y}$ is replaced by a space L spanned by the vectors $\mathbf{Y}_1 \cdots \mathbf{Y}_m$. Show how the space $M \oplus L$ can be decomposed as $M \oplus L'$, where the basis vectors of L', denoted as $\mathbf{Y}'_1 \cdots \mathbf{Y}'_m$, are orthogonal to M and to each other. Show that for any arbitrary vector $\mathbf{Z}$,

$$\mathbf{P}^{\perp}_{M\oplus L}\mathbf{Z} = \mathbf{P}^{\perp}_M\mathbf{Z} - \sum_{i=1}^{m}\mathbf{P}_{\mathbf{Y}'_i}\mathbf{Z} . \tag{6.P.1}$$

6-2 Under what conditions is $\sin^2\theta(T)$, defined by (6.2.28), equal to one? Interpret this result in the context of (6.2.34). Show that $\sin^2\theta(T|1,n) = 1$ implies $e_f(T-2|n-1) = 0$.

6-3 Assume that an autoregressive time series is described by the difference equation,

$$y(T) = 0.9y(T-1) + 0.8y(T-2) + 0.7y(T-3) + \delta(T) \tag{6.P.2}$$

where $\delta(T)$ is an uncorrelated sequence and at each iteration is equal to +1 or -1 with equal probability.

a. Compute the second-order MMSE prediction coefficients and PARCOR coefficients for this sequence.

b. Assume that a second-order prewindowed LS algorithm is used to estimate the prediction coefficients given the sequence $y(0), y(1), \cdots$. Compute the following variables, which enter the LS transversal algorithm (6.3.63), assuming that $\epsilon_b(0|n) = \epsilon_f(0|n) = 0.1$, $n = 0, 1, 2$, $w = 1$, and that $\delta(0) = -1$, $\delta(1) = 1$, $\delta(2) = 1$, $\delta(3) = -1$, and $y(T) = 0$, $T < 0$:

$$e_f^o(T),\ \mathbf{f}(T),\ e_f(T),\ \epsilon_f(T),$$

$$\mathbf{g}(T|3),\ e_b^o(T),\ \mathbf{b}(T),\ \mathbf{g}(T|2)$$

for $T = 0, 1, 2, 3$. Where unspecified, the order of the variable is 2.

c. Compute the following variables which enter the LS lattice algorithm (6.3.64):

$$e_f(T|n),\ e_b(T|n),\ k_{n+1}^f(T),\ k_{n+1}^b(T),$$

$$\epsilon_f(T|n),\ \epsilon_b(T|n),\ \gamma(T|n)$$

for $n = 0, 1, 2$, and $T = 0, 1, 2, 3$.

d. Suppose that

$$d(T) = 0.5y(T) - 0.5y(T-1)\ . \tag{6.P.3}$$

Compute the MMSE joint process transversal and lattice coefficients.

e. Using the variables in parts (b) and (c) compute the LS joint process estimation variables $\mathbf{c}(3)$, $e_c(3)$, $k_1^c(3)$, and $k_2^c(3)$.

f. Repeat parts (a)-(c), but instead of initializing ϵ_f and ϵ_b to a small positive value, use an exact initialization procedure. For part (a) this means using (6.3.78), (6.3.79), (6.5.19), and (6.5.20). For part (b) compute the recursions (6.3.64) for n=0 to $n = \min(T,2)$.

g. Verify that the quantities in part (e) for T=3 were correctly computed by using the variable definitions, where the inverse of the 2×2 sample covariance matrix at T=3 is computed directly.

6-4 List the prewindowed LS lattice recursions (6.3.64) in the order in which they would be programmed on a computer assuming that $T > N$, and that all variables computed at time $T-1$ are known. Try to minimize the number of variables which need to be stored at each iteration. (Your answer should be in the following format:
For n=1 to N compute in the following order:
1. $K_n(T)$
2. $k_n^b(T)$, ...
List the variables computed at time $T-1$ that need to be stored in addition to the same updated variables at time T.

6-5 Suppose that instead of minimizing the exponentially weighted LS criterion, we wish to choose $\mathbf{f}(T|n)$ such that the following criterion is minimized:

$$\epsilon_f(T|N) = \mu w^T \sum_{j=1}^{N} w^{N-j+1}[f_j(T|N) - \tilde{f}_j]^2 \\ + \sum_{i=0}^{T} w^{T-i}[y(i) - \mathbf{f}'(T|N)\mathbf{y}(i)]^2, \quad T > N \tag{6.P.4}$$

where f_j is the jth component of $\mathbf{f}$ and

$$\tilde{\mathbf{f}} = [\tilde{f}_1 \cdots \tilde{f}_N]'$$

might be an initial guess of the MMSE solution, assuming $y(T)$ is stationary. This has been referred to as LS with a "soft constraint" [13].

a. Show how the fast Kalman algorithm (6.3.63) can be modified so that the criterion (6.P.4) is minimized. (Hint: modify the data vector and shifted data vectors as follows,

$$\mathbf{Y}(T) = [y(T)\ y(T-1) \cdots y(0)\ \mu^{1/2}\tilde{f}_N\ \mu^{1/2}\tilde{f}_{N-1} \cdots \mu^{1/2}\tilde{f}_1]'$$

$$z^{-1}\mathbf{Y}(T) = [y(T-1)\ y(T-2) \cdots y(0)\ 0 \cdots 0\ \mu^{1/2}]'$$

$$z^{-2}\mathbf{Y}(T) = [y(T-2)\ y(T-3) \cdots y(0)\ 0 \cdots 0\ \mu^{1/2}\ 0]'$$

$$\vdots$$

$$z^{-N}\mathbf{Y}(T) = [y(T-N)\ y(T-N-1) \cdots y(0)\ \mu^{1/2}\ 0 \cdots 0]', \tag{6.P.5}$$

where all vectors have dimension $T+N+1$. Minimization of $\epsilon_f(T|N)$ defined in (6.P.4) is now the same as the regular least squares criterion, which is the minimization of ϵ_f in (6.3.8). Use the initialization routine (6.3.79) to show that after N iterations (which corresponds to $T=0$) $\mathbf{f}(N-1|N) = \mathbf{b}(N-1|N) = 0$, and $\epsilon_f(N-1) = w^N\mu$.)

b. Show how the prewindowed LS lattice algorithm (6.3.64) can be modified to minimize the criterion (6.P.4).

c. Modify the fast Kalman and lattice joint process estimation algorithms so that the following criterion is minimized:

$$\epsilon_c(T) = \mu w^T \sum_{j=1}^{N} w^{N-j+1}[c_j(T|N) - \tilde{c}_j]^2 \\ + \sum_{i=0}^{T} w^{T-i}[d(i) - \mathbf{c}'(T|N)\mathbf{y}(i)]^2. \tag{6.P.6}$$

d. Show that the soft constraint LS criterion is equivalent to the approximate initialization procedure (6.3.68).

6-6 *Decision Directed Lattice Equalizer*

Decision directed equalization was discussed in chapter 2. Recall that the output of the equalizer in this case is

$$\hat{d}(T) = \sum_{j=-L}^{L} c_j(T-1)y(T-j) \tag{6.P.7}$$

where $y(T)$ is the output of the channel at time T, and $c_j(T)$, $L \leqslant j \leqslant L$, are the transversal filter coefficients at time T. The transmitted binary sample at time T, $d(T)$, is unknown and must be estimated by thresholding $\hat{d}(T)$. The prediction error

$$e_c^o(T) = d(T) - \hat{d}(T) \tag{6.P.8}$$

is used to update the equalizer coefficients. Either an SG or LS algorithm can be used to update the equalizer coefficients. If, for instance, the fast Kalman algorithm is used, then $e_c^o(T)$ is the oblique error, which is the error computed in (6.5.14). The algorithm (6.3.63), (6.5.14), and (6.5.10) can be used without modification. Suppose, however, that we wish to use an LS lattice equalizer in decision-directed mode. In this case the recursions (6.3.64), (6.5.11), and the top component of (6.5.6) cannot be used without modification since the input $d(T)$ is determined from $\hat{d}(T)$, which is not computed by the lattice recursions. Rather than generating the oblique error $e_c^o(T|2L+1)$, the lattice recursions (6.3.64), (6.5.11), and (6.5.6) generate the current residual $e_c(T|2L+1)$. Derive the following recursions [46], which constitute an LS lattice algorithm that generates the "causal" estimates $\hat{d}(T)$:

$$K_{n+1}(T) = wK_{n+1}(T-1) + e_f^o(T|n)e_b^o(T-1|n)\,[1-\gamma(T-1|n)] \tag{6.P.9a}$$

$$k_{n+1}^f(T) = \frac{K_{n+1}(T)}{\epsilon_f(T|n)}, \quad k_{n+1}^b(T) = \frac{K_{n+1}(T)}{\epsilon_b(T-1|n)} \tag{6.P.9b}$$

$$e_f^o(T|n+1) = e_f^o(T|n) - k_{n+1}^b(T-1)e_b^o(T-1|n) \tag{6.P.9c}$$

$$e_b^o(T|n+1) = e_b^o(T-1|n) - k_{n+1}^f(T-1)e_f^o(T|n) \tag{6.P.9d}$$

$$\epsilon_f(T|n+1) = \epsilon_f(T|n) - k_{n+1}^b(T)K_{n+1}(T) \tag{6.P.9e}$$

$$\epsilon_b(T|n+1) = \epsilon_b(T-1|n) - k_{n+1}^f(T)K_{n+1}(T) \tag{6.P.9f}$$

$$\gamma(T|n+1) = \gamma(T|n) + \frac{e_b^{o^2}(T|n)}{\epsilon_b(T|n)}[1-\gamma(T|n)]^2 \tag{6.P.9g}$$

$$\hat{d}(T|n+1) = \hat{d}(T|n) + \frac{K_{n+1}^c(T-1)}{\epsilon_b(T-1|n)}e_b^o(T|n) \tag{6.P.9h}$$

$$e_c^o(T|n) = d(T) - \hat{d}(T|n) \tag{6.P.9i}$$

$$K_{n+1}^c(T) = wK_{n+1}^c(T-1) + e_c^o(T|n)e_b^o(T|n)[1 - \gamma(T|n)] \tag{6.P.9j}$$

How should the algorithm be initialized?

6-7 The algorithms derived in this chapter apply to scalar valued data. Also of interest in applications such as adaptive antenna arrays and geophysical signal processing is the case where each input sample $y(T)$ is a p-dimensional *vector*. Consider, in particular, the case where the input samples $y(T)$ are complex valued (p=2). This case arises in channel equalization where the real and imaginary parts of $y(T)$ are respectively the in-phase and quadrature signal components. Given two arbitrary vectors **u** and **v** with complex components, we define the inner product,

$$<\mathbf{u},\mathbf{v}> = \mathbf{u}^*\mathbf{v} \tag{6.P.10}$$

where the symbol "*" denotes complex conjugate transpose. Similarly, given two complex valued matrices **U** and **V**,

$$<\mathbf{U},\mathbf{V}> \equiv \mathbf{U}^*\mathbf{V}\,. \tag{6.P.11}$$

The complex forward and backward prediction error vectors are defined as

$$\mathbf{E}_f(T|n) = \mathbf{P}_{M(T|1,n)}^{\perp}\mathbf{Y}(T) = \mathbf{Y}(T) - \mathbf{S}(T|1,n)\mathbf{f}(T) \tag{6.P.12a}$$

and

$$\begin{aligned}\mathbf{E}_b(T|n) &= \mathbf{P}_{M(T|0,n-1)}^{\perp}z^{-n}\mathbf{Y}(T) \\ &= \mathbf{Y}(T-n) - \mathbf{S}(T|0,n-1)\mathbf{b}(T)\,,\end{aligned} \tag{6.P.12b}$$

where the elements of all vectors and matrices are complex in general.

a. Compute the projection $\mathbf{P}_M\mathbf{Y}$ in terms of $\mathbf{Y}$ and the basis vectors of M.

b. Verify that the projection recursions (6.2.6) and (6.2.34) are still valid and derive the following complex fast Kalman prediction algorithm:

$$e_f^o(T) = y(T) - \mathbf{f}'(T-1)\mathbf{y}(T-1) \tag{6.P.13a}$$

$$\mathbf{f}(T) = \mathbf{f}(T-1) + e_f^o(T)\mathbf{g}(T-1|N) \tag{6.P.13b}$$

$$e_f(T) = y(T) - \mathbf{f}'(T)\mathbf{y}(T-1) \tag{6.P.13c}$$

$$\epsilon_f(T) = w\epsilon_f(T-1) + e_f^*(T)e_f^o(T) \tag{6.P.13d}$$

$$[\mathbf{g}(T|N+1)]_1 = \frac{e_f^*(T)}{\epsilon_f(T)} \tag{6.P.13e}$$

$$[\mathbf{g}(T|N+1)]_{2,N+1} = \mathbf{g}(T-1|N) - [\mathbf{g}(T|N+1)]_1\mathbf{f}(T) \tag{6.P.13f}$$

$$e_b^o(T) = y(T-N) - \mathbf{b}'(T-1)\mathbf{y}(T) \tag{6.P.13g}$$

$$\mathbf{b}(T) = \frac{\mathbf{b}(T-1) + e_b^o(T)[\mathbf{g}(T|N+1)]_{1,N}}{1 - e_b^o(T)[\mathbf{g}(T|N+1)]_{N+1}} \tag{6.P.13h}$$

$$\mathbf{g}(T|N) = [\mathbf{g}(T|N+1)]_{1,n} + [\mathbf{g}(T|N+1)]_{N+1}\mathbf{b}(T)\ . \tag{6.P.13i}$$

6-8 Referring to the last problem, modify the prewindowed lattice prediction algorithm (6.3.64) so that it applies to a complex valued input sequence $\{y(T)\}$.

6-9 *Fixed-Order Prewindowed LS Algorithm*
As mentioned in section 6.3, there are numerous modifications that can be made to the fast Kalman algorithm (6.3.63) (i.e., see references [47, 29, 14, 13]). Let

$$\tilde{\mathbf{g}}(T|n) \equiv \frac{1}{1-\gamma(T|n)}\mathbf{g}(T|n)$$

and derive the following steady-state fixed-order LS algorithm [13],

$$e_f^o(T) = y(T) - \mathbf{f}'(T-1)\mathbf{y}(T) \tag{6.P.14a}$$

$$e_f(T) = e_f^o[1 - \gamma(T-1|N)] \tag{6.P.14b}$$

$$\epsilon_f(T) = w\epsilon_f(T-1) + e_f(T)e_f^o(T) \tag{6.P.14c}$$

$$1 - \gamma(T|N+1) = w\frac{\epsilon_f(T-1)}{\epsilon_f(T)}[1 - \gamma(T-1|N)] \tag{6.P.14d}$$

$$\tilde{\mathbf{g}}(T|N+1) = \begin{bmatrix} 0 \\ \tilde{\mathbf{g}}(T-1|N) \end{bmatrix} + \frac{e_f^o(T)}{\epsilon_f(T-1)}\begin{bmatrix} 1 \\ -\mathbf{f}(T-1) \end{bmatrix} \tag{6.P.14e}$$

$$\mathbf{f}(T) = \mathbf{f}(T-1) + e_f(T)\tilde{\mathbf{g}}(T-1|N) \tag{6.P.14f}$$

$$e_b^o(T) = w\epsilon_b(T-1)\tilde{\mathbf{g}}_{N+1}(T|N+1) \tag{6.P.14g}$$

$$1 - \gamma(T|N) = \frac{1 - \gamma(T|N+1)}{1 - e_b^o(T)[1 - \gamma(T|N+1)]\tilde{\mathbf{g}}_{N+1}(T|N+1)} \tag{6.P.14h}$$

$$e_b(T) = e_b^o(T)[1 - \gamma(T|N)] \tag{6.P.14i}$$

$$\epsilon_b(T) = w\epsilon_b(T-1) + e_b^o(T)e_b(T) \tag{6.P.14j}$$

$$\begin{bmatrix} \tilde{\mathbf{g}}(T|N) \\ 0 \end{bmatrix} = \tilde{\mathbf{g}}(T|N+1) + \tilde{g}_{N+1}(T|N+1)\begin{bmatrix} -\mathbf{b}(T-1) \\ 1 \end{bmatrix} \tag{6.P.14k}$$

$$\mathbf{b}(T) = \mathbf{b}(T-1) + e_b(T)\tilde{\mathbf{g}}(T|N) \tag{6.P.14l}$$

$$e_c^o(T) = d(T) - \mathbf{c}'(T-1)\mathbf{y}(T) \tag{6.P.14m}$$

$$e_c(T) = e_c^o(T)[1 - \gamma(T|N)] \tag{6.P.14n}$$

$$\mathbf{c}(T) = \mathbf{c}(T-1) + e_c(T)\tilde{\mathbf{g}}(T|N) \tag{6.P.14o}$$

Where unspecified, the order of the variable is N. How does the computational complexity of this algorithm compare with the fast Kalman joint process estimator derived in sections 6.3 and 6.5?

6-10 a. Show that given two arbitrary vectors $\mathbf{u}$ and $\mathbf{v}$, and a subspace M,

$$<\mathbf{u},P_M\mathbf{v}> = <\mathbf{v},P_M\mathbf{u}> . \tag{6.P.15}$$

b. Show that the following LS residuals can be written as

$$e_f(T|n) = y(T) - \sum_{j=1}^{n}\hat{\phi}_j g_j(T-1|n) \tag{6.P.16a}$$

$$e_b(T|n) = y(T-n) - \sum_{j=0}^{n-1}\hat{\phi}_{n-j} g_{j+1}(T|n) \tag{6.P.16b}$$

$$e_f^{\bullet}(T|n) = y(T_0+n) - \sum_{j=1}^{n}\hat{\phi}_j h_j(T-1|n) \tag{6.P.16c}$$

$$e_b^{\bullet}(T|n) = y(T_0) - \sum_{j=0}^{n-1}\hat{\phi}_{n-j} h_{j+1}(T|n) \tag{6.P.16d}$$

where $\hat{\phi}_j = <\mathbf{Y}(T),z^{-j}\mathbf{Y}(T)>$, g_j is the jth element of $\mathbf{g}$ defined by (6.3.38) and (6.4.15a), and h_j is the jth element of $\mathbf{h}$ defined by (6.4.15b).

6-11 a. Modify the fixed-order and order-recursive sliding window covariance algorithms (6.4.54) and (6.4.55) so that the *exponentially weighted* sum of the squared prediction errors

$$\epsilon_f(T|N) = \sum_{j=T-M+1}^{T} w^{T-j} e_f^2(j|N) \tag{6.P.17}$$

is minimized. (Hint: Use the inner product definition (6.1.3) and modify the inner product time updates (6.2.40) and (6.2.49). The order updates remain the same.)

b. Modify the fixed-order and order-recursive growing memory covariance algorithms (6.4.61) and (6.4.62) so that the exponentially weighted sum

$$\epsilon_f(T|N) = \sum_{j=n}^{T} w^{T-j} e_f^2(j|n) \tag{6.P.18}$$

is minimized.

c. By adding extra recursions to the algorithms in parts (a) and (b), modify the sliding window and growing memory covariance joint process estimation algorithms so that the exponentially weighted sum of the joint process estimation errors is minimized (i.e., change the subscript "f" in (6.P.17) and (6.P.18) to "c").

6-12 Assume that the growing memory covariance algorithm (6.4.61) is started at time $T=N-1$, and that the sample covariance matrix is

$$\hat{\mathbf{\Phi}}(N-1) = \mathbf{y}(N-1)\mathbf{y}'(N-1) + \delta\mathbf{I}, \tag{6.P.19}$$

where δ is a small constant that ensures that $\hat{\Phi}(N-1)$ is nonsingular, and the order is assumed to be N.

a. Derive a set of initial conditions for the variables $\mathbf{g}$, $\mathbf{h}$, γ, γ^*, and α by using the variable definitions at time $T=N-1$. (Hint: $\mathbf{g}(N-1|N) = \dfrac{\mathbf{y}(N-1)}{\mathbf{y}'(N-1)\mathbf{y}(N-1) + \delta}$.)

b. Derive a set of initial conditions for the growing memory covariance algorithm (6.4.61) assuming that the soft-constraint LS criterion (6.P.4) is used. (See the hint following problem 6-5a.)

6-13 Derive an exact order-recursive initialization routine for the fixed-order growing memory covariance algorithm (6.4.61). (Hint: See the discussion following (6.4.63)).

6-14 *Block LS Covariance Algorithm*

In speech processing applications it is often the case that a block of a few hundred speech samples are used to compute a set of LS prediction coefficients. The algorithms in this chapter can be used to generate the prediction coefficients recursively; however, if intermediate solutions are not important, much computation can saved by using a more standard matrix inversion algorithm (see for example references [6]and [16]).

a. Use the recursions in section 6.5 to derive an order-recursive algorithm that solves for the LS prediction coefficient vector $\mathbf{f}$ given a block of data $y(0), y(1), \ldots, y(M)$, $M \geqslant 2N$. (Hints: The necessary recursions are (6.3.6a), (6.3.6b), (6.4.18c), (6.4.26a), (6.4.26b), (6.4.27), (6.4.28), (6.4.29), (6.4.32), (6.4.33a), (6.4.34b), (6.4.38b), (6.4.39), (6.4.40), (6.4.45a), (6.4.46), (6.4.47), (6.4.51b), (6.4.53a). To compute the PARCOR coefficients k_j, $1 \leqslant j \leqslant N$, assume that the covariance matrix

$$\hat{\mathbf{\Phi}}(T-1|N) = \sum_{j=0}^{T-1} \mathbf{y}(j|N)\mathbf{y}'(j|N) = \mathbf{S}'(T|1,N)\mathbf{S}(T|1,N) \tag{6.P.20}$$

has been computed, and use the definition (6.4.14)).

b. Extend the algorithm in part (a) to the joint process estimation case by using the recursions (6.5.23), (6.5.33), (6.5.21), and (6.5.27).

c. Counting divides as multiplies, how many multiplies and additions do the algorithms in parts (a) and (b) require? How does this compare with the recursive fixed order growing memory algorithm in section 6.4?

6-15 *Fixed-Order Growing Memory Covariance Algorithm*

As with the prewindowed fast Kalman algorithm, numerous modifications can also be made to the fixed-order sliding window and growing memory covariance algorithms presented in section 6.4 (i.e., see references [48, 14, 28]). Derive the following set of recursions, which is a modified version of the growing memory covariance algorithm (6.4.61) [28],

$$e_f^o(T) = y(T) - \mathbf{f}'(T-1)\mathbf{y}(T-1) \tag{6.P.21a}$$

$$e_f(T) = e_f^o(T)[1 - \gamma(T-1)] \tag{6.P.21b}$$

$$\epsilon_f(T) = \epsilon_f(T-1) + e_f^o(T)e_f(T) \tag{6.P.21c}$$

$$\gamma(T|N+1) = \gamma(T-1|N) + \frac{e_f^2(T)}{\epsilon_f(T)} \tag{6.P.21d}$$

$$\tilde{\mathbf{g}}(T|N+1) = \begin{bmatrix} 0 \\ \tilde{\mathbf{g}}(T-1|N) \end{bmatrix} + \frac{e_f^o(T)}{\epsilon_f(T-1)} \begin{bmatrix} 1 \\ -\mathbf{f}(T-1) \end{bmatrix} \tag{6.P.21e}$$

$$\mathbf{f}(T) = \mathbf{f}(T-1) + e_f(T)\tilde{\mathbf{g}}(T-1|N) \tag{6.P.21f}$$

$$e_b^o(T) = \epsilon_b(T-1)\tilde{\mathbf{g}}_{N+1}(T|N+1) \tag{6.P.21g}$$

$$1 - \gamma(1,T|N) = \frac{1 - \gamma(T|N+1)}{1-[1-\gamma(T|N+1)]e_b^o(T)\tilde{\mathbf{g}}_{N+1}(T|N+1)} \tag{6.P.21h}$$

$$e_b(T) = e_b^o(T)[1 - \gamma(1,T|N)] \tag{6.P.21i}$$

$$\epsilon_b(T) = \epsilon_b(T-1) + e_b^o(T)e_b(T) \tag{6.P.21j}$$

$$\begin{bmatrix} \tilde{\mathbf{g}}(1,T|N) \\ 0 \end{bmatrix} = \tilde{\mathbf{g}}(T|N+1) + \tilde{\mathbf{g}}_{N+1}(T|N+1)\begin{bmatrix} -\mathbf{b}(T-1) \\ 1 \end{bmatrix} \tag{6.P.21k}$$

$$\mathbf{b}(T) = \mathbf{b}(T-1) + e_b(T)\tilde{\mathbf{g}}(1,T) \tag{6.P.21l}$$

$$\beta(T) \equiv \mathbf{h}'(T-1)\mathbf{y}(T) \tag{6.P.21m}$$

$$\beta^{\bullet}(T) \equiv \mathbf{g}'(1,T)\mathbf{y}(N-1) = \frac{1-\gamma(1,T)}{1-\gamma^{\bullet}(T-1)}\beta(T) \tag{6.P.21n}$$

$$1-\gamma^{\bullet}(T) = \frac{1-\gamma^{\bullet}(T-1)}{1-\beta(T)\beta^{\bullet}(T)} \tag{6.P.21o}$$

$$1-\gamma(T) = \frac{1-\gamma(1,T)}{1-\beta(T)\beta^{\bullet}(T)} \tag{6.P.21p}$$

$$\tilde{\mathbf{g}}(T) = \tilde{\mathbf{g}}(1,T) - \frac{\beta(T)}{1-\gamma^{\bullet}(T-1)}\mathbf{h}(T-1) \tag{6.P.21q}$$

$$\mathbf{h}(T) = \mathbf{h}(T-1) - \beta(T)[1-\gamma(T)]\tilde{\mathbf{g}}(T) \tag{6.P.21r}$$

$$e_c^o(T) = d(T) - \mathbf{c}'(T-1)\mathbf{y}(T) \tag{6.P.21s}$$

$$e_c(T) = e_c^o(T)[1-\gamma(T)] \tag{6.P.21t}$$

$$\mathbf{c}(T) = \mathbf{c}(T-1) + e_c(T)\tilde{\mathbf{g}}(T) \tag{6.P.21u}$$

Where unspecified, the order of all vectors is N, and $\tilde{\mathbf{g}}(T_0,T|n) = \dfrac{\mathbf{g}(T_0,T|n)}{1-\gamma(T_0,T|n)}$. How does the computational complexity of this algorithm compare with the fixed-order growing memory covariance algorithm derived in sections 6.4 and 6.5?

6-16 The geometric approach developed in sections 6.1 - 6.2 can be used to derive the normalized algorithms in section 6.6. Given arbitrary vectors $\mathbf{v}$ and $\mathbf{w}$, we first define the normalized version of $\mathbf{v}'\mathbf{P}_M^{\perp}\mathbf{w}$ as $\rho_M(\mathbf{v},\mathbf{w})$ given by (6.6.1).

a. Derive the normalized order update (6.6.2) from (6.2.52).

b. Use (6.2.40) to show that

$$\rho_{M(T)}[\mathbf{v}(T),\mathbf{Y}(T)] =$$

$$\left\{\left(1-\rho^2_{M(T)}[\mathbf{u}(T),\mathbf{v}(T)]\right)\left(1-\rho^2_{M(T)}[\mathbf{u}(T),\mathbf{Y}(T)]\right)\right\}^{1/2}\rho_{M(T-1)}[\mathbf{v}(T-1),\mathbf{Y}(T-1)]$$

$$+\ \rho_{M(T)}[\mathbf{u}(T),\mathbf{v}(T)]\rho_{M(T)}[\mathbf{u}(T),\mathbf{Y}(T)]. \tag{6.P.22}$$

c. Using the facts that

$$\bar{e}_f(T|n) = \rho_{M(T|1,n)}[\mathbf{u}(T),\mathbf{Y}(T)], \tag{6.P.23a}$$

$$\bar{e}_b(T|n) = \rho_{M(T|0,n-1)}[\mathbf{u}(T),z^{-n}\mathbf{Y}(T)], \tag{6.P.23b}$$

$$\bar{k}_n(T) = \rho_{M(T|1,n)}[\mathbf{Y}(T),z^{-1}\mathbf{E}_b(T|n)], \tag{6.P.23c}$$

$$\bar{e}_c(T|n) = \rho_{M(T|0,n-1)}[\mathbf{u}(T),\mathbf{D}(T)], \tag{6.P.23d}$$

and

$$\bar{k}_n^c(T) = \rho_{M(T|1,n)}[\mathbf{D}(T),\mathbf{E}_b(T|n)], \tag{6.P.23e}$$

derive the normalized lattice joint process estimator recursions (6.6.25), (6.6.31), (6.6.37), and (6.6.40).

6-17 Derive expressions for the unnormalized lattice variables $e_f(T|n)$, $e_b(T|n)$, $K_n(T)$, $\epsilon_f(T|n)$, $\epsilon_b(T|n)$, and $\gamma(T|n)$, in terms of $\epsilon_f(T|0)$ and the normalized lattice variables $\bar{e}_f$, $\bar{e}_b$, and $\bar{k}_j$.

6-18 Use the definitions in problem 6-7 and the normalized projection recursions in problem 6-16 to derive a complex normalized lattice joint process estimator.

6-19 *Normalized Fixed-Order Prewindowed Algorithm*
To derive a normalized fixed-order LS algorithm, the normalized vectors $\bar{\mathbf{f}}$, $\bar{\mathbf{b}}$, $\mathbf{c}$, and $\bar{\mathbf{g}}$ are defined in analogy with the normalized whitening filters $\bar{F}_{T|n}(z)$, $\bar{B}_{T|n}(z)$, and $\bar{G}_{T|n}(z)$ defined by (6.6.51). In particular,

$$\bar{\mathbf{g}}(T|n) \equiv \frac{\mathbf{g}(T|n)}{\sqrt{1-\gamma(T|n)}} .$$

For notational convenience, the operation is defined,

$$x^c \equiv (1-x^2)^{1/2}, \tag{6.P.24a}$$

and similarly,

$$x^{-c} \equiv (1-x^2)^{-1/2}. \tag{6.P.24b}$$

a. Derive the following normalized fixed-order prewindowed algorithm [13],

$$\bar{e}_f^o(T) = \frac{y(T)}{\epsilon_f^{1/2}(T-1)} - \bar{\mathbf{f}}'(T-1)\mathbf{y}(T) \tag{6.P.25a}$$

$$z(T) \equiv \bar{e}_f(T)\bar{e}_f^{-c}(T) = w^{-1/2}\bar{e}_f^o(T)[1-\gamma(T-1|N)]^{1/2} \tag{6.P.25b}$$

$$\bar{e}_f^{-c}(T) = \{1+z^2(T)\}^{1/2} \tag{6.P.25c}$$

$$\bar{e}_f(T) = \bar{e}_f^c(T)z(T) \tag{6.P.25d}$$

$$[1-\gamma(T|N+1)]^{1/2} = \bar{e}_f^c(T)[1-\gamma(T-1|N)]^{1/2} \tag{6.P.25e}$$

$$\bar{\mathbf{f}}(T) = w^{-1/2}\bar{e}_f^c(T)\bar{\mathbf{f}}(T-1) + \bar{e}_f(T)\bar{\mathbf{g}}(T-1) \tag{6.P.25f}$$

$$\epsilon_f^{1/2}(T) = w^{1/2}\epsilon_f^{1/2}(T-1)e_f^{-c}(T) \tag{6.P.25g}$$

$$\bar{\mathbf{g}}(T|N+1) = \bar{e}_f^{-c}(T)\begin{bmatrix}0\\ \bar{\mathbf{g}}(T-1|N)\end{bmatrix} + z(T)\begin{bmatrix}\epsilon_f^{-1/2}(T)\\ -\bar{\mathbf{f}}(T)\end{bmatrix} \tag{6.P.25h}$$

$$\bar{e}_b(T) = w^{1/2}\epsilon_b^{1/2}(T-1)\bar{\mathbf{g}}_{N+1}(T|N+1) \tag{6.P.25i}$$

$$\bar{e}_b^c(T) = \{1-\bar{e}_b^2(T)\}^{1/2} \tag{6.P.25j}$$

$$\epsilon_b^{1/2}(T) = w^{1/2}\epsilon_b^{1/2}(T-1)\bar{e}_b^{-c}(T) \tag{6.P.25k}$$

$$[1-\gamma(T|N)]^{1/2} = [1-\gamma(T|N+1)]^{1/2}\bar{e}_b^{-c}(T) \tag{6.P.25l}$$

$$\bar{\mathbf{b}}(T) = w^{-1/2}\bar{e}_b^{-c}(T)\bar{\mathbf{b}}(T-1) + \bar{e}_b^{-c}(T)\bar{e}_b(T)[\bar{\mathbf{g}}(T|N+1)]_{1,N} \tag{6.P.25m}$$

$$\bar{\mathbf{g}}(T|N) = \bar{e}_b^c(T)[\bar{\mathbf{g}}(T|N+1)]_{1,N} + \bar{e}_b(T)\bar{\mathbf{b}}(T) \tag{6.P.25n}$$

b. Show how the joint process estimation variables $e_c^o(T)$, $e_c(T)$, and $\mathbf{c}(T)$ can be computed.

c. Show how the normalized joint process estimation variables $\bar{e}_c^o(T)$, $\bar{e}_c(T)$, and $\bar{\mathbf{c}}(T)$ can be computed.

d. Counting divisions as multiplies, how many multiplies, additions, and square roots per iteration do the algorithms in parts a), b), and c) require?

6-20 a. Derive the normalized sliding window covariance algorithm (6.6.57).

b. Derive the normalized growing memory covariance algorithm (6.6.60).

c. Extend the algorithms in parts (a) and (b) so that they apply to the joint process estimation case.

6-21 a. Derive the normalized prewindowed whitening filter algorithm (6.6.52).

b. Derive a normalized sliding window covariance whitening filter algorithm. (Hint: See the paragraph following equation (6.6.69).)

c. Derive a normalized growing memory covariance whitening filter algorithm.

d. Extend the algorithms in parts (b) and (c) so that they compute the normalized joint process estimation filter $\bar{C}_{T_0,T|n}(z)$.

6-22 *Normalized Fixed-Order Growing Memory Covariance Algorithm*
In analogy with the normalized covariance lattice algorithms derived in section 6.6.4, normalized fixed-order covariance algorithms can also be obtained. The normalized vectors $\bar{\mathbf{f}}(T_0,T|n)$, $\bar{\mathbf{b}}(T_0,T|n)$, $\bar{\mathbf{c}}(T_0,T|n)$, $\bar{\mathbf{g}}(T_0,T|n)$, and $\bar{\mathbf{h}}(T_0,T|n)$ are defined in analogy with the normalized whitening filters $\bar{F}_{T_0,T|n}(z)$, $\bar{B}_{T_0,T|n}(z)$, $\bar{C}_{T_0,T|n}(z)$, $\bar{G}_{T_0,T|n}(z)$, and $\bar{H}_{T_0,T|n}(z)$ (see (6.6.51) and (6.6.53)). In particular,

$$\bar{\mathbf{h}}(T_0,T|n) \equiv \frac{\mathbf{h}(T_0,T|n)}{\sqrt{1-\gamma^*(T_0,T|n)}}.$$

a. Derive the following set of recursions, which constitutes a normalized version of the fixed-order growing memory covariance algorithm in problem 6-15 [28],

$$\bar{e}_f^o(T) = \frac{y(T)}{\epsilon_f^{1/2}(T-1)} - \bar{\mathbf{f}}'(T-1)\mathbf{y}(T) \tag{6.P.26a}$$

$$z(T) \equiv \bar{e}_f(T)\bar{e}_f^{-c}(T) = \bar{e}_f^o(T)[1-\gamma(T-1)]^{1/2} \tag{6.P.26b}$$

$$\bar{e}_f^{-c}(T) = (1+z^2)^{1/2}; \quad \bar{e}_f^c(T) = \frac{1}{\bar{e}_f^{-c}(T)} \tag{6.P.26c}$$

$$\bar{e}_f(T) = \bar{e}_f^c(T)z \tag{6.P.26d}$$

$$[1-\gamma(T|N+1)]^{1/2} = \bar{e}_f^c(T)[1-\gamma(T-1|N)]^{1/2} \tag{6.P.26e}$$

$$\epsilon_f^{1/2}(T) = w^{1/2}\epsilon_f^{1/2}(T-1)e_f^{-c}(T) \tag{6.P.26f}$$

$$\bar{\mathbf{f}}(T) = \bar{e}_f^c(T)\bar{\mathbf{f}}(T-1) + \bar{e}_f(T)\bar{\mathbf{g}}(T-1) \tag{6.P.26g}$$

$$\bar{\mathbf{g}}(T|N+1) = \bar{e}_f^{-c}(T)\begin{bmatrix} 0 \\ \bar{\mathbf{g}}(T-1|N) \end{bmatrix} + z(T)\begin{bmatrix} \epsilon_f^{-1/2}(T) \\ -\bar{\mathbf{f}}(T) \end{bmatrix} \tag{6.P.26h}$$

$$\bar{e}_b(T) = \epsilon_b^{1/2}(T-1)\bar{\mathbf{g}}_{N+1}(T|N+1) \tag{6.P.26i}$$

$$\bar{e}_b^c(T) = [1-\bar{e}_b^2(T)]^{1/2} \tag{6.P.26j}$$

$$\epsilon_b^{1/2}(T) = \epsilon_b^{1/2}(T-1)\bar{e}_b^{-c}(T) \tag{6.P.26k}$$

$$[1-\gamma(1,T|N)]^{1/2} = [1-\gamma(T|N+1)]^{1/2}\bar{e}_b^{-c}(T) \tag{6.P.26l}$$

$$\bar{\mathbf{b}}(T) = \bar{e}_b^{-c}(T)\bar{\mathbf{b}}(T-1) + \bar{e}_b(T)\bar{e}_b^{-c}(T)[\bar{\mathbf{g}}(T|N+1)]_{1,N} \tag{6.P.26m}$$

$$\bar{\mathbf{g}}(1,T|N) = \bar{e}_b^c(T)[\bar{\mathbf{g}}(T|N+1)]_{1,N} + \bar{e}_b(T)\bar{\mathbf{b}}(T) \tag{6.P.26n}$$

$$\bar{\beta}(T) \equiv \bar{\mathbf{h}}'(T-1)\mathbf{y}(T)[1-\gamma(1,T)]^{1/2} \tag{6.P.26o}$$

$$\bar{\beta}^c(T) = [1-\bar{\beta}^2(T)]^{1/2} \tag{6.P.26p}$$

$$[1-\gamma(T)]^{1/2} = [1-\gamma(1,T)]^{1/2}\bar{\beta}^{-c}(T) \tag{6.P.26q}$$

$$\bar{\mathbf{g}}(T) = \bar{\beta}^{-c}(T)\bar{\mathbf{g}}(1,T) - \bar{\beta}^{-c}(T)\bar{\beta}(T)\bar{\mathbf{h}}(T-1) \tag{6.P.26r}$$

$$\bar{\mathbf{h}}(T) = \bar{\beta}^c(T)\bar{\mathbf{h}}(T-1) - \bar{\beta}(T)\bar{\mathbf{g}}(T) \tag{6.P.26s}$$

Where unspecified, the order of the variable is N. x^c and x^{-c} are defined by (6.P.24).

b. Show how the unnormalized joint process estimation variables $e_c(T)$ and $\mathbf{c}(T)$ can be computed.

c. Show how the normalized joint process estimation variables $\bar{e}_c(T)$ and $\bar{\mathbf{c}}(T)$ can be computed.

d. Counting divides as multiplies, how much computation per iteration do the algorithms in parts a) through c) require?

REFERENCES

1. M.L. Honig and D.G. Messerschmitt, "Comparison of LS and Stochastic Gradient Lattice Predictor Algorithms Using Two Performance Criteria," *IEEE Trans. on Acoustics, Speech, and Signal Processing* **ASSP-32**(2) p. 441 (April 1984).
2. R.S. Medaugh and L.J. Griffiths, "A Comparison of Two Fast Linear Predictors," *Proc. 1981 IEEE ICASSP*, (April 1981).
3. R. S. Medaugh and L. J. Griffiths, "Further Results of a Least-Squares and Gradient Adaptive-Lattice Algorithm Comparison," *Proc. IEEE ICASSP*, pp. 1412-1415 (May 1982).
4. B. Friedlander, "Lattice Filters for Adaptive Processing," *Proceedings of the IEEE* **70**(8) pp. 829-867 (August 1982).
5. L. Ljung, M. Morf, and D. Falconer, "Fast calculation of gain matrices for recursive estimation schemes," *Int. J. Control* **27**(1) pp. 1-19 (1978).
6. M. Morf, B. Dickinson, T. Kailath , and A. Vieira, "Efficient Solutions of Covariance Equations for Linear Prediction," *IEEE Trans. on ASSP* **ASSP-25**(5) pp. 429-435 (Oct. 1977).
7. D.D. Falconer and L. Ljung, "Application of Fast Kalman Estimation to Adaptive Equalization," *IEEE Trans. Comm.* **COM-26** pp. 1439-1446 (Oct. 1978).
8. D. Godard, "Channel Equalization Using Kalman Filter for Fast Data Transmission," *IBM J. of Res. and Dev.*, pp. 267-273 (May 1974).
9. M. Morf, D.T. Lee, J.R. Nickolls, and A. Vieira, "A Classification of Algorithms for ARMA Models and Ladder Realizations ," *Proc. 1977 IEEE Conf. ASSP*, pp. 13-19 (April 1977).
10. M. Morf, A. Vieira , and D.T. Lee, "Ladder Forms for Identification and Speech Processing," *Proc. 1977 IEEE Conference Decision and Control*, pp. 1074-1078 (Dec. 1977).
11. D.T.L. Lee, M. Morf, and B. Friedlander, "Recursive Square Root Ladder Estimation Algorithms," *IEEE Trans. on ASSP* **ASSP-29**(3) pp. 627-641 (June 1981).
12. C. Samson, "A Unified Treatment of Fast Kalman Algorithms for Identification," *Int. Journal of Control* **35**(5) pp. 909-934 (May 1982).

13. J. M. Cioffi and T. Kailath, "Fast, Recursive-Least Squares, Transversal Filters for Adaptive Filtering," *IEEE Trans. on Acoustics, Speech, and Signal Processing* **ASSP-32**(2) (April 1984).

14. N. Kalouptsidis, G. Carayannis, and D. Manolakis, "A Fast Covariance Type Algorithm for Sequential LS Filtering and Prediction," *IEEE Trans. on Automatic Control,* to appear

15. M. L. Honig, "Recursive Fixed-Order Covariance Least Squares Algorithms," *Bell System Technical Journal* **62**(10, Part 1) pp. 2961-2992 (Dec. 1983).

16. S. Marple, "Efficient Least-Squares FIR System Identification," *IEEE Trans. on ASSP* **ASSP-29**(1) pp. 62-73 (Feb. 1981).

17. N. Kalouptsidis, D. Manolakis, and G. Carayannis, "A Family of Computationally Efficient Algorithms for Multichannel Signal Processing - A Tutorial Review," *Signal Processing* **5**(1) pp. 5-19 (Jan. 1983).

18. B. Porat and T. Kailath, "Normalized Lattice Algorithms for Least-Squares FIR System Identification," *IEEE Trans. on Acoustics, Speech, and Signal Processing* **31**(1) pp. 122-128 (Feb. 1983).

19. V. Solo, "Finite Data Lattice Algorithms for Instrumental Variable Recursions," *IEEE Trans. on Acoustics, Speech, and Signal Processing* **ASSP-31**(5) pp. 1202-1210 (Oct. 1983).

20. G.C. Goodwin and K.S. Sin, *Adaptive Filtering Prediction and Control,* Prentice-Hall, Englewood Cliffs N.J. (1984).

21. M. J. Shensa, "Recursive Least-Squares Lattice Algorithms: A Geometrical Approach," *IEEE Trans. Automatic Control* **AC-26** pp. 695-702 (June 1981).

22. M. Morf, C. H. Muravchik , and D. T. Lee, "Hilbert Space Array Methods for Finite Rank Process Estimation and Ladder Realizations for Adaptive Signal Processing," *Proc. IEEE Int. Conf. on Acoustics, Speech, and Signal Processing,* pp. 856-859 (March 1981).

23. B. Porat, B. Friedlander , and M. Morf, "Square Root Covariance Ladder Algorithms," *IEEE Trans. on Automatic Control* **AC-27**(4) pp. 813-829 (August 1982).

24. J. M. Cioffi and T. Kailath, "Fast, Fixed-Order, Least-Squares Algorithms for Adaptive Filtering," *Proc. IEEE 1983 Int. Conf. on Acoustics, Speech, and Signal Processing,* (April, 1983).

25. A. Ben-Israel and T. N. E. Greville, *Generalized Inverses; Theory and Applications,* John Wiley & Sons, New York (1974).

26. C. L. Lawson and R. J. Hanson, *Solving Least Squares Problems,* Prentice Hall, Inc., Englewood Cliffs, New Jersey (1974).

27. A.H. Gray, Jr. and J.D. Markel, "A Normalized Digital Filter Structure," *IEEE Trans. ASSP* **ASSP-23** pp. 268-277 (June 1975).

28. J. M. Cioffi and T. Kailath, "Windowing Methods and their Efficient Transversal-Filter Implementation for the RLS Adaptive-Filtering

Criterion," *IEEE Trans. on ASSP*, (to appear).

29. G. Carayannis, D. Manolakis, and N. Kalouptisidis, "A Fast Sequential Algorithm for Least Squares Filtering and Prediction" *IEEE Trans. on ASSP* **ASSP-31(6)** pp. 1394-1402 (Dec. 1983).
30. D.T.L. Lee and M. Morf, "A Novel Innovations Based Time-Domain Pitch Detector," *Proc. 1980 IEEE ICASSP*, (April 1980).
31. L.R. Rabiner and R.W. Schafer, *Digital Processing of Speech Signals*, Prentice-Hall, Englewood Cliffs, N.J. (1978).
32. M. Morf and D.T.L. Lee, "Fast Algorithms for Speech Modeling," Technical Report M308-1, Stanford Univ. Information Systems Laboratory, Stanford, Ca. (Dec. 1978).
33. B. Gold and L. Rabiner, "Parallel Processing Techniques for Estimating Pitch Periods of Speech in the Time Domain," *J. Acoust. Soc. Am.* **46** pp. 442-448 (Aug. 1969).
34. D. T. Lee, B. Friedlander, and M. Morf, "Recursive Ladder Algorithms for ARMA Modeling," *IEEE Trans. on Automatic Control* **AC-27**(4) pp. 753-764 (August 1982).
35. A. Benveniste and C. Chaure, "AR and ARMA Identification Algorithms of Levinson Type: An Innovations Approach," *IEEE Trans. on Aut. Control* **AC-26**(6) pp. 1243-1261 (Dec. 1981).
36. B. Friedlander, "System Identification Techniques for Adaptive Signal Processing," *Circuits Systems Signal Processing* **1**(1) pp. 1-41 (1982).
37. B. Friedlander and M. Morf, "Least Squares Algorithms for Adaptive Linear-Phase Filtering," *IEEE Trans. on Acoustics, Speech, and Signal Processing* **ASSP-30**(3) pp. 381-390 (June 1982).
38. S. L. Marple, Jr., "A Fast Least Squares Linear Phase Adaptive Filter," *Proc. 1984 Int. Conf. Acoustics, Speech, and Signal Processing*, (21.8) (March 1984).
39. B.Friedlander, "Instrumental Variable Methods for ARMA Spectral Estimation," *IEEE Transactions on Acoustics, Speech, and Signal Processing* **ASSP-31**(2) p. 404 (April 1983).
40. N. Kalouptsidis, G. Carayannis, and D. Manolakis, "Fast Design of Multichannel FIR Least-Squares Filters with Optimum Lag," *IEEE Transaction on Acoustics, Speech, and Signal Processing* **ASSP-32**(1) pp. 48-59 (Feb. 1984).
41. G. Carayannis, D. Manolakis, and N. Kalouptsidis, "Efficient Algorithms and Structures for Lagged Least Squares (LS) FIR Filters in the Case of Prewindowed Signals," *Proc. 1984 Int. Conf. Acoustics, Speech, and Signal Processing*, (43.5.1) (March 1984).
42. M. R. Gevers and V. J. Wertz, "A D-Step Predictor in Lattice and Ladder Form," *IEEE Trans. on Automatic Control* **AC-28**(4) pp. 465-475 (April 1983).

43. F. Ling and J. G. Proakis, "A Generalized Multichannel Least Squares Lattice Algorithm Based on Sequential Processing Stages," *IEEE Transactions on Acoustics, Speech, and Signal Processing* **ASSP-32**(2) pp. 381-389 (April 1984).

44. H. Lev-Ari and T. Kailath, "Lattice Filter Parametrization and Modeling of Nonstationary Processes," *IEEE Transactions on Information Theory* **Vol. IT-30**(No. 1) pp. pp. 2-16 (Jan. 1984).

45. A.W. Naylor and G.R. Sell, *Linear Operator Theory in Engineering and Science,* Holt, Rinehart and Winston, Inc., New York (1971).

46. E. Satorius and J. Pack, "Application of Least Squares Lattice Algorithms to Adaptive Equalization," *IEEE Trans. Comm.* **COM-29** pp. 136-142 (Feb. 1981).

47. D. Manolakis, G. Carayannis, and N. Kalouptsidis, "On the Computational Organization of Fast Sequential Algorithms," *Proc. 1984 Int. Conf. Acoustics, Speech, and Signal Processing*, (Paper 43.7.1) (March 1984).

48. D. Manolakis, F. Ling, and J. G. Proakis, "Efficient Least-Squares Algorithms for Finite Memory Adaptive Filtering," *1984 Conf. on Information Sciences and Systems*, (March 1984).

7

CONVERGENCE

This chapter will attempt to characterize the performance of the numerous adaptation algorithms which have been derived in this book. The performance of an adaptive algorithm will be measured in terms of its convergence properties. Our approach will be to assume that the input given data samples are a realization of a wide-sense stationary random process with known statistics, and study how the adaptive filter solution approaches the MMSE solution. While this approach is unrealistic in the sense that inputs are actually nonstationary in many instances, this simplifying assumption is a concession to tractability and gives useful insight.

The convergence of an adaptation algorithm can be measured in two ways. The first way is to determine analytically or empirically how the output error (prediction or joint process) decreases with time. This is usually done by calculating the output mean-square error (MSE). The second method is to determine analytically or empirically how the filter coefficients approach their optimum value vs. time. Since the coefficients are actually fluctuating, even asymptotically for a fixed step-size algorithm, this is usually done by calculating the mean value of the filter coefficients. The two measures are of course closely related to one another.

As to which method of measuring convergence is most appropriate, it depends on the application. In applications such as channel equalization and echo cancellation, the objective of the adaptive algorithm is to rapidly minimize the output estimation error. The size of the estimation error is therefore a direct measure of the performance of the adaptive filter. A specific measure of performance for these cases is the speed at which the output mean squared estimation error approaches its asymptotic value as a function of the algorithm step size, or exponential weighting factor, and the input statistics. The actual achieved MSE will be somewhat larger than the minimum MSE (MMSE) attainable with a fixed coefficient filter due to statistical fluctuations of the filter coefficients.

In applications such as short-term spectral estimation of speech, the objective of the adaptive algorithm is to rapidly obtain an accurate estimate of the prediction coefficients. A specific measure of performance for applications such as this is the speed at which the mean values of the filter coefficients approach their asymptotic values, and the asymptotic standard deviations of the filter coefficients.

This chapter attempts to characterize the performance measures mentioned above for some of the recursive algorithms derived in previous chapters. This is not an easy task. A rigorous and comprehensive theoretical treatment of the convergence properties of almost any one of the adaptive algorithms presented in this book has yet to appear. However, approximate analyses are available that give very useful insight into how each algorithm will perform in a given environment. In all cases, checks with computer simulation indicate that the analytical results in this chapter are sufficiently accurate for design purposes.

The material in this chapter is arranged in order of increasing complexity. We start with a convergence analysis of the stochastic gradient (SG) transversal algorithm in section 7.1. By making approximations, it is possible to obtain closed form expressions for the output mean squared estimation error and the mean values of the filter coefficients as functions of time.

The convergence properties of an adaptive lattice filter using an SG algorithm are much harder to characterize because of the nonlinear dependence each coefficient has on preceding coefficients. Nevertheless, the convergence properties of only one stage of an SG adaptive lattice algorithm can be analyzed in a similar manner to the SG transversal algorithm, if the input forward and backward prediction errors are assumed to be stationary processes. This is done in section 7.2, and a similar analysis is performed for the LS adaptation algorithm in section 7.3. Of course, in practice the inputs to a given lattice stage are not stationary due to the adaptation of the previous stages, even for a stationary input signal to the whole filter. However, approximate stationarity is achieved after the preceding stages have converged in some stochastic sense.

Once an understanding of how one lattice stage adapts is obtained, the next step is to determine how one adapting lattice stage affects the convergence of successive lattice stages. This is done in an approximate manner in section 7.4, resulting in simple dynamic convergence models for SG lattice algorithms. Unfortunately, these models must be evaluated on a computer; however, they can be used to predict output MSE and the mean values of filter coefficients as functions of time, given only second-order statistical information about the input sequence. Furthermore, the lattice adaptation model requires much less computation than the multiple simulation runs of the actual algorithm needed to estimate statistical properties.

Dynamic convergence models from which output MSE and mean coefficient values can be computed are also derived for both transversal and lattice prewindowed LS algorithms.

Some of the results on convergence properties are illustrated by explicitly comparing the convergence properties of two-stage SG transversal and lattice algorithms in section 7.5. It is shown that the orthogonalization, or decoupling, property of the lattice filter tends to speed up convergence relative to the SG transversal algorithm, although this is not universally the case. The two-stage comparison is given to provide insight into the considerably more difficult N-stage case.

The joint process estimation case is considered for all the adaptation algorithms in section 7.6. Section 7.7 contains a further discussion of LS algorithms and gives examples of where LS algorithms clearly perform better than SG algorithms. It is shown that in adaptive filtering applications where the input signal to noise ratio is quite high, the "algebraic", rather than the statistical properties of the input are of most importance. In these cases an LS algorithm can estimate the filter coefficients (transversal *or* lattice) in a very short time that depends only on the number of filter coefficients, whereas SG algorithms are much slower due to their reliance on statistical averages. On the other hand, if the input is very noisy, an LS algorithm must also average out the effects of the noise, and hence in these cases a recursive LS algorithm does not converge much faster than an appropriately defined SG lattice algorithm.

The adaptation properties studied in this chapter are illustrated in section 7.8, which contains computer simulated curves of averaged output squared prediction error and averaged spectral estimates vs. time using the algorithms discussed.

7.1. SG TRANSVERSAL ALGORITHM

In this section the convergence properties of the SG transversal algorithm are discussed in some detail, using the earlier discussion in section 3.2 as a starting point. Because the SG transversal algorithm has been widely used in various applications, some of which were mentioned in chapter 3, numerous methods of analysis of its convergence properties have appeared in the literature. Alternative methods to that used here [1] are given in [2, 3, 4, 5] and in [6] where a time-varying adaptive step size is considered. Although more rigorous treatments of the SG transversal convergence have appeared, they require considerably more mathematical sophistication than assumed here, and do not really offer a great deal of additional insight into the performance of the algorithm in different environments.

Throughout the subsequent discussion, the input sequences $y(T)$ and $d(T)$ are assumed to be stationary random processes with known statistics. Although this assumption is not valid in many applications, the resulting analysis will give insight into adaptation of the filter for a nonstationary input where the statistical variations of the input are slow relative to the adaptation speed of the filter. This assumption is necessary to get tractable results, and also allows us to compare the performance of the adaptation algorithm against the ideal MMSE solution of section 3.1.

The objective of the SG transversal algorithm is to drive the filter coefficients to those values which minimize the short-term output MSE. Assuming that at time $T = 0$ the filter coefficients are at suboptimal values and that the algorithm parameters are properly set, the output MSE at $T = 0$ should be greater than the minimum MSE and should subsequently decay towards some asymptotic value. As pointed out in chapter 3, the filter coefficients will not exactly converge to the MMSE solution if the algorithm step size is constant. Rather, because a noisy estimate of the error gradient is used to adapt the filter, the coefficients will have some asymptotic variance that causes the asymptotic MSE to be somewhat greater than the minimum mean square error obtained from an optimum fixed coefficient filter.

The SG transversal algorithm is repeated here for convenience,

$$\begin{aligned} \mathbf{c}(T) &= \mathbf{c}(T-1) + \beta e(T)\mathbf{y}(T) \\ &= \mathbf{c}(T-1) + \beta\Big[d(T) - \mathbf{y}'(T)\mathbf{c}(T-1)\Big]\mathbf{y}(T) , \end{aligned} \tag{7.1.1}$$

where, as in previous chapters, $\mathbf{c}(T)$ is the vector of filter coefficients at time T, $\mathbf{y}(T)$ is the N-element vector of observed data samples, $d(T)$ is the desired filter output, and β is the algorithm step size. Where unspecified, all vectors are assumed to contain N elements, where N is the order of the filter. The filter residual $e(T)$ is the same as the joint process residual e_c. The subscript is not necessary in this case and will be omitted for convenience. In the case of linear prediction, the vector $\mathbf{c}$ would be replaced by the vector $\mathbf{f}$, and the residual would be denoted as e_f.

7.1.1. Mean Coefficient Vector

Before the output MSE of the filter as a function of time is investigated, it is simple and instructive to examine the mean-value trajectories of the coefficients. Recall from chapter 3 that an estimate of the mean coefficient vector is obtained by taking expected values of both sides of (7.1.1),

$$\begin{aligned} E[\mathbf{c}(T)] &= E[\mathbf{c}(T-1)] + \beta E\Big\{\Big[d(T) - \mathbf{y}'(T)\mathbf{c}(T-1)\Big]\mathbf{y}(T)\Big\} \\ &= E\Big[\Big(\mathbf{I} - \beta\mathbf{y}(T)\mathbf{y}'(T)\Big)\mathbf{c}(T-1)\Big] + \beta E[d(T)\mathbf{y}(T)] . \end{aligned} \tag{7.1.2}$$

In general, evaluation of the first expectation on the right hand side is quite difficult due to the dependence of the filter coefficient vector on the data sequence, and due to possible correlations present in the data sequence. In particular, the filter coefficient vector $\mathbf{c}(T-1)$ is a nonlinear function of all of the past data values. This function can be explicitly evaluated by iterating difference equation (7.1.1) (problem 7-1). Obtaining bounds on the expected value of this function let alone computing its exact value, requires a great deal of effort. If β is very small, however, $\mathbf{c}(T)$ will not fluctuate greatly about its mean value. Furthermore, the smaller the step size, the more "smoothing" will take place, and the less effect recent data values will have on the estimate $\mathbf{c}(T)$.

To facilitate the analysis it is therefore assumed, as in chapter 3, that $\mathbf{c}(T)$ is approximately independent of the input sequence $\{y(j)\}$, $j = 0, 1, 2, \ldots, T$. This "independence assumption" has been the subject of much concern in recent years since, strictly speaking, the coefficient vector is strongly dependent upon the input data samples even when they are in turn independent from sample to sample. The more rigorous analysis referred to earlier attempts to analyze the convergence properties of the SG transversal algorithm without this independence assumption. By assuming certain regularity conditions on the statistical properties of the input sequence, it is possible to rigorously prove that the mean value of the filter coefficient vector does converge, provided that the algorithm step size is small enough [7]. The analytical results based upon the independence assumption have also been found to agree quite well with computer simulations of the SG transversal algorithm. Equation (7.1.2) is therefore rewritten as

$$E[\mathbf{c}(T)] \approx (1 - \beta\mathbf{\Phi})E[\mathbf{c}(T-1)] + \beta\mathbf{p} , \tag{7.1.3}$$

where $\mathbf{\Phi} \equiv E[\mathbf{y}(T)\mathbf{y}'(T)]$ is the $N \times N$ autocorrelation matrix for the sequence $y(T)$ and the cross-correlation vector $\mathbf{p} \equiv E[d(T)\mathbf{y}(T)]$. Since $\mathbf{\Phi}$ is symmetric and assumed to be positive definite, it admits the factorization, which is equivalent to (3.1.24),

$$\mathbf{\Phi} = \mathbf{V\Lambda V}' , \tag{7.1.4}$$

where $\mathbf{\Lambda}$ is the diagonal matrix of eigenvalues,

$$\mathbf{\Lambda} = \text{diag}\,[\lambda_1 \ \lambda_2 \cdots \lambda_N] , \tag{7.1.5}$$

and $\mathbf{V}$ is the orthonormal matrix whose jth column is the eigenvector $\mathbf{v}_j$ of $\mathbf{\Phi}$ associated with the jth eigenvalue λ_j . The vectors $\mathbf{v}_j$ are the same as those used in section 3.1.3 and satisfy the property (3.1.23), which is equivalent to saying $\mathbf{V}^{-1}=\mathbf{V}'$.

The following transformation of variables, or coordinate rotation, will be useful,

$$\begin{aligned} \mathbf{q}(T) &\equiv \mathbf{c}(T) - \mathbf{c}_{opt} , \\ \tilde{\mathbf{q}}(T) &\equiv \mathbf{V}'\mathbf{q}(T) , \end{aligned} \tag{7.1.6}$$

and

$$\tilde{\mathbf{y}}(T) = \mathbf{V}'\mathbf{y}(T) , \tag{7.1.8}$$

where the MMSE coefficient vector $\mathbf{c}_{opt} = \mathbf{\Phi}^{-1}\mathbf{p}$. The jth component of the transformed vectors is therefore obtained by premultiplying the original vector with the jth eigenvector $\mathbf{v}_j'$. The components of the transformed data vector are uncorrelated. In particular,

$$E[\tilde{\mathbf{y}}(T)\tilde{\mathbf{y}}'(T)] = \mathbf{\Lambda} . \tag{7.1.9}$$

Multiplying (7.1.2) on the left by $\mathbf{V}'$ gives

$$E[\tilde{\mathbf{q}}(T)] \approx [\mathbf{I} - \beta\mathbf{\Lambda}]E[\tilde{\mathbf{q}}(T-1)] . \tag{7.1.10}$$

Since $\mathbf{\Lambda}$ is diagonal, (7.1.10) can be written in component form as

$$E[\tilde{q}_j(T)] \approx (1 - \beta\lambda_j)^T \tilde{q}_j(0) , \tag{7.1.11}$$

where $\tilde{q}_j$ is the jth component of $\tilde{\mathbf{q}}$. The mean coefficient vector error, $E[\mathbf{q}(T)] = \mathbf{V}E[\tilde{\mathbf{q}}(T)]$, therefore converges exponentially to zero according to N normal modes provided that β satisfies the constraint (3.1.26), where $\lambda_{\max}$ is the largest eigenvalue of $\mathbf{\Phi}$. From (7.1.11) the time constant associated with the jth normal mode (time for $E[q_j(T)]$ to decay to $\frac{1}{e}q_j(0)$) is

$$\tau_j \approx \frac{1}{\ln(1 - \beta\lambda_j)} \approx \frac{1}{\beta\lambda_j} , \tag{7.1.12}$$

assuming β is small. The mean value of the tap vector can therefore be approximated as

$$E[\mathbf{c}(T)] \approx \mathbf{c}_{opt} + \mathbf{V}(\mathbf{I} - \beta\mathbf{\Lambda})^T \tilde{\mathbf{q}}(0) \; . \qquad (7.1.13)$$

The second term on the right side of the equation is the error term which decays to zero assuming β is small enough. Recall from chapter 3 the graphical interpretation of (7.1.12) and (7.1.13). In particular, the convergence of the mean coefficient vector is limited by the smallest eigenvalue λ_{min}, which produces the largest time constant τ_{min}. It therefore takes more iterations to make adjustments in $\mathbf{c}$ along the direction of the eigenvector corresponding to λ_{min} than along any other direction in the N-dimensional coefficient vector space. Recall also that due to the relation (3.1.30), which bounds the extreme eigenvalues λ_{min} and λ_{max}, there is a convenient relation between the unevenness of the input signal spectrum and the rate of convergence of $E[\mathbf{c}(T)]$. The more uneven the signal spectrum, the larger the ratio of largest to minimum eigenvalue of $\mathbf{\Phi}$ is likely to be, thereby causing the SG transversal algorithm to converge more slowly according to (7.1.13).

As discussed in section 3.2.2.3, in many cases the step size β is normalized by an estimate of the input signal variance. We will use a slightly different method of performing this normalization here. In particular, choose the step size to be

$$\beta(T) = \frac{1}{\sigma^2(T)} = \frac{1}{(1-\alpha)\sigma^2(T-1) + y^2(T)} \; , \qquad (7.1.14)$$

where α is a constant close to zero, so that as T tends to infinity,

$$E\left[\frac{1}{\beta(T)}\right] \rightarrow \frac{E[y^2(T)]}{\alpha} \; .$$

In this case α controls both the speed of convergence (averaging time) of the variance estimate, and also the scaling factor for the variance estimate. It will now be shown that this normalized step-size results in a much reduced dependence of convergence speed on input signal variance.

In order to approximately ascertain the effect of this normalization on the convergence of the mean coefficient vector, the step size β can be replaced by $E[\beta(T)] \approx 1/E[\sigma^2(T)]$ in equations (7.1.12) and (7.1.13). This is a reasonable approximation as long as the constant α is small enough to "smooth out" the statistical fluctuations in the step size $\beta(T)$, so that it can be regarded as virtually independent of the data samples. If the step size at time T=0 is initialized at the asymptotic value α/ϕ_0, where $\phi_0 = E[y^2(T)]$, then $E[\beta(T)] \approx \alpha/\phi_0$ for all T, and the time constant (7.1.13) becomes

$$\begin{aligned} \tau_j &\approx \frac{\phi_0}{\alpha\lambda_j} \\ &= \frac{\sum_{j=1}^{N} \lambda_j}{N\alpha\lambda_j} \end{aligned} \qquad (7.1.15)$$

where the fact that

$$\sum_{j=1}^{N} \lambda_j = \text{trace}\, \mathbf{\Phi} = N\phi_0$$

has been used (problem 7-2). The normal mode time constants are therefore proportional to the ratio λ_{av}/λ_j , where $\lambda_{av} = \phi_0$ is the "average" eigenvalue of $\mathbf{\Phi}$. A change in the input signal variance therefore causes a much less dramatic change in convergence speed, and furthermore, the stability constraint (3.1.26) is satisfied as long as $0 < \alpha < 2\lambda_{av}/\lambda_{\max}$.

The optimal selection of β, which results in the fastest convergence of the output MSE, is considered later in this section.

7.1.2. Output MSE

Having described the behavior of the mean coefficient vector, the filter's output MSE is now investigated. Assuming for the time being that the coefficient vector $\mathbf{c}(T)$ is known, then the output MSE conditioned on the coefficient vector at time T is

$$\begin{aligned} \epsilon(T) &= E[d^2(T)] - 2\mathbf{c}'(T)\mathbf{p} + \mathbf{c}'(T)\mathbf{\Phi}\mathbf{c}(T) \\ &= E[d^2(T)] - \mathbf{c}_{opt}'\mathbf{\Phi}\mathbf{c}_{opt} + 2\mathbf{c}'(T)[\mathbf{\Phi}\mathbf{c}_{opt}-\mathbf{p}] \\ &\quad + [\mathbf{c}(T)-\mathbf{c}_{opt}]'\mathbf{\Phi}[\mathbf{c}(T)-\mathbf{c}_{opt}] \, . \end{aligned} \tag{7.1.16}$$

Using the fact that $\mathbf{c}_{opt} = \mathbf{\Phi}^{-1}\mathbf{p}$ gives

$$\epsilon(T) = \epsilon_{\min} + \mathbf{q}'(T)\mathbf{\Phi}\mathbf{q}(T) \ , \tag{7.1.17}$$

where

$$\epsilon_{\min} = E[d^2(T)] - \mathbf{p}'\mathbf{\Phi}^{-1}\mathbf{p} \tag{7.1.18}$$

is the minimum MSE attainable with a fixed coefficient filter. If we now attempt to average (7.1.17) over the filter coefficient vector, then using the transformation (7.1.6) and (7.1.7) gives an expression for the excess output MSE due to adaptation,

$$\epsilon_{ex}(T) = E[\tilde{\mathbf{q}}'(T)\mathbf{\Lambda}\tilde{\mathbf{q}}(T)] = \sum_{j=1}^{N} \lambda_j E[\tilde{\mathbf{q}}_j^2(T)] \, . \tag{7.1.19}$$

If the expectation on the right hand side is ignored, then this expression is exactly the same as (3.1.29), which gives the output MSE for the deterministic gradient algorithm. In the case of the SG algorithm, we must ensemble average over the filter coefficients. In order to compute this quantity in terms of second-order input statistics, the $N \times N$ error covariance matrix for the transformed filter coefficient error vector is defined,

$$\mathbf{Q}(T) \equiv E[\tilde{\mathbf{q}}(T)\tilde{\mathbf{q}}'(T)] \, . \tag{7.1.20}$$

Using this definition (7.1.19) can be rewritten as

$$\epsilon_{ex}(T) = \text{trace}\{\mathbf{\Lambda}\ \mathbf{Q}(T)\} \, . \tag{7.1.21}$$

In order to study the evolution of $\mathbf{Q}(T)$, the MMSE coefficient vector $\mathbf{c}_{opt}$ is subtracted from both sides of (7.1.1) and the coordinate transformation (7.1.6-8) is used to get (problem 7-3)

$$\begin{aligned}\tilde{\mathbf{q}}(T) &= \tilde{\mathbf{q}}(T-1) + \beta e(T)\tilde{\mathbf{y}}(T) \\ &= \left[\mathbf{I} - \beta\left(\tilde{\mathbf{y}}(T)\tilde{\mathbf{y}}'(T)\right)\right]\tilde{\mathbf{q}}(T-1) + \beta e_{opt}(T)\tilde{\mathbf{y}}(T)\ ,\end{aligned} \tag{7.1.22}$$

where the sequence of estimation errors that results from using the MMSE coefficient vector is denoted as

$$e_{opt}(T) \equiv d(T) - \mathbf{c}_{opt}'\mathbf{y}(T)\ . \tag{7.1.23}$$

It is now straightforward to derive the following equation (problem 7-3), which describes the evolution of the the coefficient error matrix $[\mathbf{c}(T) - \mathbf{c}_{opt}][\mathbf{c}(T) - \mathbf{c}_{opt}]'$,

$$\begin{aligned}\tilde{\mathbf{q}}(T)\tilde{\mathbf{q}}'(T) &= \left[\mathbf{I} - \beta\left(\tilde{\mathbf{y}}(T)\tilde{\mathbf{y}}'(T)\right)\right]\tilde{\mathbf{q}}(T-1)\tilde{\mathbf{q}}'(T-1)\left[\mathbf{I} - \beta\left(\tilde{\mathbf{y}}(T)\tilde{\mathbf{y}}'(T)\right)\right] \\ &+ 2\beta\left[\mathbf{I} - \beta\left(\tilde{\mathbf{y}}(T)\tilde{\mathbf{y}}'(T)\right)\right]\left(\tilde{\mathbf{q}}(T-1)\tilde{\mathbf{y}}'(T)\right)e_{opt}(T) + \beta^2 e_{opt}^2(T)\left(\tilde{\mathbf{y}}(T)\tilde{\mathbf{y}}'(T)\right).\end{aligned} \tag{7.1.24}$$

In order to compute output MSE, this equation must be averaged to obtain a difference equation for the coefficient error covariance matrix $\mathbf{Q}$. Several approximations will made to obtain this average. Using the independence assumption, the expected value of the first term on the right hand side is

$$\begin{aligned}&E\left\{\left(\mathbf{I} - \beta\left(\tilde{\mathbf{y}}(T)\tilde{\mathbf{y}}'(T)\right)\right)\tilde{\mathbf{q}}(T-1)\tilde{\mathbf{q}}'(T-1)\left(\mathbf{I} - \beta\left(\tilde{\mathbf{y}}(T)\tilde{\mathbf{y}}'(T)\right)\right)\right\} \approx \\ &\mathbf{Q}(T-1)(\mathbf{I} - 2\beta\boldsymbol{\Lambda}) + \beta^2 E\left[\left(\tilde{\mathbf{y}}(T)\tilde{\mathbf{y}}'(T)\right)\mathbf{Q}(T-1)\left(\tilde{\mathbf{y}}(T)\tilde{\mathbf{y}}'(T)\right)\right].\end{aligned} \tag{7.1.25}$$

Designating $\tilde{y}_j$ as the jth component of $\tilde{\mathbf{y}}(T)$, and $\tilde{q}_j$ as the jth component of $\tilde{\mathbf{q}}(T-1)$, the l,mth element of the last term of (7.1.25) is approximated as

$$\left\{E\left[\left(\tilde{\mathbf{y}}(T)\tilde{\mathbf{y}}'(T)\right)\mathbf{Q}(T-1)\left(\tilde{\mathbf{y}}(T)\tilde{\mathbf{y}}'(T)\right)\right]\right\}_{l,m}$$

$$\begin{aligned}
&= E\left([\tilde{y}_l\tilde{\mathbf{y}}'(T)]\ \mathbf{Q}\ [\tilde{y}_m\tilde{\mathbf{y}}(T)]\right) \\
&= \sum_{k=1}^{N}\sum_{j=1}^{N} E[\tilde{q}_k\tilde{q}_j]E[\tilde{y}_l\tilde{y}_k\tilde{y}_j\tilde{y}_m] \\
&\approx \sum_{j=1}^{N} E[\tilde{q}_j^2]\ E[\tilde{y}_l^2]\ E[\tilde{y}_j^2]\ \delta_{lm} \\
&= \sum_{j=1}^{N} E[\tilde{q}_j^2]\lambda_l\lambda_j\delta_{lm} \\
&= \delta_{lm}\lambda_l \text{ trace } [\boldsymbol{\Lambda}\mathbf{Q}(T-1)]\ ,
\end{aligned} \tag{7.1.26}$$

where δ_{lm} is the Kronecker delta. The fourth-order statistics in (7.1.25) have been approximated by second-order statistics. The resulting matrix specified by (7.1.26) is diagonal. The off-diagonal elements of this matrix have been approximated by zero since correlations between pairs of components of the transformed coefficient error vector, $\tilde{q}_k$ and $\tilde{q}_j$, are assumed to be weak, and similarly, the contributions to the off-diagonal elements from the fourth-order statistics of the uncorrelated random variables $\tilde{y}_j$, $j = 1, \ldots, N$ are assumed to be small. However, if the statistics of the input sequence $y(T)$ are known to be Gaussian, then the fourth-order statistics in (7.1.25) can be evaluated *exactly* in terms of second-order statistics, yielding a somewhat different answer than that given in (7.1.26). The resulting expression for output MSE in this case is left as a problem (problem 7-6).

Examining the second term of the right hand side of (7.1.24) and again using the independence assumption gives

$$\begin{aligned}
&E\left\{2\beta\left(\mathbf{I} - \beta\left(\tilde{\mathbf{y}}(T)\tilde{\mathbf{y}}'(T)\right)\right)\left[\tilde{\mathbf{q}}(T-1)\tilde{\mathbf{y}}'(T)\right]e_{opt}(T)\right\} \\
&\approx 2\beta\ E[\tilde{\mathbf{q}}(T-1)]\left\{E\left[e_{opt}(T)\tilde{\mathbf{y}}'(T)\right] - \beta\ E\left[\left(\tilde{\mathbf{y}}'(T)\tilde{\mathbf{y}}(T)\right)e_{opt}(T)\tilde{\mathbf{y}}'(T)\right]\right\}.
\end{aligned} \tag{7.1.27}$$

Because $E[\mathbf{c}(T)]$ is converging to $\mathbf{c}_{opt}$, it follows from (7.1.6) and (7.1.7) that $E[\tilde{\mathbf{q}}(T-1)]$ is converging to zero, and hence must be small for moderately large T. In addition, from the principle of orthogonality,

$$E\left[e_{opt}(T)\tilde{\mathbf{y}}(T)\right] = 0\ , \tag{7.1.28}$$

so that the term given by (7.1.27) should be small relative to (7.1.26), and hence we approximate (7.1.27) by zero.

Finally, examining the last term of (7.1.24) gives

$$E\left[\beta^2 e_{opt}^2(T)\tilde{\mathbf{y}}(T)\tilde{\mathbf{y}}'(T)\right] \approx \beta^2\epsilon_{\min}\mathbf{\Lambda} \ . \qquad \mathbf{(7.1.29)}$$

This is again an approximation because e_{opt} is a linear function of the input sequence, and hence exact evaluation of this term requires knowledge of the fourth-order statistics of the input sequence. Combining (7.1.24-29) gives

$$\mathbf{Q}(T) \approx \mathbf{Q}(T-1)(\mathbf{I}-2\beta\mathbf{\Lambda}) + \beta^2\mathbf{\Lambda}\ \text{trace}\ [\mathbf{\Lambda}\mathbf{Q}(T-1)] + \beta^2\epsilon_{\min}\mathbf{\Lambda} \ . \qquad (7.1.30)$$

Since the off-diagonal elements of (7.1.30) are zero, the vector $\mathbf{s}(T)$ is defined whose jth component is $E[\tilde{q}_j^2(T)]$. From (7.1.30) it follows that

$$\mathbf{s}(T) \approx \mathbf{A}\ \mathbf{s}(T-1) + \beta^2\epsilon_{\min}\boldsymbol{\lambda} \ , \qquad (7.1.31)$$

where $\boldsymbol{\lambda}$ is the N-dimensional vector of eigenvalues, and the $N \times N$ matrix $\mathbf{A}$ has elements

$$A_{lm} = (1 - 2\beta\lambda_l)\delta_{lm} + \beta^2\lambda_l\lambda_m \ , \qquad (7.1.32)$$

and can be written as

$$\mathbf{A} = \beta^2\boldsymbol{\lambda}\boldsymbol{\lambda}' + (\mathbf{I} - 2\beta\mathbf{\Lambda}) \ . \qquad (7.1.33)$$

Referring to (7.1.19), which relates the excess MSE to the elements of the vector $\mathbf{s}(T)$, it follows from (7.1.31) that the dynamics of the output MSE are governed by the eigenvalues of the matrix $\mathbf{A}$, which are denoted as μ_j, $1 \leqslant j \leqslant N$. Since $\mathbf{A}$ is symmetric, and assuming β is chosen small enough so that the elements of $\mathbf{A}$ are positive, then $\mathbf{A}$ can be factored in the same way that the autocorrelation matrix $\mathbf{\Phi}$ was factored in (7.1.4), that is,

$$\mathbf{A} = \mathbf{U}\ \mathbf{M}\ \mathbf{U}' \ , \qquad (7.1.34)$$

where $\mathbf{M}$ is the diagonal matrix of eigenvalues of $\mathbf{A}$, and the jth column of $\mathbf{U}$ is the orthonormal eigenvector associated with μ_j. The transformed vector $\tilde{\mathbf{s}}(T) \equiv \mathbf{U}'\mathbf{s}(T)$ can then be substituted for $\mathbf{s}(T)$ in (7.1.31), yielding the difference equation, which has no cross-coupling between components,

$$\tilde{\mathbf{s}}(T) = \mathbf{U}\ \tilde{\mathbf{s}}(T-1) + \beta^2\epsilon_{\min}\mathbf{U}'\boldsymbol{\lambda} \ . \qquad (7.1.35)$$

The resulting decoupled equation has the same interpretations as were given to the transformed difference equation for the mean coefficient vector (7.1.10). At each successive iteration, adjustments to the components of $\mathbf{s}$ are made in the directions of the eigenvectors of $\mathbf{A}$, denoted as $\mathbf{u}_j$, $j = 1, \ldots, N$, and the amount of adjustment in each direction is determined by the associated eigenvalue μ_j. The output MSE of the filter, as described by (7.1.19) and (7.1.31), decays according to a weighted sum of the components of $\mathbf{s}$.

Because $\mathbf{A}$ has real elements and is symmetric, it has real eigenvalues. The solution to (7.1.35), and therefore (7.1.31), will remain bounded for all T if and only if $|\mu_j| < 1$ for $1 \leqslant j \leqslant N$. Because the eigenvalues μ_j depend on the step size β, a stability requirement for β, which ensures the convergence of the coefficient variances and hence from (7.1.19) the output MSE, can be derived by finding values of β which ensure that all eigenvalues of $\mathbf{A}$ have magnitudes less than unity. Summing the lth row of $\mathbf{A}$ gives

$$\sum_{m=1}^{N} \left[(1 - 2\beta\lambda_l)\lambda_{lm} + \beta^2 \lambda_l \lambda_m \right] = 1 - \beta\lambda_l \left(2 - \beta \sum_{m=1}^{N} \lambda_m \right), \quad (7.1.36)$$

which is less than unity if

$$0 < \beta < \frac{2}{\sum_{m=1}^{N} \lambda_m} = \frac{2}{N\,\phi_0} . \quad (7.1.37)$$

This condition ensures that all elements of the matrix $\mathbf{A}$, as specified by (7.1.33), are positive. A symmetric matrix that has all positive elements, and whose row sums are strictly less than unity must have eigenvalues with absolute values less than unity (problem 7-4). The condition (7.1.37) is therefore a stability requirement for the step size β. This stability requirement is significantly more severe than the requirement (3.1.26), which is necessary for the convergence of the mean coefficient vector. This is intuitively reasonable since the requirement (3.1.26) does not account for the variance of the coefficients, which contributes to the output MSE. The constraints (3.1.26) and (7.1.37) in fact imply that there is a range of step sizes for which the mean coefficient values converge, but the output MSE diverges. The maximum value of β allowed by (7.1.37) decreases as the number of coefficients N increases.

Suppose now that the step size is normalized by an estimate of the input signal variance. As before, if we assume that $E[\beta(T)] \approx \alpha/\phi_0$, then the stability constraint (7.1.37) becomes

$$0 < \alpha < \alpha_{\max} = \frac{2}{N} . \quad (7.1.38)$$

This constraint is quite useful, since it depends only on the number of filter coefficients. Insofar as a reasonable estimate of the variance of the input is obtained, we can precompute $\alpha_{\max}$ and by selecting $\alpha < \alpha_{\max}$ be assured that the adaptive algorithm will converge.

Theoretically, it is possible to obtain from (7.1.31) and (7.1.19), a step size β_{opt} that causes the fastest convergence of the output MSE. In general, this step size depends in a complicated way on the eigenvalues λ_j, $j = 1, \ldots, N$, and therefore cannot be easily computed from the input sequence. Later on, however, a specific example will be considered where this optimal step-size can be easily computed.

7.1.3. Asymptotic Coefficient Variance and Output MSE

Having characterized the output MSE as a function of time, we now compute the asymptotic variances of the coefficients and the asymptotic output MSE. Substituting (7.1.33) into (7.1.31) and using (7.1.19) gives

$$\mathbf{s}(T) \approx (\mathbf{I} - 2\beta\mathbf{\Lambda})\mathbf{s}(T-1) + \beta^2[\epsilon_{ex}(T) + \epsilon_{\min}]\boldsymbol{\lambda} . \quad (7.1.39)$$

If β satisfies the constraint (7.1.37), then $\lim_{T\to\infty} \mathbf{s}(T) = \lim_{T\to\infty} \mathbf{s}(T-1)$, and taking limits of both sides of (7.1.39) as T approaches infinity gives

$$\lim_{T\to\infty} 2\beta\Lambda\mathbf{s}(T) \approx \beta^2\left[\lim_{T\to\infty} \epsilon_{ex}(T) + \epsilon_{\min}\right]\boldsymbol{\lambda}\ , \qquad \textbf{(7.1.40)}$$

or

$$\lim_{T\to\infty} \mathbf{s}(T) = \frac{\beta}{2}\,\mathbf{1}\left[\lim_{T\to\infty} \epsilon_{ex}(T) + \epsilon_{\min}\right], \qquad (7.1.41)$$

where **1** is the vector whose components are unity. Premultiplying (7.1.41) by $\boldsymbol{\lambda}'$ gives

$$\lim_{T\to\infty} \epsilon_{ex}(T) = \frac{\beta}{2}\left(\sum_{j=1}^{N} \lambda_j\right)\left[\lim_{T\to\infty} \epsilon_{ex}(T) + \epsilon_{\min}\right]. \qquad (7.1.42)$$

Substituting $N\phi_0$ for $\sum_{j=1}^{N}\lambda_j$, and solving for $\epsilon_{ex}(\infty)$ gives an expression for the asymptotic excess MSE due to adaptation,

$$\lim_{T\to\infty} \epsilon_{ex}(T) = \frac{\beta\epsilon_{\min}N\phi_0}{2-\beta\,N\phi_0}\ . \qquad (7.1.43)$$

Substituting (7.1.43) into (7.1.41) gives an expression for the asymptotic (transformed) coefficient variance,

$$\lim_{T\to\infty} E[\tilde{q}_j^{\,2}(T)] = \frac{\beta\epsilon_{\min}}{2-\beta\,N\phi_0}\ . \qquad (7.1.44)$$

If the normalized step size (7.1.14) is used, then substituting $E[\beta(T)] \approx \alpha/\phi_0$ for β in (7.1.43) indicates that the output MSE is dependent only upon α and the number of filter coefficients.

Equations (7.1.12) and (7.1.44) indicate that as β increases, the convergence time of the filter decreases; however, the variance of the coefficients and hence output MSE increases. In the case of a stationary input, it is therefore desirable to keep β large initially for fast convergence and subsequently decrease its value to minimize the output MSE. A time varying optimal step size sequence $\beta(T)$ which approximately minimizes output MSE at each iteration has been derived [6]. However, suboptimal strategies are also available. In particular, a relatively large constant step size can be used initially to rapidly decrease the output MSE, and subsequently, a smaller constant step size can be used to reduce asymptotic coefficient variance and output MSE, and simultaneously track slowly varying changes in the input statistics. This *gear shifting* technique is most popular in applications such as channel equalization, where a rapid reduction in output MSE is desired so that data transmission can begin, and the channel impulse response and hence the input statistics to the equalizer change very slowly.

7.1.4. Example: White Input Signal

To illustrate the previous results, suppose that the input process $\{y(T)\}$ used to adapt an SG transversal joint process estimator is white (successive samples are uncorrelated). This is precisely the situation encountered in echo cancellation of data signals. Recall from chapter 2 that in this case the input to the canceler,

$y(T)$, is the uncorrelated data sequence. The desired signal $d(T)$ is the output of the channel, and the objective of the adaptive filter is to synthesize a replica of the channel impulse response. The autocorrelation matrix in this case is

$$\mathbf{\Phi} = \phi_0 \mathbf{I} , \tag{7.1.45}$$

where $\phi_0 = E[y^2(T)]$ and $\mathbf{I}$ is the identity matrix. All eigenvalues of $\mathbf{\Phi}$ are therefore equal to ϕ_0 and the eigenvectors are the columns of the identity matrix. The transformed variables $\tilde{\mathbf{q}}$ and $\tilde{\mathbf{y}}$ from (7.1.7) and (7.1.8) are therefore the same as $\mathbf{q}$ and $\mathbf{y}$, respectively. Assuming an unnormalized step size, the mean value of each filter coefficient convergences exponentially to zero with time constant

$$\tau_j \approx \frac{1}{\beta\phi_0} . \tag{7.1.46}$$

Using the definition (7.1.33), the matrix $\mathbf{A}$ in this case is

$$\mathbf{A} = \beta^2\phi_0^2 \mathbf{1}\mathbf{1}' + (1 - 2\beta\phi_0)\mathbf{I} . \tag{7.1.47}$$

Substituting into (7.1.31) gives,

$$\mathbf{s}(T) = \beta^2\phi_0^2 \mathbf{1}\ [\mathbf{1}'\mathbf{s}(T-1)] + (1 - 2\beta\phi_0)\mathbf{s}(T-1) + \beta^2\epsilon_{\min}\boldsymbol{\lambda} , \tag{7.1.48}$$

where the jth component of $\mathbf{s}$ is in this case the variance of the coefficient c_j. Premultiplying both sides of (7.1.48) by $\boldsymbol{\lambda}' = \phi_0\mathbf{1}'$ gives

$$\begin{aligned} \epsilon_{ex}(T) &= \beta^2\phi_0^2 N\epsilon_{ex}(T-1) + (1 - 2\beta\phi_0)\epsilon_{ex}(T-1) + \beta^2\epsilon_{\min}\phi_0^2 N \\ &= \left(\beta^2\phi_0^2 N - 2\beta\phi_0 + 1\right)\epsilon_{ex}(T-1) + \beta^2\phi_0^2 N\epsilon_{\min} . \end{aligned} \tag{7.1.49}$$

Iterating (7.1.49) gives an expression for the excess output MSE as a function of time,

$$\epsilon_{ex}(T) = \left(1 - 2\beta\phi_0 + \beta^2\phi_0^2 N\right)^T \left[\epsilon_{ex}(0) - \frac{\beta N\phi_0\epsilon_{\min}}{2 - \beta N\phi_0}\right] + \frac{\beta N\phi_0\epsilon_{\min}}{2 - \beta N\phi_0} . \tag{7.1.50}$$

The first term is a transient term, which tends to zero, and the second term is the asymptotic excess MSE due to adaptation, which agrees with (7.1.37)

The optimal (constant) step size that results in the fastest possible convergence to the asymptotic MSE can be computed by minimizing the factor that multiplies $\epsilon_{ex}(T-1)$ in (7.1.49). This occurs when $\beta = 1/N\phi_0$. (Notice from (7.1.43) that for stability β must be less than $2/N\phi_0$.) Substituting the optimal step size into (7.1.50) gives an expression for excess output MSE,

$$\epsilon_{ex}(T) = \left(1 - \frac{1}{N}\right)^T [\epsilon_{ex}(0) - \epsilon_{\min}] + \epsilon_{\min} . \tag{7.1.51}$$

The output MSE therefore exhibits exponential convergence to its asymptotic value, which is twice the minimum value $\epsilon_{\min}$. As the number of coefficients N increases, the speed of convergence decreases. As mentioned previously, in practice the step size is often decreased after convergence in order to further decrease the asymptotic excess MSE.

For correlated inputs, the computation of output MSE produced by the SG transversal algorithm is considerably more difficult than for the example just considered, since the computation of eigenvalues and eigenvectors of the autocorrelation matrix requires more effort. Nevertheless, once these quantities are known, computation of the approximate mean coefficient trajectories and output MSE becomes relatively straightforward. In general, the output MSE consists of the sum of exponentially decaying terms corresponding to each normal mode. This situation is to be contrasted with the lattice algorithms to be analyzed in the next few sections, for which no closed form expressions currently exist for either the mean coefficient trajectories or output MSE. A comparison of the formulas given in this section with output MSE obtained from computer simulation of the SG transversal algorithm is given in reference [1] (see also section 7.8).

7.2. SG LATTICE SINGLE-STAGE CONVERGENCE

In the last section the convergence properties of the SG transversal algorithm were investigated resulting in relatively simple analytical expressions describing the convergence of the output MSE. We now attempt to derive analogous results for both SG and LS adaptive lattice algorithms. As mentioned in chapter 4, in most applications lattice algorithms (both SG and LS) converge substantially faster than the SG transversal algorithm. Lattice algorithms are considerably more difficult to analyze than the SG transversal algorithm, however, due to the highly nonlinear nature of the adaptation.

In this section the convergence properties of a single stage of an SG multi-stage lattice predictor are investigated assuming that both inputs to that stage are stationary. This is equivalent to assuming that the previous stages have fixed coefficients, as would approximately be true if they had already adapted. In this way, by making approximations analogous to those that were made in the last section, explicit expressions for the mean value of the nth coefficient and the $(n+1)$st order output MSE can be obtained as functions of the second-order residual statistics at the input to the nth stage of the predictor. The effects of one adapting stage on successive adapting stages are considered in section 7.4. In this section only SG lattice algorithms are considered. Analogous results for the prewindowed LS lattice algorithm are derived in the next section. As in the last section, a number of simplifying assumptions are made in order to obtain simple results that give insight into the convergence process.

7.2.1. Stochastic Gradient Lattice Algorithms

The lattice order recursions (4.2.9) and (4.2.10) are repeated here for convenience,

$$e_f(T|n) = e_f(T|n-1) - k_n^b(T)e_b(T-1|n-1) \tag{7.2.1a}$$

$$e_b(T|n) = e_b(T-1|n-1) - k_n^f(T)e_f(T|n-1) \ , \tag{7.2.1b}$$

where the notation is identical to that used in chapter 4. The value of $k_n^b(T)$ which minimizes the mean squared forward prediction error $E[e_f^2(T|n)]$ is

$$k_{n,opt}(T) = \frac{E[e_f(T|n-1)e_b(T-1|n-1)]}{E[e_b^2(T-1|n-1)]} . \tag{7.2.2}$$

Observe that dependence of $k_{n,opt}$ upon the time interval T is indicated. This is due to the adaptation of $k_1^b(T)$, $k_1^f(T), \ldots, k_{n-1}^b(T)$, $k_{n-1}^f(T)$, which cause $e_f(T|n-1)$ and $e_b(T|n-1)$ to be nonstationary. If the coefficients are constant, then

$$E[e_f^2(T|n)] = E[e_b^2(T|n)], \quad 1 \leqslant n \leqslant N , \tag{7.2.3}$$

where N is the order of the filter, so that the optimal value of k_n^f is the same as $k_{n,opt}$ given by (7.2.2). Throughout this section it is therefore assumed that $k_n(T) = k_n^b(T) = k_n^f(T)$.

Two SG lattice algorithms will be considered. The first algorithm is obtained by adapting $k_n(T)$ to minimize $E[e_f^2(T|n)$, and is given by (problem 4-32)

$$k_n(T+1) = [1-\beta_u e_b^2(T-1|n-1)]k_n(T)+\beta_u e_f(T|n-1)e_b(T-1|n-1), \tag{7.2.4}$$

where β_u is the adaptation step size. The subscript u stands for "unnormalized". This is to distinguish it from the step size β_n used in the second "normalized" algorithm. The second algorithm considered attempts to recursively estimate the expectations $E[e_f(T|n-1)e_b(T-1|n-1)]$ and $E[e_f(T|n-1)]$ by time averages. In this way, using (7.2.3), an estimate for $k_{n,opt}$ is obtained, i.e.,

$$K_n(T+1) = (1-\beta_n)K_n(T) + e_f(T|n-1)e_b(T-1|n-1) \tag{7.2.5a}$$

$$F(T+1|n) = (1-\beta_n)F(T|n) + e_f^2(T|n-1) \tag{7.2.5b}$$

and

$$k_n(T) = \frac{K_n(T)}{F(T|n)} . \tag{7.2.5c}$$

It is easily verified that (7.2.5) can be rewritten as

$$k_n(T+1) = \left[1 - \frac{e_b^2(T|n-1)}{F(T|n)}\right]k_n(T) + \frac{e_f(T|n-1)e_b(T-1|n-1)}{F(T|n)} . \tag{7.2.6}$$

The algorithm (7.2.5) is therefore equivalent to (7.2.4) where β_u is replaced by the normalized step size $1/F(T|n)$. Also notice that the updates (7.2.5) are basically the same as the prewindowed LS updates (6.3.49), (6.3.50), and (6.3.64b) where the gain $\gamma(T-1|n)$ (i.e., $\sin^2\theta(T|1,n)$) is set to zero. The "unnormalized" and "normalized" SG algorithms given respectively by (7.2.4) and (7.2.6) are the only SG adaptive lattice algorithms considered here. The other SG algorithms discussed in chapter 4 can be analyzed in exactly the same way as the algorithms considered here.

The objective of the SG lattice algorithm is to adapt the PARCOR coefficients to the set of values which minimizes the short term mean squared output. In the case of a stationary input, it is therefore desirable that the coefficients be driven to their fixed optimal values as rapidly as possible. In analogy with the SG transversal algorithm, because a noisy estimate of the error gradient is used to adapt the coefficients, each PARCOR coefficient has some nonzero variance even after convergence which increases the output MSE over the minimum MSE attainable with a fixed coefficient filter.

From (7.2.4) if it is assumed that β_u is small enough so that the mean value of $k_n(T)$ converges, then

$$E_\infty[k_n(T)e_b^2(T-1|n-1)] = E_\infty[e_f(T|n-1)e_b(T-1|n-1)] , \tag{7.2.7}$$

where $E_\infty[x(T)] \equiv \lim_{T\to\infty} E[x(T)]$ denotes the asymptotic mean value of the sequence $x(T)$, assuming that it exists. If $k_n(T)$ and $e_b(T-1|n-1)$ are approximately independent, it follows that $E_\infty[k_n(T)] \approx \lim_{T\to\infty} k_{n,opt}(T)$. (Note that $k_{n,opt}$ is time varying due to the adaptation of previous stages, and hence the reference to the *asymptotic* value of $k_{n,opt}$.) This "independence assumption" is analogous to the independence assumption made in the last section for the SG transversal algorithm. Referring to (7.2.4), $k_n(T)$ and $e_b(T-1|n-1)$ are independent only if the backward error sequence $e_b(T|n-1)$ is independent from sample to sample. Nevertheless, the justification for using this assumption as an approximation is the same as for the SG transversal algorithm. Specifically, the step size β_u is assumed to be small enough so that $k_n(T)$ varies relatively slowly and is consequently insensitive to the statistical variations of the input sequences $e_f(T|n-1)$ and $e_b(T|n-1)$. Similarly, assuming that β_n is sufficiently small, from (7.2.6) it follows that

$$E_\infty\left[\frac{k_n(T)e_f^2(T|n-1)}{F(T|n)}\right] = E_\infty\left[\frac{e_f(T|n-1)e_b(T-1|n-1)}{F(T|n)}\right] . \tag{7.2.8}$$

As with the unnormalized algorithm, $k_n(T)$ and $e_f(T|n-1)$ are usually dependent, so that evaluation of the expected values in (7.2.8) is extremely difficult in general. Also, if X and Y are two random variables, then $E\left[\frac{X}{Y}\right] \neq \frac{E[X]}{E[Y]}$ (in general), and hence the estimate of $k_{n,opt}$ obtained using either (7.2.4) and (7.2.6) can be biased; however, this bias is small in most applications and decreases with the adaptive step size.

7.2.2. Single-Stage Convergence Time

We first attempt to evaluate the mean value of the nth PARCOR coefficient as a function of time using first (7.2.4) and then (7.2.5). Iterating both sides of the time update (7.2.4) and taking expected values of both sides gives

$$E[k_n(T+1)] = k_n(0)E\left\{\prod_{j=1}^{T}[1 - \beta_u e_b^2(j-1|n-1)]\right\} \tag{7.2.9}$$

$$+ \beta_u E\left\{\sum_{j=1}^{T-1} e_f(j|n-1)e_b(j-1|n-1) \prod_{l=j+1}^{T} [1 - \beta_u e_b^2(l-1|n-1)]\right\}$$

$$+ \beta_u E[e_f(T|n-1)e_b(T-1|n-1)] .$$

Evaluation of the right hand terms is again quite difficult due to correlations present in both sequences $e_f(T|n-1)$ and $e_b(T|n-1)$. To greatly simplify the discussion, we therefore again invoke the independence assumption. In particular, it is assumed that $k_n(T+1)$ is approximately independent of $e_f(j|n-1)$ and $e_b(j|n-1)$ for $j = 0, 1, \ldots, T$. Intuitively, this assumption should improve as the prediction order n increases since each stage of the lattice attempts to whiten its two input signals. Taking expected values of both sides of (7.2.4) and iterating therefore gives

$$E[k_n(T+1)] - k_{n,opt} \approx \left\{1 - \beta_u E[e_b^2(T-1|n-1)]\right\}^T [k_n(0) - k_{n,opt}] . \tag{7.2.10}$$

The mean value of $k_n(T)$ therefore decays exponentially towards $k_{n,opt}$ with time constant

$$\tau_n \approx -\frac{1}{\ln\{1 - \beta_u E[e_b^2(T|n-1)]\}} \approx \frac{1}{\beta_u E[e_b^2(T|n-1)]} . \tag{7.2.11}$$

The first stage time constant is therefore $\tau_1 \approx 1/\beta_u \phi_0$ where ϕ_m, $m = 0, 1, 2, \cdots$ is the autocorrelation sequence of the input signal. If the input is uncorrelated, then $E[e_b^2(T|n)] \approx \phi_0$, $n = 0, 1, \ldots, N$, and the trajectories of the mean values of the PARCOR coefficients are the same as the trajectories of the mean values of the transversal coefficients using the SG algorithm (7.1.1). A disadvantage of this unnormalized algorithm (7.2.4) that is approximately eliminated by the normalized algorithm (7.2.6) is the dependence of adaptation speed upon the input signal variance.

In order to estimate an equivalent time constant for the normalized algorithm, the fixed step size β_u in (7.2.4) must be replaced by the normalized step size $\frac{1}{F(T|n)}$. If it is assumed that the input signal variance is known and that the estimate $F(T|n)$ is initialized at the value $\frac{1}{\beta_n} E[e_f^2(T|n-1)]$, then from (7.2.5b), $E[F(T|n)] = \frac{1}{\beta_n} E[e_f^2(T|n-1)]$ for all T. Furthermore, if β_n is very small so that $F(T|n)$ does not fluctuate greatly about its mean value, then the mean coefficient time constant becomes

$$\begin{aligned} \tau_n &\approx -\frac{1}{\ln\left\{1 - E\left[\frac{e_f^2(T|n-1)}{F(T|n)}\right]\right\}} \approx -\frac{1}{\ln\left\{1 - \frac{E[e_f^2(T|n-1)]}{E[F(T|n)]}\right\}} \\ &\approx -\frac{1}{\ln(1-\beta_n)} = \frac{1}{\beta_n} , \end{aligned} \tag{7.2.12}$$

implying that in this case the single stage time constant is dependent only upon β_n. In practice the input signal variance is generally not known, so that $E[F(T|n)]$ varies with T. In order to approximately determine the effect of this time-varying variance estimate on the mean coefficient trajectory, (7.2.5a) and (7.2.5b) can be iterated giving the expression,

$$k_n(T+1) = \frac{(1-\beta_n)^T K_n(0) + \sum_{j=1}^{T}(1-\beta_n)^{T-j} e_f(j|n-1)e_b(j-1|n-1)}{(1-\beta_n)^T F(0|n) + \sum_{j=1}^{T}(1-\beta_n)^{T-j} e_f^2(j|n-1)} \quad (7.2.13)$$

Multiplying through by the denominator, taking expected values of both sides, and invoking the independence assumption gives an expression for the mean coefficient trajectory,

$$E[k_n(T)] \approx \frac{\beta_n(1-\beta_n)^T K_n(0)+[1-(1-\beta_n)^T]E[e_f(T|n-1)e_b(T-1|n-1)]}{\beta_n(1-\beta_n)^T F(0|n)+[1-(1-\beta_n)^T]E[e_f^2(T|n-1)]}. \quad (7.2.14)$$

It is easily verified that as $T\to\infty$, $E[k_n(T)]$ converges (approximately) to $k_{n,opt}$. The trajectory of $E[k_n(T)]$ is not exponential; however, defining τ_n as the time it takes $E[k_n(T)]$ to reach the value $k_{n,opt} + [k_n(0) - k_{n,opt}]g$ where $0<g<1$, it follows that

$$\beta_n(1-\beta_n)^{\tau_n} K_n(0) + [1 - (1-\beta_n)^{\tau_n}]E[e_f(T|n-1)e_b(T-1|n-1)] \approx$$

$$[(1-g)k_{n,opt}+gk_n(0)]\left\{\beta_n(1-\beta_n)^{\tau_n} F(0|n) + [1 - (1-\beta_n)^{\tau_n}]E[e_f^2(T|n-1)]\right\}.$$

Dividing both sides by $E[e_f^2(T|n-1)]$ and using (7.2.2) and (7.2.5c) gives

$$(1-\beta_n)^{\tau_n} \approx \quad (7.2.15)$$

$$\frac{[(1-g)k_{n,opt} + gk_n(0)] - k_{n,opt}}{\beta_n \frac{K_n(0)}{E[e_f^2(T|n-1)]} - k_{n,opt} - [(1-g)k_{n,opt} + gk_n(0)]\left\{\beta_n \frac{F(0|n)}{E[e_f^2(T|n-1)]} - 1\right\}}$$

$$= \frac{g[k_n(0)-k_{n,opt}]}{(1-g)\beta_n \frac{F(0|n)}{E[e_f^2(T|n-1)]}[k_n(0)-k_{n,opt}] + g[k_n(0)-k_{n,opt}]}$$

$$= \frac{1}{\frac{1-g}{g}\beta_n \frac{F(0|n)}{E[e_f^2(T|n-1)]} + 1},$$

so that

$$\tau_n \approx \frac{1}{\beta_n}\ln\left[\beta_n\left(\frac{1-g}{g}\right)\frac{F(0|n)}{E[e_f^2(T|n-1)]} + 1\right].$$

In this case the "time constant" τ_n depends upon the *normalized* input signal variance

$$\frac{E[e_f^2(T|n-1)]}{F(0|n)}.$$

If

$$F(0|n) = \frac{1}{\beta_n} E[e_f^2(T|n-1)\ ,$$

then this time constant becomes the same as the time constant given by (7.2.12). Equation (7.2.15) indicates that as $F(0|n)$ decreases, so does τ_n. Recall, however, that τ_n describes the convergence of the *mean value* of the nth stage coefficient, assuming both inputs are *stationary*. This assumption is certainly not true while the first $n-1$ stages are adapting. Also, although the mean value of $k_n(T)$ may converge very fast, the variance of $k_n(T)$ and hence the output MSE may take much longer to converge. Because $F(T|n)$ appears in the denominator of (7.2.6), $F(0|n)$ cannot be set arbitrarily close to zero. Empirical results indicate that values of $F(0|n)$ below a given threshold, which depends upon the input statistics and number of stages, are likely to cause the algorithm to diverge.

7.2.3. Single-Stage Asymptotic Output MSE

We now attempt to characterize the asymptotic output MSE of the nth lattice stage, assuming that the $(n-1)$st order residuals are stationary. The result will be an expression for the nth stage MSE in terms of the $(n-1)$st stage MSE. This gives us an iterative way in which to compute the asymptotic output MSE at all stages of the filter, assuming that the signals at each stage of the lattice are stationary. This is equivalent to assuming that all of the PARCOR coefficients have converged in an average sense. In fact it will be shown that the asymptotic output MSE can be approximated, provided that the *variances* of all of the lattice coefficients have converged to their asymptotic values. This is analogous to the SG transversal analysis in the last section, which expressed output MSE as a weighted sum of the variances of the (transformed) filter coefficients. After describing the asymptotic MSE, which applies to all stages of the lattice, the dynamics of the output MSE of a single stage is subsequently considered in section 7.2.4.

The nth stage lattice coefficient is first rewritten as

$$k_n(T) = E[k_n(T)] + \tilde{k}_n(T)\ , \tag{7.2.16}$$

where $\tilde{k}_n(T)$ represents the fluctuations of $k_n(T)$ about its mean value and $E[\tilde{k}_n(T)] = 0$. Squaring both sides of (7.2.1a) and taking asymptotic expected values of both sides assuming that $\tilde{k}_n(T)$ is independent of $e_f(T|n-1)$ and $e_b(T|n-1)$ gives

$$\begin{aligned} E_\infty[e_f^2(T|n)] &\approx E_\infty[e_f^2(T|n-1)] + \{E_\infty[k_n(T)]\}^2 E_\infty[e_b^2(T-1|n-1)] \\ &+ E_\infty[\tilde{k}_n^2(T)] E[e_b^2(T-1|n-1)] - 2E_\infty[k_n(T)] E_\infty[e_f(T|n-1)e_b(T-1|n-1)]\ . \end{aligned} \tag{7.2.17}$$

Assuming that $E[k_n(T)]$ converges to $k_{n,opt}$, i.e., $E_\infty[k_n(T)] \approx k_{n,opt}$, which is specified by (7.2.2) and (7.2.3), then (7.2.17) can be rewritten as

$$E_\infty[e_f^2(T|n)] \approx \{1 - k_{n,opt}^2 + E_\infty[\tilde{k}_n^2(T)]\}\ E[e_f^2(T|n-1)]\ , \tag{7.2.18}$$

where $E_\infty[\tilde{k}_n^2(T)] \equiv \mathrm{var}_\infty k_n \approx E_\infty[k_n^2(T)] - k_{n,opt}^2$ denotes the asymptotic variance of $k_n(T)$. The asymptotic variance of $k_n(T)$ therefore contributes asymptotic excess MSE at the output of the nth stage of the lattice that is approximated by $[\text{var}_\infty k_n]\cdot E_\infty[e_f^2(T|n-1)]$. The expression (7.2.18) for asymptotic MSE can be iterated to give the nth stage asymptotic MSE in terms of the input signal variance ϕ_0 and the variances of the PARCOR coefficients.

Completion of the description therefore requires an expression for $\text{var}_\infty k_n$ for each of the two algorithms considered. The asymptotic variance of k_n using (7.2.4) is obtained by squaring (7.2.4) and taking expected values of both sides using the independence assumption. The result is

$$\text{var}_\infty k_n \approx \beta_u \frac{E_\infty\left\{\left[k_{n,opt}\, e_b^2(T-1|n-1) - e_f(T|n-1)e_b(T-1|n-1)\right]^2\right\}}{2E_\infty[e_b^2(T|n-1)] - \beta_u E_\infty[e_b^4(T|n-1)]} . \tag{7.2.19}$$

If $e_f(T|n-1)$ and $e_b(T-1|n-1)$ are further assumed to be jointly Gaussian, the fourth-order statistics in (7.2.19) can be evaluated in terms of second-order statistics. In particular, if X_1, X_2, X_3, and X_4 are Gaussian, then it can be shown that

$$E(X_1X_2X_3X_4) = E(X_1X_2)E(X_3X_4) + E(X_1X_3)E(X_2X_4) + E(X_1X_4)E(X_2X_3) , \tag{7.2.20}$$

from which it follows that (problem 7-8)

$$\text{var}_\infty k_n \approx \beta_u \frac{(1 - k_{n,opt}^2)E_\infty[e_b^2(T|n-1)]}{2 - 3\beta_u E_\infty[e_b^2(T|n-1)]} . \tag{7.2.21}$$

It is interesting to note that as $|k_{n,opt}| \rightarrow 1$, (7.2.21) predicts that $\text{var}_\infty k_n \rightarrow 0$. Referring to the discussion on lattice coefficient sensitivity in chapter 5, this property is desirable in applications such as spectral estimation where in general the closer $|k_n|$ is to one, the more accuracy is needed to represent k_n in order to stay within a given spectral deviation.

Equations (7.2.18) and (7.2.21) can be used to compute asymptotic output MSE for an N-stage lattice filter when the adaptation algorithm (7.2.4) is used. An expression for the asymptotic variance of k_n when (7.2.6) is used remains to be computed. An expression for the asymptotic variance of $k_n(T)$ using (7.2.6) is analogous to (7.2.19) and is again obtained by assuming $k_n(T)$ is independent of $e_b(j-1|n-1)$ and $e_f(j|n-1)$, $0 \leqslant j \leqslant T$,

$$\text{var}_\infty k_n \approx \frac{E_\infty\left[k_{n,opt}\frac{e_f^2(T|n-1)}{F(T|n)} - \frac{e_f(T|n-1)e_b(T-1|n-1)}{F(T|n)}\right]^2}{2E_\infty\left[\frac{e_f^2(T|n-1)}{F(T|n)}\right] - E_\infty\left[\frac{e_f^4(T|n-1)}{F^2(T|n)}\right]} . \tag{7.2.22}$$

If β_n is small enough so that $\lim_{T\to\infty} F(T|n) \approx \frac{1}{\beta_n}E_\infty[e_f^2(T|n-1)]$, then (7.2.22) can be rewritten as

$$\mathrm{var}_\infty k_n \approx \frac{E_\infty\left\{\left[k_{n,opt}e_f^2(T|n-1) - e_f(T|n-1)e_b(T-1|n-1)\right]^2\right\}}{2\left\{\frac{E_\infty[e_f^2(T|n-1)]}{\beta_n}\right\}E_\infty[e_f^2(T|n-1)] - E_\infty[e_f^4(T|n-1)]} . \qquad \textbf{(7.2.23)}$$

Coefficient variance produced by the unnormalized algorithm (7.2.19) is therefore approximately equal to coefficient variance produced by the normalized algorithm (7.2.23) provided that

$$\beta_u = \frac{\beta_n}{E_\infty[e_f^2(T|n-1)]} . \qquad (7.2.24)$$

Equation (7.2.23) was obtained by assuming that $E[F(T|n)]$ is approximately constant. In order to approximately account for the statistical fluctuations of $F(T|n)$, we square both sides of (7.2.5c), multiply through by the denominator, and take expected values of both sides using the independence assumption to get

$$\mathrm{var}_\infty k_n \approx \frac{E_\infty[K_n^2(T)]}{E_\infty[F^2(T|n)]} - k_{n,opt}^2 . \qquad (7.2.25)$$

Evaluation of the right hand side assuming that $e_f(T|n-1)$ and $e_b(T-1|n-1)$ are jointly Gaussian is left as a problem (problem 7-8). The result is

$$\mathrm{var}_\infty k_n \approx \frac{\beta_n\left\{1 - k_{n,opt}^2 + 2\sum_{m=1}^{\infty}(1-\beta_n)^m\,[\rho_m^2(1-2k_{n,opt}^2) + \nu_m\nu_{-m}]\right\}}{2 + \beta_n\left[1 + 4\sum_{m=1}^{\infty}(1-\beta_n)^m\rho_m^2\right]} , \qquad (7.2.26)$$

where

$$\rho_m = \frac{E_\infty[e_f(j|n-1)e_f(j+m|n-1)]}{E_\infty[e_f^2(j|n-1)]}$$

and

$$\nu_m = \frac{E_\infty[e_f(j|n-1)e_b(j+m-1|n-1)]}{E_\infty[e_f^2(j|n-1)]} .$$

As the order n increases, the sequence $e_f(T|n-1)$ becomes uncorrelated and $\nu_m \approx 0$, $m \neq 0$. In this case (7.2.26) simplifies to

$$\mathrm{var}_\infty k_n \approx \frac{\beta_n}{2+\beta_n}(1 - k_{n,opt}^2) \approx \frac{1}{2}\beta_n(1 - k_{n,opt}^2) , \qquad (7.2.27)$$

which for small β_n is essentially equivalent to (7.2.21) and (7.2.24).

Checks with computer simulation [8] have indicated that (7.2.18), (7.2.21), and (7.2.27) are accurate when an uncorrelated Gaussian noise sequence is used as the input to a multi-stage lattice. In general, these formulas become less accurate as the input becomes more correlated. It has been empirically observed that the dependence between the coefficients and input data tends to make coefficient variance smaller and output MSE somewhat larger than what the formulas in this section predict. An intuitive explanation for this observation is that as the input becomes more correlated, the short term fluctuations of the data samples decrease, causing the coefficients to fluctuate less. On the other hand, highly correlated inputs tend to produce biased coefficient estimates, which tend to increase the output MSE.

7.2.4. Transient Behavior of Single-Stage Output MSE

Thus far only the asymptotic behavior of the output MSE of a filter stage has been investigated. We now investigate how single-stage MSE varies with time. For convenience, only the unnormalized algorithm (7.2.4) is considered. By making the same approximations as were made in the last section, the analysis in this section can also be applied to the normalized algorithm (7.2.6).

Squaring (7.2.1a) and taking expected values of both sides, using the independence assumption and (7.2.3), gives

$$\epsilon(T|n) \equiv E[e_f^2(T|n)] \approx \left\{1 + E[k_n^2(T)] - 2k_{n,opt}E[k_n(T)]\right\}\epsilon_{n-1} . \qquad (7.2.28)$$

The $n-1$st order MSE, $\epsilon_{n-1} \equiv E[e_f^2(T|n-1)]$, is assumed to be constant. Substituting (7.2.4) for $k_n(T)$, assuming that $e_f(T|n-1)$ and $e_b(T-1|n-1)$ are jointly Gaussian and again using the independence assumption, the following difference equation for $\epsilon(T|n)$ can be obtained after some manipulation (problem 7-9),

$$\epsilon(T|n) \approx [\alpha(3\alpha - 2) + 1]\epsilon(T-1|n) + 2\alpha(1 - \alpha)\epsilon_{\min} , \qquad (7.2.29)$$

where $\alpha \equiv \beta_u \epsilon_{n-1}$ and $\epsilon_{\min} = (1 - k_{n,opt}^2)\epsilon_{n-1}$. Equation (7.2.29) implies that $E_\infty[e_f^2(T|n)] \approx \dfrac{2(1-\alpha)}{2-3\alpha}\epsilon_{\min}$, which agrees with (7.2.18) and (7.2.21). The condition

$$0 < \alpha(3\alpha - 2) + 1 < 1 ,$$

is required for stability, or equivalently,

$$0 < \alpha < \frac{2}{3} . \qquad (7.2.30)$$

This stability requirement for output MSE is stricter than the stability requirement for coefficient mean values ($0 < \alpha < 2$) obtained from (7.2.10). This is analogous to the stability results obtained for the SG transversal algorithm in section 7.1, which state that the maximum step size that ensures the convergence of the output MSE is less than the maximum step size that ensures convergence of the mean values of the coefficients. This is again due to the variance of the coefficients, which in both cases make significant contributions to the output MSE. Examining (7.2.29), the fastest convergence rate occurs when $3\alpha^2 - 2\alpha + 1$ is minimized, i.e., when $\alpha = \dfrac{1}{3}$. In this case the asymptotic output MSE is approximately $\dfrac{4}{3}\epsilon_{\min}$. It is interesting to note that the optimal (constant) step size is one half the maximum allowable step size. This is also true for the example discussed in section 7.1.4.

It is emphasized that the stability requirement (7.2.30) was derived for only one stage of an adaptive lattice under the assumption that the inputs to that stage are stationary and jointly Gaussian. As more stages are added to the lattice, the fluctuations of the added filter coefficients will tend to increase the output MSE and gradient noise in successive coefficient estimates. The latter effect may tend to drive a filter coefficient far from its optimal value. This effect can propagate through the lattice causing successive coefficients to diverge. One would therefore expect that for the N-stage case a stricter stability requirement than (7.2.30), which depends upon the filter order, is needed. The stability constraint,

$$0 < \alpha < \frac{2}{3N\phi_0}, \tag{7.2.31}$$

has been empirically observed to be sufficient [8]. In analogy with the LS lattice algorithm, which is stable if the exponential weighting factor w is less than unity, the normalized algorithm (7.2.6) has been empirically observed to be stable if the step size β_n satisfies $0 < \beta_n < 1$.

7.3. LS LATTICE SINGLE-STAGE CONVERGENCE

The same type of single-stage analysis that was applied to SG lattice algorithms in the last section is now applied to the LS lattice. For convenience, only the prewindowed lattice algorithm (6.3.65) is considered. The techniques in this section can be similarly applied to the LS covariance algorithms derived in chapter 6; however, for the stationary input case considered here, the corresponding results should be nearly identical to those obtained here. The LS lattice algorithm is even harder to analyze than the SG lattice algorithms, because of the additional state variables entering the LS algorithm (i.e., $\gamma(T|n)$, $n = 1, \ldots, N-1$), and because the algorithm recursions are somewhat more complicated than the recursions entering the SG algorithms. Nevertheless, results similar to those obtained in the last section can be obtained in the LS case by making some additional approximations. The results for mean coefficient time constants and output MSE obtained here are in fact nearly identical to those obtained in the last section for the normalized SG lattice algorithm. This is partially due to the nature of the approximations made in the analysis, as well as more fundamentally the similarity in algorithm performance in certain environments.

The reader is cautioned that many of the approximations to be made in this section and the next are quite poor in certain situations. In particular, in some applications where the input does not look like Gaussian noise, the approximate "mean-value" type of analysis employed here can be misleading when applied to LS algorithms. In applications such as channel equalization, where only a very small amount of noise is added by the channel relative to the signal strength, the fast convergence of LS algorithms is due to *algebraic* properties, to be discussed in section 7.7, rather than on the fast convergence of certain time averages, as is the case for the SG lattice algorithms. Throughout this section we therefore implicitly assume that the input to the filter has a high noise content, which the adaptive algorithm must average out in order to obtain good estimates of the MMSE filter coefficients. Because of this averaging or smoothing, many of the algorithm parameters vary slowly and can be treated as essentially independent of the other adapting parameters.

7.3.1. Single-Stage Time Constant

The prewindowed LS lattice algorithm recursions (6.3.65) are first used to obtain single-stage time constants for the coefficient means. Taking expected values of both sides of (6.3.65a) gives

$$E[K_{n+1}(T)] = wE[K_{n+1}(T-1)] + E\left[\frac{e_f(T|n)e_b(T-1|n)}{1-\gamma(T-1|n)}\right]$$
$$\approx wE[K_{n+1}(T-1)] + \frac{E[e_f(T|n)e_b(T-1|n)]}{1-E[\gamma(T-1|n)]} . \tag{7.3.1}$$

To evaluate $E[\gamma(T|n)]$, we first take expected values of both sides of (6.3.64g),

$$E[\gamma(T|n)] \approx E[\gamma(T|n-1)] + \frac{E[e_b^2(T|n-1)]}{E[\epsilon_b(T|n-1)]} . \tag{7.3.2}$$

By assumption the previous stages have converged and hence $E[\epsilon_b(T|n-1)] = E_\infty[\epsilon_b(T|n-1)]$. Taking asymptotic expected values of both sides of (6.3.51) (assuming exponential weighting) gives

$$E_\infty[\epsilon_b(T|n-1)] \approx \frac{1}{1-w}\,\frac{E[e_b^2(T|n-1)]}{1-E[\gamma(T|n-1)]} ,$$

or

$$\frac{E[e_b^2(T|n-1)]}{E[\epsilon_b(T|n-1)]} \approx (1-w)\{1-E[\gamma(T|n-1)]\} . \tag{7.3.3}$$

Substituting (7.3.3) into (7.3.2) gives an order recursion for $E[\gamma(T|n)]$,

$$1-E[\gamma(T|n)] \approx w\{1-E[\gamma(T|n-1)]\} . \tag{7.3.4}$$

Using the boundary condition $\gamma(T|0)=0$, (7.3.4) can be rewritten as

$$E[\gamma(T|n)] \approx 1-w^n . \tag{7.3.5}$$

Substituting (7.3.5) into (7.3.1) gives a difference equation for $E[K_{n+1}(T)]$,

$$E[K_{n+1}(T)] \approx wE[K_{n+1}(T-1)] + w^{-n}E[e_f(T|n)e_b(T-1|n)] \tag{7.3.6}$$

$$= w^T\left\{K_n(0) - \frac{w^{-n}}{1-w}E[e_f(T|n)e_b(T-1|n)]\right\} + \frac{w^{-n}}{1-w}E[e_f(T|n)e_b(T-1|n)] .$$

Similarly, taking expected values of both sides of (6.3.51) and using (7.3.5) gives

$$E[\epsilon_b(T|n)] \approx wE[\epsilon_b(T-1|n)] + w^{-n}E[e_b^2(T|n)]$$
$$= w^T\left\{\epsilon_b(0|n) - \frac{w^{-n}}{1-w}E[e_b^2(T|n)]\right\} + \frac{w^{-n}}{1-w}E[e_b^2(T|n)] \,. \tag{7.3.7}$$

Using (7.3.6) and (7.3.7), and the approximation $E[k_n^b(T)] \approx \dfrac{E[K_n(T)]}{E[\epsilon_b(T-1|n-1)]}$, as in the last section we can compute the time τ_n^b it takes $E[k_n^b(T)]$ to reach the value $k_n^b(0) + g\left(E_\infty[k_n^b(T)] - k_n^b(0)\right)$ where $0 < g < 1$ (problem 7-11),

$$\tau_n^b \approx -\frac{1}{\ln w}\ln\left[w^n(1-w)\frac{g}{1-g}\frac{\epsilon_b(0|n-1)}{E[e_b^2(T|n-1)]} + 1\right]. \tag{7.3.8}$$

Assuming that $E[e_b^2(T-1|n-1)] = E[e_f^2(T|n-1)]$ implies that the same time constant also applies to $k_n^f(T)$. The only difference between (7.3.8) and the analogous time constant formula for the SG lattice algorithm (7.2.15) is the added factor w^n in brackets. In general w is slightly less than unity so that τ_n^b will approximately equal the analogous SG lattice "time constant". This similarity is not surprising since the fundamental difference between the LS lattice predictor and the SG lattice predictor given by (7.2.5) is the added LS gain γ. If $\gamma(T|n-1)$ is assumed to be constant, as is approximately the case if the previous stages have converged, the algorithms are basically the same.

7.3.2. Single-Stage Output MSE

Single-stage output MSE is now approximated assuming both inputs are stationary. We wish to compute both the asymptotic value of $E[e_f^2(T|n+1)]$, where $e_f(T|n+1)$ is the signal in the lattice at time T, in addition to the asymptotic value of $E[e_f^{o^2}(T|n+1)]$, which denotes the oblique mean squared prediction error. In particular, $e_f(T|n+1)$ is computed from regression coefficients based upon the data $y(T), y(T-1), \ldots, y(0)$. As the exponential weighting factor decreases, the more recent values of $e_f^2(T|n+1)$ are weighted more heavily, and hence $E_\infty[e_f^2(T|n+1)]$ decreases. For example, as $w \rightarrow 0$,

$$\frac{K_{n+1}(T)}{\epsilon_b(T-1|n)} \rightarrow \frac{e_f(T|n)e_b(T-1|n)}{e_b^2(T-1|n)} \,,$$

and from (6.3.64a), $e_f^2(T|n+1) \rightarrow 0$. In contrast, $e_f^o(T|n+1)$ is computed from regression coefficients calculated at time $T-1$, and hence is the causal LS prediction error obtained by estimating the value of $y(T)$ given $y(T-1), y(T-2), \ldots, y(0)$. As w decreases, the second-order statistics used to predict $y(T)$ are effectively estimated from fewer samples and hence $E_\infty[e_f^{o^2}(T|n+1)]$ increases. In practice "output MSE" refers to $E[e_f^{o^2}(T|n)]$. Also notice that the oblique error, rather than the error which appears in the lattice, is analogous to the error that appears in the SG transversal algorithm. In the case of the SG lattice, the errors that appear in the lattice may correspond to either e_f and e_b or e_f^o and e_b^o, depending on whether the lattice coefficients are updated before or after the errors at the output of each stage are computed. The analysis of SG lattice MSE in the last section assumed that the nth coefficient is updated *after* the nth order error is computed so that the resulting MSE is greater than the MMSE obtained with a fixed coefficient filter. This is illustrated further by the examples in section 7.8.

From (6.3.41) the oblique forward error is given by

$$e_f^o(T|n) = \frac{e_f(T|n)}{1 - \gamma(T-1|n)} , \tag{7.3.9}$$

so that

$$E_\infty[e_f^{o^2}(T|n)] \approx \frac{E_\infty[e_f^2(T|n)]}{E_\infty\{[1 - \gamma(T-1|n)]^2\}} .$$

Assuming the variance of $\gamma(T|n)$ is small after convergence, so that $E_\infty[\gamma^2] \approx \{E_\infty[\gamma]\}^2$, then (7.3.9) implies that

$$E_\infty[e_f^{o^2}(T|n)] \approx \frac{1}{w^{2n}} E_\infty[e_f^2(T|n)] . \tag{7.3.10}$$

In order to compute the output MSE, the "lattice MSE" $E[e_f^2(T|n)]$ must also be computed.

To approximate $E_\infty[e_f^2(T|n+1)]$, asymptotic expected values of both sides of (6.3.50) are taken to get

$$E_\infty[\epsilon_f(T|n)] = \frac{1}{1-w} E_\infty\left[\frac{e_f^2(T|n)}{1 - \gamma(T-1|n)}\right] \approx \frac{1}{w^n(1-w)} E[e_f^2(T|n)] . \tag{7.3.11a}$$

Similarly, (6.3.51) implies that

$$E_\infty[\epsilon_b(T|n)] \approx \frac{1}{w^n(1-w)} E[e_b^2(T|n)] . \tag{7.3.11b}$$

Taking asymptotic expected values of both sides of (6.3.64e) gives

$$E_\infty[\epsilon_f(T|n+1)] \approx E_\infty[\epsilon_f(T|n)] - \frac{E_\infty[K_{n+1}^2(T)]}{E_\infty[\epsilon_b(T-1|n)]} . \tag{7.3.12}$$

Substituting (7.3.11) into (7.3.12) and assuming $E[e_f^2(T|n)] = E[e_b^2(T-1|n)]$ gives an order-recursive expression for the MSE that appears in the lattice,

$$E_\infty[e_f^2(T|n+1)] \approx w\left\{1 - w^{2n}(1-w)^2 \frac{E_\infty[K_{n+1}^2(T)]}{(E[e_b^2(T-1|n)])^2}\right\} E[e_f^2(T|n)] . \tag{7.3.13}$$

In order to compute $E[K_{n+1}^2(T)]$, we square both sides of (6.3.64a) and take asymptotic expected values of both sides using the independence assumption. The result is

$$E_\infty[K_{n+1}^2(T)] \approx w^2 E_\infty[K_{n+1}^2(T-1)] + 2wE_\infty[K_{n+1}(T)]\frac{E[e_f(T|n)e_b(T-1|n)]}{1-E[\gamma(T-1|n)]} + \frac{E[e_f^2(T|n)e_b^2(T-1|n)]}{E\left\{[1-\gamma(T-1|n)]^2\right\}} . \quad \textbf{(7.3.14)}$$

Substituting

$$E_\infty[K_{n+1}(T)] \approx \frac{1}{w^n(1-w)}E[e_f(T|n)e_b(T-1|n)] ,$$

which is obtained from (7.3.6), and $E_\infty\left\{[1-\gamma(T-1|n)]^2\right\} \approx w^{2n}$ into (7.3.14) gives

$$E_\infty[K_{n+1}^2(T)] \approx \frac{2w}{w^{2n}(1-w)(1-w^2)}\{E[e_f(T|n)e_b(T-1|n)]\}^2 + \frac{1}{w^{2n}(1-w^2)}E[e_f^2(T|n)e_b^2(T-1|n)] . \quad (7.3.15)$$

Substituting (7.3.15) into (7.3.13) and using (7.2.3) gives

$$E_\infty[e_f^2(T|n+1)] \approx w\left\{1-\frac{2w}{1+w}k_{n+1,opt}^2-\left(\frac{1-w}{1+w}\right)\frac{E[e_f^2(T|n)e_b^2(T-1|n)]}{(E[e_f^2(T|n)])^2}\right\}E[e_f^2(T|n)] .$$

(7.3.16)

If it is further assumed that $e_f(T|n)$ and $e_b(T-1|n)$ are jointly Gaussian, then the fourth-order statistics in (7.3.16) can be evaluated in terms of second-order statistics,

$$E_\infty[e_f^2(T|n+1)] \approx \frac{2w}{1+w}(w - k_{n+1,opt}^2)E_\infty[e_f^2(T|n)] . \quad (7.3.17)$$

Equation (7.3.17) gives an order recursive expression for the $(n+1)$st order lattice MSE in terms of the nth order lattice MSE. As $w \to 0$, (7.3.17) predicts that the MSE approaches zero. This is consistent with the discussion at the beginning of this section. Checks with computer simulation [8] have indicated that (7.3.17) is accurate when a Gaussian input sequence is used, and the exponential weight w is close to one. If w is very small (i.e., less than 0.8), however, $K_{n+1}(T)$, $\epsilon_f(T|n)$, and $\epsilon_b(T|n)$ start to track the immediate variations in e_f and e_b and hence the independence assumptions used to derive (7.3.17) are no longer approximately true. Substitution of (7.3.17) into (7.3.10) yields the corresponding expression for the (oblique) output MSE.

Also of interest is the asymptotic variance of the filter coefficients. In analogy with the computation of coefficient variance for the normalized SG lattice algorithm in the last section, this is approximated as

$$\mathrm{var}_\infty k_{n+1}^b \approx \frac{E_\infty[K_{n+1}^2(T)]}{E_\infty[\epsilon_b^2(T|n)]} - k_{n+1,opt}^2 . \qquad \textbf{(7.3.18)}$$

Squaring (6.3.51) (assuming exponential weighting) and taking asymptotic expectations assuming $\epsilon_b(T-1|n)$ is uncorrelated with $\dfrac{e_b^2(T|n)}{1-\gamma(T|n)}$ gives

$$E_\infty[\epsilon_b^2(T|n)] \approx \frac{1}{1-w^2}\left\{2wE_\infty[\epsilon_b(T|n)]\frac{E[e_b^2(T|n)]}{1-E[\gamma(T|n)]} + \frac{E[e_b^4(T|n)]}{E\{[1-\gamma(T|n)]^2\}}\right\}. \qquad (7.3.19)$$

Using (7.3.11b), (7.3.5), and assuming as a first-order approximation that $e_b(T|n)$ is Gaussian gives

$$E_\infty[\epsilon_b^2(T|n)] \approx \frac{3-w}{w^{2n}(1-w)(1-w^2)}\{E[e_b^2(T|n)]\}^2 . \qquad (7.3.20)$$

Similarly, assuming e_f and e_b are jointly Gaussian in (7.3.15) gives

$$E_\infty[K_{n+1}^2(T)] \approx \frac{1}{w^{2n}(1-w^2)}\left[\frac{2}{1-w}k_{n+1,opt}^2 + 1\right]\{E[e_f^2(T|n)]\}^2 . \qquad (7.3.21)$$

Substituting (7.3.21) and (7.3.20) into (7.3.18) gives an expression for the asymptotic coefficient variance,

$$\mathrm{var}_\infty k_{n+1}^b \approx \frac{1-w}{3-w}(1 - k_{n+1,opt}^2) . \qquad (7.3.22)$$

Since w corresponds to $(1-\beta_n)$ in the last section, this formula is identical to the formula for coefficient variance, (7.2.27), which results from using the SG algorithm (7.2.5). (Also notice that the method used to derive the expression for coefficient variance (7.2.26) using the SG algorithm (7.2.5) can also be applied in this case.)

A comparison of the formulas in this section with their counterparts in section 7.2 illustrates the similarities between the LS lattice and SG lattice. In particular, calculated single-stage time constants, coefficient variance, and output MSE for both the LS and SG lattice algorithms are essentially identical. The first-order techniques used in this section therefore cannot be used to predict the performance advantage that the LS lattice algorithm offers relative to the SG lattice algorithm in some applications. On the other hand, this analysis indicates that when the input sequence is Gaussian noise, the LS lattice does not offer a significant performance advantage relative to the SG lattice algorithm. This is because the difference in performance between the two algorithms is primarily due to the weighting factors $\gamma(T|n)$, $1 \leqslant n \leqslant N$, in the LS lattice algorithm, which tend to remain small (i.e., near zero) for stationary Gaussian inputs. A further comparison of SG and LS lattice algorithms is given in section 7.8.

7.3.3. Summary

In this section and the last, the convergence properties of a single stage of an adaptive lattice filter have been analyzed assuming that the inputs are stationary. The mean trajectories of the filter coefficients and the asymptotic variance of each filter coefficient and its effect upon the output MSE have been approximated for both SG lattice algorithms and the prewindowed LS lattice algorithm. The approximations that were used are:

1. The filter coefficients are independent of the input data sequence, which is analogous to the independence assumption used to analyze the SG transversal algorithm in section 7.1,
2. The average $E\left[\frac{X(T)}{Y(T)}\right]$ can be replaced by $\frac{E[X(T)]}{E[Y(T)]}$ for the cases considered, and
3. The error sequences e_b and e_f are jointly Gaussian.

The second assumption was used because the variables $X(T)$ and $Y(T)$ are, for most of the cases considered, time averages of second-order input statistics. The variances of $X(T)$ and $Y(T)$ therefore become relatively small as T increases. The third assumption was used to approximate fourth-order statistics by second-order statistics. Since e_f and e_b are linear combinations of the input, a loose application of the Central Limit Theorem indicates that in some cases the joint distribution of e_f and e_b may in fact be close to a Gaussian distribution.

The results in this section and the last can be used to gain insight into the relationships between convergence time, output MSE, the step size β (or w), and second-order input signal statistics. The time constants also give some idea of the speed of convergence of the entire filter, if the worst case assumption is made that the first $(n-1)$ stages have to adapt before the nth stage can begin its adaptation. A loose "upper bound" on the convergence time of the filter can therefore be obtained by adding up the time constants for each stage of the filter. Intuitively, however, it is clear that this estimate of convergence time is very pessimistic, and hence in the next section a simple model for the adaptation of a multi-stage filter is presented.

7.4. MULTI-STAGE ADAPTATION MODELS

Intuitively, it would seem likely that the nth coefficient of a lattice filter will start adapting in the direction of its asymptotic optimum value before the first $(n-1)$ stages have completed their adaptation. In fact, it can be shown that (problem 7-12)

$$\left.\frac{\partial}{\partial k_j} k_{n,opt}\right|_{*} = 0, \quad 1 \leqslant j \leqslant n-1 \tag{7.4.1}$$

where "*" refers to the condition

$$k_m = k_{m,opt}, \quad 1 \leqslant m \leqslant n-1. \tag{7.4.2}$$

The nth stage optimal PARCOR coefficient $k_{n,opt}$ is therefore to first order insensitive to incremental changes in k_1 through k_{n-1} when the latter are in the region of their optimum values. This indicates that coefficient k_n will start adapting in the direction of $k_{n,opt}$ before the previous stages converge.

The time constant calculations in the last two sections are unsatisfactory for predicting the behavior of the multi-stage adaptation, since the effect of previous stages' adaptation must be taken into account. Our goal is to compute the mean values of the N PARCOR coefficients and the output MSE as functions of time. Both of these functions could be obtained by averaging the results of multiple simulations of the algorithms. However, since multiple simulations are expensive and do not provide much insight, in this section a simpler model for the multi-stage adaptation is developed. Adaptation models for the SG lattice algorithms considered in section 7.2 are derived first. Simple modifications result in an adaptation model for the prewindowed LS lattice algorithm. A similar type of adaptation model is subsequently presented for the LS transversal (fast Kalman) algorithm.

7.4.1. SG Lattice Adaptation Models

An approximate analytical technique for estimating the adaptation of the nth PARCOR coefficient is to simply ignore the statistical fluctuation of $k_1(T)$ through $k_{n-1}(T)$ about their mean values, since those fluctuations should have little effect on the *mean value* of $k_n(T)$. It can then be assumed that $k_1(T)$ through $k_{n-1}(T)$ are following their deterministic mean value trajectories. These mean value trajectories plus the input statistics can be used to compute a set of second-order statistics for $e_f(T|n-1)$ and $e_b(T-1|n-1)$ vs. time, which in turn can be used to predict the mean value trajectory of $k_n(T)$. Proceeding one stage at a time starting with k_1, the mean value trajectories of all the PARCOR coefficients can thereby be approximated. The resulting model, which unfortunately must be represented by a computer program rather than analytically, is nevertheless much simpler and less expensive than a simulation. Furthermore, by plotting quantities such as $k_{n,opt}$ vs. time, much insight can be gained.

In order to describe the model in more detail, the errors $e_f(T|n)$ and $e_b(T|n)$ are first rewritten as linear combinations of $y(T), y(T-1), \ldots, y(T-n)$. In particular,

$$e_f(T|n) = y(T) - \sum_{j=1}^{n} f_j(T|n)\, y(T-j) = \bar{\mathbf{f}}'(T|n)\mathbf{y}(T|N+1) \tag{7.4.3a}$$

and

$$e_b(T|n) = y(T-n) - \sum_{j=0}^{n-1} b_j(T|n)\, y(T-j) = \bar{\mathbf{b}}'(T|n)\mathbf{y}(T|N+1) \tag{7.4.3b}$$

where

$$\bar{\mathbf{f}}(T|n) = [1\ -f_1(T|n) \cdots -f_n(T|n)\ 0 \cdots 0]', \tag{7.4.4a}$$

$$\bar{\mathbf{b}}(T|n) = [-b_1(T|n) \cdots -b_n(T|n)\ \ 1\ 0 \cdots 0]', \tag{7.4.4b}$$

and

$$\mathbf{y}(T|N+1) = [y(T)\, y(T-1) \cdots y(T-N)]' , \tag{7.4.4c}$$

and all vectors have dimension $N+1$. (Notice that $\bar{\mathbf{f}}$ and $\bar{\mathbf{b}}$ are different from the vectors $\mathbf{f}$ and $\mathbf{b}$ used previously.) The residuals $e_f(T|n)$ and $e_b(T|n)$ can therefore be advantageously represented by the coefficient vectors $\bar{\mathbf{f}}(T|n)$ and $\bar{\mathbf{b}}(T|n)$. The lattice recursions (7.2.1), with $k_n(T)$ replaced by its mean-value, can then be rewritten as

$$\bar{\mathbf{f}}(T|n) = \bar{\mathbf{f}}(T|n-1) - E[k_n(T)][z^{-1}\bar{\mathbf{b}}(T-1|n-1)] \tag{7.4.5}$$

and

$$\bar{\mathbf{b}}(T|n) = [z^{-1}\bar{\mathbf{b}}(T-1|n-1)] - E[k_n(T)]\bar{\mathbf{f}}(T|n-1) , \tag{7.4.6}$$

where $z^{-1}\bar{\mathbf{b}}(T-1|n-1)$ represents $\bar{\mathbf{b}}(T-1|n-1)$ shifted "down" one element, i.e.,

$$[z^{-1}\bar{\mathbf{b}}(T-1|n)]_j = [\bar{\mathbf{b}}(T-1|n)]_{j-1} , \quad 2 \leqslant j \leqslant N+1 ,$$

$$[z^{-1}\bar{\mathbf{b}}(T-1|n)]_1 = 0 .$$

The vector $z^{-1}\bar{\mathbf{b}}(T-1|n-1)$ must be used instead of $\bar{\mathbf{b}}(T-1|n-1)$ since from (7.4.3b) $e_b(T-1|n-1)$ is a linear combination of $y(T-1), y(T-2), \ldots, y(T-n)$. Given $\bar{\mathbf{f}}(T|n-1)$, $\bar{\mathbf{b}}(T-1|n-1)$, and the expected value of $k_n(T)$, then (7.4.5) and (7.4.6) can be used to compute the nth order coefficient vectors at time T. Once the trajectories of the nth order coefficient vectors $\bar{\mathbf{f}}$ and $\bar{\mathbf{b}}$ are known, the second-order statistics of e_f and e_b can be estimated as

$$\begin{aligned} E[e_f^2(T|n)] &= E\left\{\left[y(T) - \sum_{j=1}^{n} f_j(T|n)y(T-j)\right]^2\right\} \\ &\approx \sum_{j=1}^{n+1}\sum_{m=1}^{n+1} [\bar{\mathbf{f}}(T|n)]_j [\bar{\mathbf{f}}(T|n)]_m \phi_{j-m} \end{aligned} \tag{7.4.7}$$

and

$$\begin{aligned} &E[e_f(T|n)e_b(T-1|n)] = \\ &E\left\{\left[y(T)-\sum_{j=1}^{n} f_j(T|n)y(T-j)\right]\left[y(T-n-1)-\sum_{j=1}^{n} b_j(T-1|n)y(T-j)\right]\right\} \\ &\approx \sum_{j=1}^{n+1}\sum_{m=2}^{n+2} [\bar{\mathbf{f}}(T|n)]_j [z^{-1}\bar{\mathbf{b}}(T-1|n)]_m \phi_{j-m} \end{aligned} \tag{7.4.8}$$

where $\phi_j = E[y(T)y(T-j)]$. Finally, the trajectory of the $(n+1)$st PARCOR coefficient mean follows from (7.2.4) or (7.2.5) where each random element is replaced by its mean. For example, in the case of the normalized step size algorithm the approximation

$$E[k_n(T)] \approx \frac{E[K_n(T)]}{E[F(T|n)]} \tag{7.4.9}$$

is made. The complete adaptation model for the SG lattice algorithm (7.2.5) is therefore specified by the mean order updates (7.4.5) and (7.4.6), the computation of MSE and the cross-correlation of e_f and e_b given by (7.4.7) and (7.4.8), and the mean updates for the filter coefficients (7.4.9), where $E[K_n(T)]$ and $E[F(T|n)]$ are obtained by taking expected values of both sides of (7.2.5a) and (7.2.5b). The mean values of the PARCOR coefficients and the second-order statistics of the forward and backward errors are computed in an order-recursive fashion starting at the first stage. In particular, $E[k_1(T)]$ is first computed from $E[e_f^2(T|0)] = E[e_b^2(T|0)] = \phi_0$ and $E[e_f(T|0)e_b(T-1|0)] = \phi_1$. This enables the computation of the second-order statistics of $e_f(T|1)$ and $e_b(T|1)$, which in turn enables the computation of $E[k_2(T)]$, and so forth. It should be noted that since reducing the step size β will reduce the statistical fluctuations present in each coefficient, the accuracy of the model should improve as β decreases.

To illustrate the model, figure 7-1 shows a plot of $E[k_4(T)]$, using the SG algorithm (7.2.5), for a particular case of second-order input statistics and $\beta_n = 0.025$. Shown in addition to $E[k_4(T)]$ as generated by the model is $k_{4,opt}(T)$, the value towards which $E[k_4(T)]$ is converging at time T. Because $k_{4,opt}(0)$ is initially greater than $k_4(0)$, the trajectory of $E[k_4(T)]$ starts off in the wrong direction! Once $k_{4,opt}$ reaches its asymptotic value, however, $E[k_4(T)]$ converges monotonically to $k_{4,opt}$. This type of behavior, where the trajectory of the mean value of the coefficient changes direction, cannot occur with the SG transversal algorithm. Also shown in figure 7-1 is $E[k_4(T)]$, as obtained by averaging the results of 200 simulations of the algorithm. The input was obtained by passing a white Gaussian noise sequence through a ten pole filter.

Some of the difference between the model curve and the simulated curve can be attributed to the different asymptotic values towards which they converge.

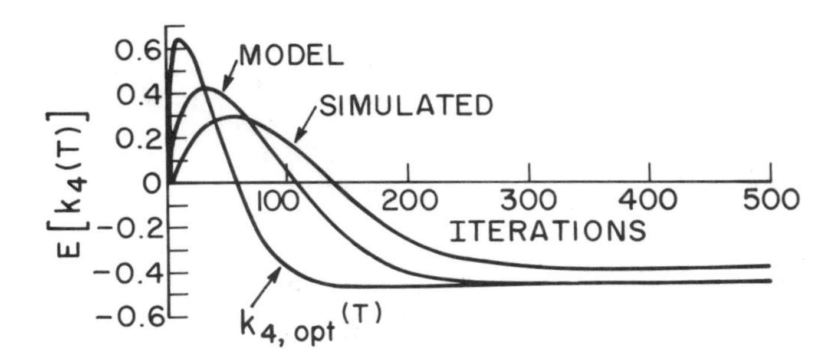

Figure 7-1. Mean value trajectory of the PARCOR coefficient $k_4(T)$ using the SG lattice algorithm (7.2.5). Also shown is the trajectory of $k_{4,opt}(T)$ computed from the model.

The value of $E[k_n(T)]$ as generated by the model converges to the optimum value of k_n for the given input statistics under the assumption that $k_1, \ldots, k_{n-1}$ are fixed at their optimal values. In practice, however, k_1 through k_{n-1} have some nonzero variance not accounted for in the model. The bias shown in the simulated curve is therefore partially due to the fact that the variances of k_1 through k_{n-1} perturb the statistics of $e_f(T|n-1)$ and $e_b(T-1|n-1)$, which in turn perturb the value of $k_{n,opt}$ from the MMSE value computed by the model. Correlations between the coefficients and input data also contribute to coefficient bias. An interesting empirical observation is that the simulated asymptotic expected value of $|k_n|$ always appears to be less than $|E_\infty[k_n]|$ as generated by the model. This is intuitively satisfying if we view the effect of previous coefficient variations as partially whitening the input to the current stage.

7.4.2. LS Lattice Adaptation Model

The adaptation model for the LS lattice is again obtained by ignoring the effect of statistical fluctuations of the coefficients $k_n^b(T)$ and $k_n^f(T)$, $1 \leqslant n \leqslant N$. Given second-order information about the input sequence, $k_{n+1}^b(T)$ and $k_{n+1}^f(T)$ can be replaced by their mean value trajectories and a set of simple deterministic iterative equations can be obtained. The residuals e_f and e_b are again represented by the coefficient vectors $\bar{\mathbf{f}}(T|n)$ and $\bar{\mathbf{b}}(T|n)$ defined by (7.4.4). The order updates for $\bar{\mathbf{f}}(T|n)$ and $\bar{\mathbf{b}}(T|n)$ are therefore the same as for the SG case except that $E[k_n^f(T)]$ and $E[k_n^b(T)]$ are substituted for $E[k_n(T)]$. The updates that are used to compute $E[k_n^f(T)]$ and $E[k_n^b(T)]$ are

$$E[\gamma(T|n)] \approx E[\gamma(T|n-1)] + \frac{E[e_b^2(T|n-1)]}{E[\epsilon_b(T|n-1)]} \tag{7.4.10}$$

$$E[\epsilon_b(T|n)] \approx wE[\epsilon_b(T-1|n)] + \frac{E[e_b^2(T|n)]}{1 - E[\gamma(T|n)]} \tag{7.4.11}$$

$$E[\epsilon_f(T|n)] \approx wE[\epsilon_f(T-1|n)] + \frac{E[e_f^2(T|n)]}{1 - E[\gamma(T-1|n)]} \tag{7.4.12}$$

$$E[K_{n+1}(T)] \approx wE[K_{n+1}(T-1)] + \frac{E[e_f(T|n)e_b(T-1|n)]}{1 - E[\gamma(T-1|n)]} \tag{7.4.13}$$

$$E[k_{n+1}^f(T)] \approx \frac{E[K_{n+1}(T)]}{E[\epsilon_f(T|n)]} \quad \text{and} \quad E[k_{n+1}^b(T)] = \frac{E[K_{n+1}(T)]}{E[\epsilon_b(T-1|n)]}\,, \tag{7.4.14}$$

where $E[e_f(T|n)e_b(T-1|n)]$ and $E[e_f^2(T|n)]$ are computed respectively from (7.4.8) and (7.4.7), and

$$E[e_b^2(T|n)] \approx \sum_{j=1}^{n+1}\sum_{m=1}^{n+1} [\bar{\mathbf{b}}(T|n)]_j [\bar{\mathbf{b}}(T|n)]_m \phi_{j-m}\,. \tag{7.4.15}$$

Equations (7.4.5), (7.4.6) (with $E[k_{n+1}(T)]$ replaced by its forward and backward counterparts), (7.4.7), and (7.4.8) combined with (7.4.10) through (7.4.15) complete the adaptation model for the LS lattice filter. Initialization of the convergence model is accomplished in exactly the same way as for the LS lattice algorithm. The mean cost functions can be initialized at some small value δ, and the model recursions are computed from $n=0$ to $n=N$ at each iteration T. Figure 7-2 compares the mean value trajectory of $k_4^b(T)$ obtained from the model and by simulation for the same input statistics used to illustrate the SG lattice model in figure 7-1. Also shown is the trajectory of

$$k_{4,opt}^b(T) = \frac{E[e_f(T|n)e_b(T-1|n)]}{E[e_b^2(T-1|n)]} \,. \tag{7.4.16}$$

The exponential weighting constant in figure 7-1 is $w = 1-\beta_n$, where β_n was the adaptation constant used in figure 7-1. The speeds of convergence for the two algorithms are therefore similar. In analogy with the single-stage lattice results presented in sections 7.2 and 7.3, plots of mean coefficient trajectories and output MSE vs. time obtained from the LS lattice model are nearly identical to the same curves obtained from the SG lattice adaptation model. A further discussion and comparison of SG and LS algorithms is given in sections 7.7 and 7.8.

7.4.3. Fast Kalman Adaptation Model

The fast Kalman algorithm, or prewindowed LS transversal algorithm, is given by (6.3.63). Because the LS lattice and fast Kalman algorithms both minimize the same cost criterion, if output MSE is the only item of interest, the adaptation model for the LS lattice algorithm would suffice for both algorithms. In addition, the formulas for asymptotic output MSE given in section 7.3 should also apply to the fast Kalman algorithm. In order to describe the mean values of the transversal coefficients as functions of time, however, a convergence model similar to the convergence models already presented is needed.

To obtain a convergence model for the fast Kalman algorithm, the residuals $e_f(T)$ and $e_b(T)$ are again represented by their respective prediction coefficient

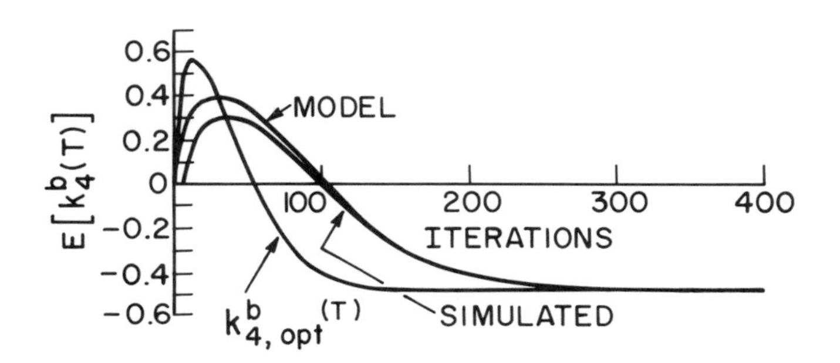

Figure 7-2. Mean value trajectory of the LS coefficient $k_4^b(T)$. Also shown is the trajectory of $k_{4,opt}^b(T)$ computed from the LS lattice adaptation model.

vectors $\bar{\mathbf{f}}(T)$ and $\bar{\mathbf{b}}(T)$ defined by (7.4.4). By definition, the oblique errors $e_f^o(T)$ and $e_b^o(T)$ are represented by the vectors $\bar{\mathbf{f}}(T-1)$ and $\bar{\mathbf{b}}(T-1)$. Since the filter coefficients are assumed to be adapting deterministically, the inverse covariance matrices $\hat{\mathbf{\Phi}}^{-1}(T|N)$ and $\hat{\mathbf{\Phi}}^{-1}(T|N+1)$ (defined by (6.3.34) and (6.3.35)) are also treated as deterministic entities. From the definition (6.3.38) it follows that each element of $\mathbf{g}(T-1|N)$ and $\mathbf{g}(T|N+1)$ is a linear combination of

$$y(T-1), y(T-2), \ldots, y(T-N) ,$$

and

$$y(T), y(T-1), \ldots, y(T-N) ,$$

respectively. The vector elements $[\mathbf{g}(T-1|N)]_j$, $1 \leqslant j \leqslant N$, and $[\mathbf{g}(T|N+1)]_j$, $1 \leqslant j \leqslant N+1$, can therefore be represented by the coefficient vectors

$$\tilde{\mathbf{g}}_j(T-1|N) = [0 \; g_{j,1}(T-1|N) \; g_{j,2}(T-1|N) \cdots g_{j,N}(T-1|N)]' \tag{7.4.17a}$$

and

$$\tilde{\mathbf{g}}_j(T|N+1) = [g_{j,1}(T|N+1) \; g_{j,2}(T|N+1) \cdots g_{j,N+1}(T|N+1)]' . \tag{7.4.17b}$$

In particular,

$$[\mathbf{g}(T-1|N)]_j = \tilde{\mathbf{g}}_j{}'(T-1|N) \; \mathbf{y}(T|N+1)$$

and

$$[\mathbf{g}(T|N+1)]_j = \tilde{\mathbf{g}}_j{}'(T|N+1) \; \mathbf{y}(T|N+1) .$$

The vectors $\tilde{\mathbf{g}}_j(T-1|N)$ and $\tilde{\mathbf{g}}_j(T|N+1)$ therefore approximate the jth row of the matrices $E[\hat{\mathbf{\Phi}}^{-1}(T-1|N)]$ and $E[\hat{\mathbf{\Phi}}^{-1}(T|N+1)]$, respectively.

Substituting $\tilde{\mathbf{g}}_j{}'\mathbf{y}(T|N+1)$ for $[\mathbf{g}]_j$ and $\bar{\mathbf{f}}'(T-1|N) \; \mathbf{y}(T|N+1)$ for $e_f^o(T|N)$, the mean-forward prediction coefficient vector update can be evaluated from (6.3.63b) using second-order statistics,

$$E[\mathbf{f}(T)]_j = E[\mathbf{f}(T-1)]_j + \sum_{l=1}^{N+1} \sum_{m=1}^{N} [\bar{\mathbf{f}}(T-1)]_l g_{j,m}(T-1|N) \phi_{l-m-1}, \; 1 \leqslant j \leqslant N , \tag{7.4.18a}$$

and

$$\bar{\mathbf{f}}(T) = [\,1 \;\; E[\mathbf{f}'(T)] \;\; 0 \cdots 0]' . \tag{7.4.18b}$$

The mean cost function update is similarly given by

$$E[\epsilon_f(T)] \approx wE[\epsilon_f(T-1)] + \sum_{l=1}^{N+1} \sum_{m=1}^{N+1} [\bar{\mathbf{f}}(T)]_l [\bar{\mathbf{f}}(T-1)]_m \phi_{l-m} . \tag{7.4.19}$$

The coefficient vectors $\tilde{\mathbf{g}}_j(T|N+1)$ can be updated by using (6.3.63e) and (6.3.63f), where the forward prediction vector $\mathbf{f}$ is replaced by its mean value. Replacing $[\mathbf{g}]_j$ by $\tilde{\mathbf{g}}_j$ and e_f by $\bar{\mathbf{f}}$ gives

$$\tilde{\mathbf{g}}_1(T|N+1) = \frac{1}{E[\epsilon_f(T)]}\,\overline{\mathbf{f}}(T) \tag{7.4.20a}$$

and

$$\tilde{\mathbf{g}}_j(T|N+1) = \tilde{\mathbf{g}}_{j-1}(T-1|N) - [\overline{\mathbf{f}}(T)]_j \tilde{\mathbf{g}}_1(T|N+1) \tag{7.4.20b}$$

for $2 \leqslant j \leqslant N+1$.

In order to compute $\overline{\mathbf{b}}(T)$, equation (6.3.61a) is used to write

$$\tilde{\mathbf{g}}_{N+1}(T|N+1) = \frac{1}{E[\epsilon_b(T)]}\,[\overline{\mathbf{b}}(T)] \;. \tag{7.4.21}$$

Since $[\overline{\mathbf{b}}(T)]_{N+1} = 1$, it follows that the $(N+1)$st element of $\tilde{\mathbf{g}}_{N+1}(T|N+1)$ is

$$g_{N+1,N+1}(T|N+1) = \frac{1}{E[\epsilon_b(T)]} \tag{7.4.22}$$

and from (7.4.21) it follows that

$$\overline{\mathbf{b}}(T) = \frac{1}{g_{N+1,N+1}(T|N+1)}\,\tilde{\mathbf{g}}_{N+1}(T|N+1) \;. \tag{7.4.23}$$

Finally, an update for the vectors $\tilde{\mathbf{g}}_j(T|N)$, $1 \leqslant j \leqslant N$, follows from equation (6.3.63i),

$$\tilde{\mathbf{g}}_j(T|N) = \tilde{\mathbf{g}}_j(T|N+1) + [\overline{\mathbf{b}}(T)]_j \tilde{\mathbf{g}}_{N+1}(T|N+1) \;. \tag{7.4.24}$$

The vectors $\tilde{\mathbf{g}}_j(T-1|N)$, $1 \leqslant j \leqslant N$, defined by (7.4.17a) and used at the Tth iteration must be replaced by the shifted vector $z^{-1}\tilde{\mathbf{g}}_j(T|N)$ at the $(T+1)$st iteration.

An adaptation model for the LS transversal filter is given by (7.4.18), (7.4.19), (7.4.20), (7.4.23), and (7.4.24). These recursions are computed recursively in time, in a similar fashion as the fast Kalman algorithm. Initialization of this adaptation model is essentially equivalent to the initialization of the fast Kalman algorithm. The elements of all vectors are set to zero, and the mean cost function can be set to some small value δ. The mean values of the prediction coefficients at each iteration are computed from (7.4.18), and the filter MSE can be estimated by using (7.4.7). Figure 7-3 shows mean value trajectories of the coefficient $f_{10}(T|10)$ obtained from the model and by computer simulation for the same input statistics used to illustrate the SG and LS lattice models. The adaptation constant w was also the same as that used in figure 7-2.

Although many approximations have been used to generate the models presented in this section, checks against computer simulated results indicate that these models can be used to obtain a reasonable assessment of algorithm performance given second-order information about the input. In addition, the approximations that have been made are useful for gaining qualitative insight into the convergence properties of the adaptive algorithms considered. This is further illustrated in the next section, where the convergence properties of two-stage SG transversal and lattice algorithms are compared.

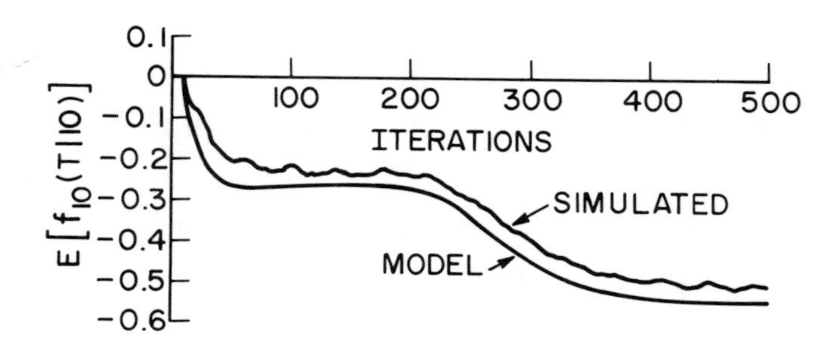

Figure 7-3. Mean value trajectory of the coefficient $f_{10}(T|10)$ computed by simulation and from the fast Kalman adaptation model.

7.5. TWO-STAGE SG COMPARISON

The results presented in the last few sections are now illustrated by examining the convergence behavior of a two-stage SG lattice algorithm as a function of the second-order input statistics. This behavior is then compared with that exhibited by a two-stage SG transversal algorithm. Although the general n-stage case is significantly more complicated, the basic ideas used to discuss the two-stage case carry through.

If it is assumed that each stage (coefficient) of an SG lattice algorithm does not start to converge until its inputs are stationary (i.e., approximately when $k_{n,opt}(T)$ is close to its asymptotic value), the filter will converge stage by stage. If the algorithm step size is normalized as in (7.2.5), then equation (7.2.12) states that the time constants for each stage are approximately dependent only upon the step size β_n and hence are independent of the input signal statistics. We have already seen, however, that the convergence speed of the nth stage is significantly influenced by the behavior of the first $(n-1)$ stages. In particular, $E[k_n(T)]$ is continually moving towards $k_{n,opt}(T)$ which *does* depend upon the input signal statistics. The trajectory of $k_{n,opt}(T)$ before it reaches its asymptotic value may therefore significantly influence the trajectory of $E[k_n(T)]$ and hence the convergence time of the filter.

In general the trajectories of $k_{n,opt}$, $n > 2$, are quite complicated (as figures 7-1 and 7-2 will testify), so that it is quite difficult to analytically determine the dependence of $k_{n,opt}(T)$, $n > 2$, upon the input statistics. For $n=2$, however, the problem simplifies considerably. To illustrate the previous discussion, consider a two-stage lattice for which $k_1(0) = k_2(0) = 0$. The dependence of the trajectories of $E[k_1(T)]$ and $E[k_2(T)]$ upon the input statistics (i.e., ϕ_0, ϕ_1, and ϕ_2) will now be determined. Since a stationary input is assumed, $E[k_1(T)]$ will converge to $k_{n,opt}$ with a time constant given by (7.2.12), which is independent of the input statistics. We therefore concentrate on the behavior of $E[k_2(T)]$. Assuming again that $k_1(T)$ is following its mean value trajectory,

a)

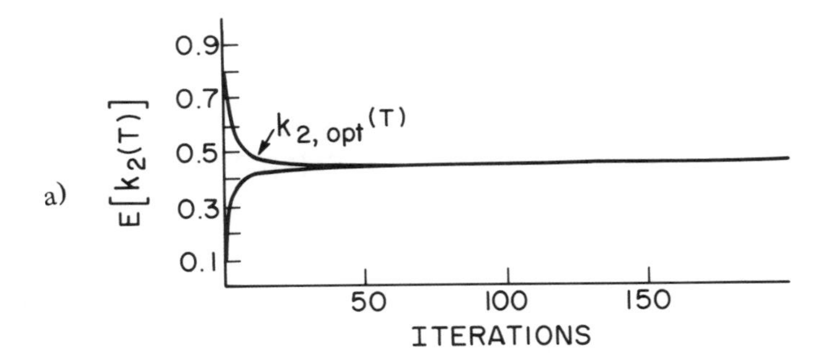

b)

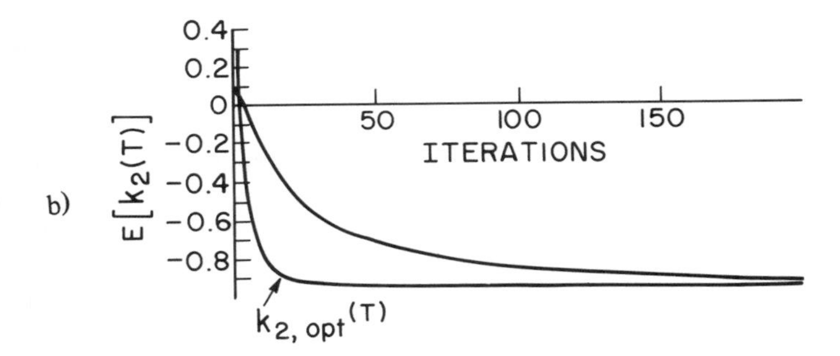

Figure 7-4. Trajectories of $E[k_2(T)]$ and $k_{2,opt}(T)$ in a two-stage adaptive lattice with different input statistics. Figure 7-4a shows a "fast mode" of the filter and figure 7-4b shows a "slow mode."

$$k_{2,opt}(T) = \frac{E[e_f(T|1)e_b(T-1|1)]}{E[e_f^2(T|1)]}$$

$$= \frac{E\Big[\Big(y(T) - k_1(T)y(T-1)\Big)\Big(y(T-2) - k_1(T)y(T-1)\Big)\Big]}{E\Big([y(T) - k_1(T)y(T-1)]^2\Big)} \tag{7.5.1}$$

$$\approx \frac{\phi_2 - 2E[k_1(T)]\phi_1 + E[k_1^2(T)]\phi_0}{\phi_0 - 2E[k_1(T)]\phi_1 + E[k_1^2(T)]\phi_0} ,$$

which can be rewritten as

$$k_{2,opt}(T) \approx 1 - \frac{\phi_2 - \phi_0}{E[e_f^2(T|1)]} . \tag{7.5.2}$$

This implies that

$$k_{2,opt}(T) - k_{2,opt}(\infty) = (\phi_0 - \phi_2)\left[\frac{1}{E_\infty[e_f^2(T|1)]} - \frac{1}{E[e_f^2(T|1)]}\right] \geqslant 0 \qquad \textbf{(7.5.3)}$$

since $0 < |\phi_2| < \phi_0$ and $E_\infty[e_f^2(T|1)] \leqslant E[e_f^2(T|1)]$, and hence

$$k_{2,opt}(T) \geqslant k_{2,opt}(\infty) \approx E_\infty[k_2(T)] \qquad (7.5.4)$$

for all T, independent of the input signal statistics. Since $k_{2,opt}(T)$ is monotonically decreasing, by changing the value of $k_{2,opt}(0) = \frac{\phi_2}{\phi_0}$ (assuming $k_1(0) = k_2(0) = 0$), as shown in figure 7-4, the convergence time of $E[k_2(T)]$ can be significantly altered along with the convergence time for the output MSE. The initial slope of the trajectory of $E[k_2(T)]$ is much greater in figure 7-4a than in figure 7-4b. Figure 7-5 shows the same type of behavior for a smaller value of $\frac{\phi_1}{\phi_0}$. Figures 7-4a and 7-5a represent a fast "mode" of the filter corresponding to $k_2(0) = 0 < k_{2,opt}$ while figures 7-4b and 7-5b represent a slower "mode" corresponding to $k_2(0) > k_{2,opt}$.

It is instructive to compare this behavior with that exhibited by the two-stage adaptive transversal filter. For N=2 the mean value of the prediction coefficient vector converges towards its optimal value according to two (fast and slow) normal modes. From (7.1.12) the time constant associated with each normal mode is $\tau_j \approx \frac{1}{\beta\lambda_j}$, j=1, 2 , where λ_j is the jth eigenvalue of the 2×2 autocorrelation matrix. Specifically, $\lambda_1 = \phi_0 + \phi_1$ and $\lambda_2 = \phi_0 - \phi_1$. Assuming the transversal coefficients $c_1(0)$ and $c_2(0)$ are fixed, each mode can be excited to different degrees by changing the value of $\frac{\phi_2}{\phi_0}$ (see problem 7-5). This is similar to the two-stage lattice behavior just discussed. Furthermore, for both the two-stage lattice and two-stage transversal filters, as $|\phi_1|$ (and hence the eigenvalue spread) increases, the difference between convergence times associated with each "mode" becomes greater.

The two-stage lattice therefore exhibits a similar type of dependence upon the input signal statistics as the two-stage transversal filter. On the other hand the difference between slow and fast "modes" as shown in figures 7-4 and 7-5 is less than the difference between normal modes in a two-stage transversal filter assuming the same value of ϕ_1. Although this discussion has implied that the convergence properties of a two-stage lattice can be influenced by the input statistics, it has been empirically observed that for larger filter orders the convergence properties of an adaptive lattice are insensitive to the input statistics (i.e., see reference [9] and the simulations in section 7.8). The effect of signal statistics on the convergence speed of the SG transversal filter has been found to be much more significant than the analogous effect on adaptive lattice filters. Because of slow convergence modes, or time constants, associated with small eigenvalues of the autocorrelation matrix, the SG transversal filter in general converges significantly more slowly than the SG lattice filter. This behavior is illustrated by the computer simulated results in section 7.8. By using the results in section 7.1 and this section, however, one can derive conditions

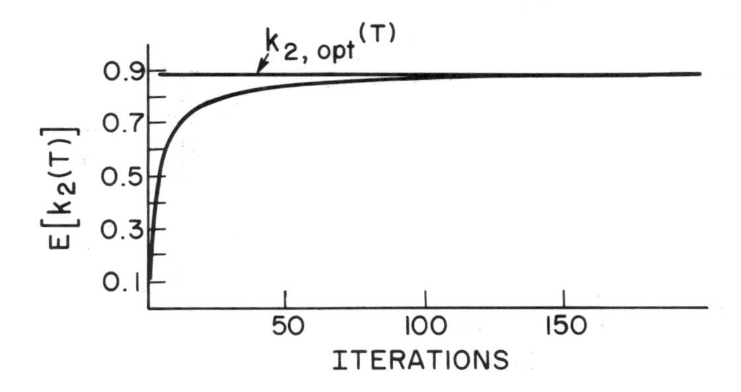

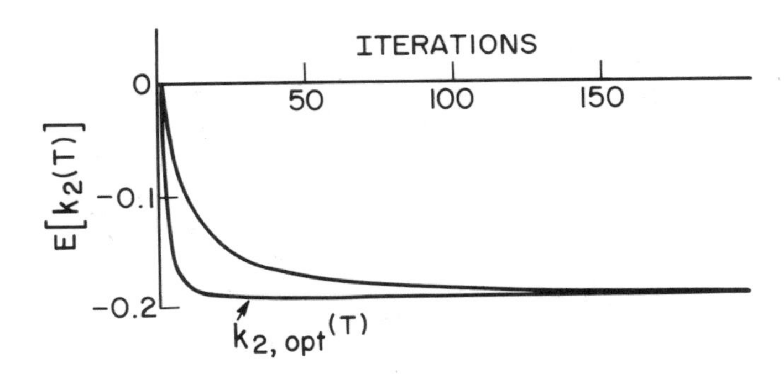

Figure 7-5. Trajectories of $E[k_2(T)]$ and $k_{2,opt}(T)$ in a two-stage adaptive lattice with different input statistics. The value of $\frac{\phi_1}{\phi_0}$ used in this case is less than that used to generate the curves in figure 7-4.

under which the two-stage SG transversal filter will converge faster than the two-stage SG lattice filter (problem 7-10).

7.6. JOINT PROCESS ESTIMATION

The analysis in section 7.1 applies to an SG transversal joint process estimator as well as the SG transversal predictor; however, the lattice results presented so far apply only to lattice predictors. In this section it is shown how the results of preceding sections can be extended in order to apply to the joint process estimation case. To begin, adaptation models for the SG lattice joint process estimators described in chapter 4 and shown in figure 4-12 are derived. An adaptation model for the LS lattice joint process estimator is subsequently derived. As before, the data sequence $\{y(T)\}$ is used to estimate some desired sequence $\{d(T)\}$.

7.6.1. SG Lattice Joint Process Estimation Algorithms

The two joint process estimator structures shown in figure 4-12 are equivalent assuming that the filter coefficients are fixed. In the adaptive mode, however, the coefficients that multiply the backward residuals of the lattice may be adapted in two ways. The algorithm corresponding to figure 4-12a is

$$e_c(T|N) = d(T) - \mathbf{k}^{c\prime}(T)\mathbf{e}_b(T) \tag{7.6.1a}$$

$$B(T+1|j) = (1-\beta)B(T|j) + e_b^2(T|j-1) \tag{7.6.1b}$$

$$k_j^c(T+1) = k_j^c(T) + \frac{1}{B(T|j)} e_b(T|j-1)e_c(T|N) \tag{7.6.1c}$$

where

$$\mathbf{k}^{c\prime}(T) = [k_1^c(T)\ k_2^c(T) \cdots k_N^c(T)]\ , \tag{7.6.2}$$

$$\mathbf{e}_b{}'(T) \equiv [e_b(T|0)\ e_b(T|1) \cdots e_b(T|N-1)]\ , \tag{7.6.3}$$

$k_j^c(T)$ is the jth joint process lattice coefficient, and $e_c(T|N)$ is the Nth-order joint process error, all at time T. The algorithm corresponding to figure 4-12b is

$$e_c(T|n+1) = e_c(T|n) - k_{n+1}^c(T)e_b(T|n) \tag{7.6.4a}$$

$$C_{n+1}(T+1) = (1-\beta)C_{n+1}(T) + e_c(T|n-1)e_b(T|n-1) \tag{7.6.4b}$$

$$\begin{aligned} k_{n+1}^c(T+1) &= \frac{C_{n+1}(T+1)}{B(T+1|n+1)} \\ &= \left[1 - \frac{e_b^2(T|n)}{B(T|n+1)}\right] k_{n+1}^c(T) + \frac{e_c(T|n)e_b(T|n)}{B(T|n+1)}\ . \end{aligned} \tag{7.6.4c}$$

The method for adapting the coefficients in (7.6.1) is similar to the SG transversal algorithm, and the second method (7.6.4) is analogous to an SG lattice prediction algorithm. Although the second method has received more attention and is analogous to the LS lattice joint process estimation algorithm, convergence models for both methods are described. Both adaptation techniques (7.6.1) and (7.6.4) are analogous to the normalized algorithm (7.2.5) used to adapt the lattice PARCOR coefficients in the sense that the constant step size β is effectively normalized by an estimate of the variance of the backward residual. Unnormalized adaptation techniques, which are analogous to (7.2.4), can also be used. In addition, it is possible to use different step sizes to adapt each different coefficient.

The optimal value of $\mathbf{k}^c(T)$, obtained by minimizing $E[e_c^2(T|N)]$, is

$$\mathbf{k}_{opt}^c = \{E[\mathbf{e}_b(T)\mathbf{e}_b{}'(T)]\}^{-1}E[d(T)\mathbf{e}_b(T)]\ . \tag{7.6.5}$$

If the lattice coefficients are optimal, then the backward errors of different orders are uncorrelated, so that the matrix $E[\mathbf{e}_b(T)\mathbf{e}_b{}'(T)]$ is diagonal and hence

$$[\mathbf{k}^c(T)]_{j,opt} = \frac{E[d(T)e_b(T|j-1)]}{E[e_b^2(T|j-1)]} \, . \tag{7.6.6}$$

Similarly, minimizing $E[e_c^2(T|n)]$ with respect to k_n^c gives

$$k_{n,opt}^c(T) = \frac{E[e_c(T|n-1)e_b(T|n-1)]}{E[e_b^2(T|n-1)]} \, . \tag{7.6.7}$$

If the PARCOR coefficients are fixed at their optimal values, then the principle of orthogonality implies that the two expressions in (7.6.6) and (7.6.7) are equivalent. Since the coefficients k_j or k_j^c, $1 \leqslant j < n$, are time-varying, e_c and e_b will be nonstationary causing $k_{n,opt}^c$ to vary with time.

7.6.2. SG Lattice Joint Process Estimator Adaptation Models

To model the lattice joint process estimation algorithm (7.6.1), equation (7.6.1c) is rewritten as

$$\begin{aligned} \mathbf{k}^c(T+1) &= \mathbf{k}^c(T) + \boldsymbol{\beta}(T)\mathbf{e}_b(T)\,[d(T) - \mathbf{e}_b{}'(T)\mathbf{k}^c(T)] \\ &= \Big[\mathbf{I} - \boldsymbol{\beta}(T)\mathbf{e}_b(T)\mathbf{e}_b{}'(T)\Big]\mathbf{k}^c(T) + \boldsymbol{\beta}(T)d(T)\mathbf{e}_b(T) \, , \end{aligned} \tag{7.6.8}$$

where $\boldsymbol{\beta}$ is a diagonal (normalized) step size matrix, i.e.,

$$\boldsymbol{\beta}(T) \equiv \text{diag}\left[\frac{1}{B(T|1)} \;\; \frac{1}{B(T|2)} \;\cdots\; \frac{1}{B(T|N)}\right] . \tag{7.6.9}$$

Taking expectations of both sides of (7.6.8) assuming $\mathbf{e}_b(T)$ and $\mathbf{k}^c(T)$ are approximately independent gives

$$\begin{aligned} E[\mathbf{k}^c(T+1)] \approx &\Big\{\mathbf{I} - E[\boldsymbol{\beta}(T)]\,E[\mathbf{e}_b(T)\mathbf{e}_b{}'(T)]\Big\}E[\mathbf{k}^c(T)] \\ &+ E[\boldsymbol{\beta}(T)]\,E[d(T)\mathbf{e}_b(T|N)] \, . \end{aligned} \tag{7.6.10}$$

Assuming that the lattice PARCOR coefficients are following trajectories close to their mean value trajectories, the convergence model for the lattice predictor previously discussed can be used to compute the second-order statistics of $\mathbf{e}_b(T)$. Specifically,

$$E[\mathbf{e}_b(T)\mathbf{e}_b{}'(T)]_{lm} = E[e_b(T|l-1)e_b(T|m-1)] \tag{7.6.11}$$

$$= E\left[\left(y(T-l+1) - \sum_{j=0}^{l-2} b_{j+1}(T|l-1)y(T-j)\right)\left(y(T-m+1) - \sum_{j=0}^{m-2} b_{j+1}(T|m-1)y(T-j)\right)\right]$$

$$\approx \phi_y(l-m) - \sum_{j=0}^{m-2} b_{j+1}(T|m-1)\phi_y(l-j-1) - \sum_{j=0}^{l-2} b_{j+1}(T|l-1)\phi_y(m-j-1)$$

$$+ \sum_{q=0}^{l-2}\sum_{j=0}^{m-2} b_{q+1}(T|l-1)b_{j+1}(T|m-1)\phi_y(q-j)$$

and

$$\{E[d(T)\mathbf{e}_b(T)]\}_m = E[d(T)e_b(T|m-1)]$$

$$= E\left[d(T)\left[y(T-m+1) - \sum_{j=0}^{m-2} b_{j+1}(T|m-1)y(T-j)\right]\right] \tag{7.6.12}$$

$$= \phi_{dy}(m-1) - \sum_{j=0}^{m-2} b_{j+1}(T|m-1)\phi_{dy}(j) ,$$

where

$$\phi_y(j) = E[y(T)y(T-j)] ,$$

$$\phi_{dy}(j) = E[d(T)y(T-j)] ,$$

and the mean backward prediction coefficients $b_j(T|m-1)$, $1 \leqslant j \leqslant m-1$, $1 \leqslant m \leqslant N$, are computed from the convergence model for the lattice predictor. The step size

$$\{E[\boldsymbol{\beta}(T+1)]\}_{nn} \approx \frac{1}{E[B(T+1|n)]},\ 1 \leqslant n \leqslant N ,$$

is computed by taking expected values of both sides of (7.6.1b), using the value of $E[e_b^2(T|n-1)]$ calculated from the lattice predictor model. The output mean square error can be estimated at time T by using the mean coefficient values computed from (7.6.10). In particular,

$$E[e_c^2(T|N)] = E\left\{\left[d(T) - \mathbf{k}^{c\prime}(T)\mathbf{e}_b(T|N)\right]^2\right\}$$

$$\approx E[d^2(T)] - 2E[\mathbf{k}^{c\prime}(T)]\, E[d(T)\mathbf{e}_b(T)] \tag{7.6.13}$$

$$+ E[\mathbf{k}^{c\prime}(T)]E[\mathbf{e}_b(T)\mathbf{e}_b{}'(T)]E[\mathbf{k}^c(T)]$$

where $E[\mathbf{k}^c(T)]$, $E[\mathbf{e}_b(T)\mathbf{e}_b{}'(T)]$, and $E[d(T)\mathbf{e}_b(T)]$ are given respectively by (7.6.10), (7.6.11), and (7.6.12). Combining (7.6.10) through (7.6.13) with the appropriate SG lattice predictor model completes the convergence model for the SG lattice joint process estimator shown in figure 4-12a. The procedure for computing the mean values of the joint process coefficients $k_j^c(T)$, $j = 1, \ldots, N$, and the output MSE is similar to the prediction case. The SG lattice predictor model is first used to compute the backward coefficients $b_j(T|n)$, $j = 1, \ldots, N$, $1 \leqslant n \leqslant N$. These are used in (7.6.11) and (7.6.12) along with the second-order input statistics to compute the cross-correlations that are subsequently needed to compute the update (7.6.10) for $\mathbf{k}^c$ and the expression for output MSE (7.6.13). This adaptation model requires significantly more computation than the adaptation models presented so far, because the matrix $E[\mathbf{e}_b(T)\mathbf{e}_b{}'(T)]$ has to be computed at each iteration. The convergence model is initialized by setting each variable to zero. figure 7-6 illustrates the model by comparing the mean value trajectory for the coefficient

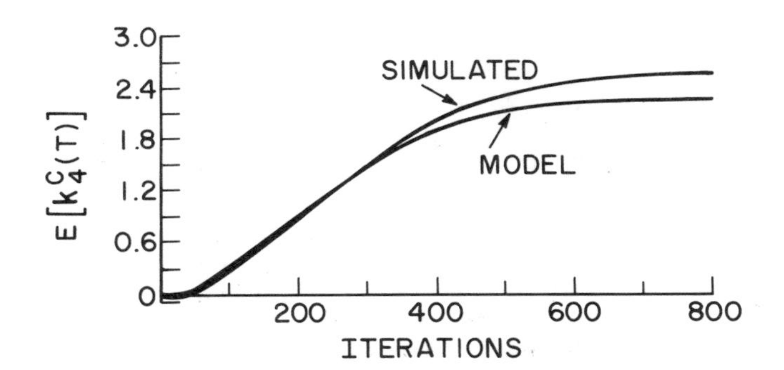

Figure 7-6. Mean value of lattice joint process coefficient $k_4^c(T|10)$ computed from multiple simulations of the algorithm (7.6.1), and from the adaptation model.

$k_4^c(T)$ computed from the model with the mean value trajectory obtained by averaging 200 separate simulations of the algorithm with a step size β=0.01. The inputs $d(T)$ and $y(T)$ used in the simulations were correlated Gaussian noise sequences.

A convergence model for the second SG lattice joint process estimation algorithm (7.6.4), which corresponds to the filter structure shown in figure 4-12b, is derived in an analogous fashion as the model for the SG lattice predictor algorithm (7.2.5). We first note from (7.6.1a) that $e_c(T|n)$ is a linear combination of $d(T), y(T-1), y(T-2), \ldots, y(T-n)$, i.e.,

$$e_c(T|n) = \bar{\mathbf{c}}'(T|n)\mathbf{y}_d(T) \tag{7.6.14}$$

where

$$\bar{\mathbf{c}}'(T|n) \equiv [1 \; -c_1(T|n) \cdots -c_n(T|n) \; 0 \; 0 \cdots 0] \tag{7.6.15}$$

and

$$\mathbf{y}_d{}'(T) \equiv [d(T) \; y(T) \; \cdots \; y(T-n) \cdots y(T-N+1)] \,. \tag{7.6.16}$$

All vectors $\bar{\mathbf{c}}(T|n)$, $1 \leqslant n \leqslant N$, have dimension $N+1$. The order recursion for the joint process error (7.6.4a) can be rewritten with $k_{n+1}^c(T)$ replaced by its mean value as

$$\bar{\mathbf{c}}(T|n+1) = \bar{\mathbf{c}}(T|n) - E[k_{n+1}^c(T)]\bar{\mathbf{b}}(T|n) \tag{7.6.17}$$

where $\bar{\mathbf{b}}$ is defined by (7.4.4b). Taking expected values of both sides of (7.6.4c) assuming $k_n^c(T)$ is approximately independent of $e_b(j|n)$, $0 \leqslant j \leqslant T$, gives

$$E[k_{n+1}^c(T+1)] \approx \left\{1 - \frac{E[e_b^2(T|n)]}{E[B(T|n)]}\right\}E[k_{n+1}^c(T)] + \frac{E[e_c(T|n)e_b(T|n)]}{E[B(T|n)]} \,. \tag{7.6.18}$$

The backward residual variance $E[e_b^2(T|n)]$, and hence $E[B(T|n)]$, $1 \leqslant n \leqslant N$, can be computed from the appropriate SG lattice predictor model. The second-order statistics of e_c and e_b can be computed,

$$E[e_c^2(T|n)] = E\Big[\overline{\mathbf{c}}'(T|n)\mathbf{y}_d(T)\mathbf{y}_d'(T)\overline{\mathbf{c}}(T|n)\Big]$$
$$\approx E[\overline{\mathbf{c}}'(T|n)]\, E[\mathbf{y}_d(T)\mathbf{y}_d'(T)]\, E[\overline{\mathbf{c}}(T|n)]\ , \tag{7.6.19}$$

and

$$E[e_c(T|n)e_b(T|n)] = E\Big[\overline{\mathbf{c}}'(T|n)\mathbf{y}_d(T)\mathbf{y}_d'(T)\overline{\mathbf{b}}(T|n)\Big]$$
$$\approx E[\overline{\mathbf{c}}'(T|n)]\, E[\mathbf{y}_d(T)\mathbf{y}_d'(T)]\, E[\overline{\mathbf{b}}(T|n)]\ . \tag{7.6.20}$$

Combining (7.6.17-20) with the appropriate SG lattice predictor model completes the convergence model for the SG lattice joint process estimation algorithm (7.6.4). The model recursions (7.6.17) and (7.6.18) are computed in exactly the same manner as the the model recursions for the SG lattice predictor. In particular, $E[e_b^2(T|0)] = \phi_y(0)$ and $E[e_c(T|n)e_b(T|0)] = \phi_{dy}(0)$ are used to compute $E[k_1^c(T)]$, which is used in (7.6.17) to compute $\overline{\mathbf{c}}(T|1)$. Equations (7.6.19) and (7.6.20) can then be used to compute $E[e_c^2(T|1)]$ and $E[e_c(T|n)e_b(T|n)]$, and so forth until the mean values of all N joint process coefficients in addition to the output MSE $E[e_c^2(T|N)]$ have been computed. This convergence model requires less computation per iteration than the previous lattice joint process convergence model since there are no matrices that need to be stored and updated at each iteration. Figure 7-7 compares the mean value trajectory of the coefficient $k_4^c(T)$ computed from the model with the mean value trajectory obtained by averaging 200 separate simulations of the algorithm (7.2.5) and (7.6.4). The input statistics were the same as those used to generate figure 7-6. Also shown is the time varying optimal value $k_{n,opt}^c(T)$, given by (7.6.7).

A comparison of figures 7-6 and 7-7 suggests that the speeds of convergence for both algorithms (7.6.1) and (7.6.4) are roughly the same. Intuitively, this seems reasonable since in both cases the lattice predictor is attempting to orthogonalize the set of inputs driving the joint process coefficients. If it is therefore assumed that the backward error covariance matrix $E[\mathbf{e}_b(T)\mathbf{e}_b'(T)]$ is diagonal, and that $E[d(T)e_b(T|n)] \approx E[e_c(T|n)e_b(T|n)]$, which is exactly true if the coefficients are fixed at their optimal values, then equations (7.6.10) and (7.6.18) imply that the mean values of the filter coefficients that result from using either (7.6.4) or (7.6.1) are approximately equal. Although this is not the case when the lattice prediction coefficients are adapting, the off-diagonal elements of the backward error covariance matrix converge to zero and hence the effect of cross-coupling between the joint process coefficients in (7.6.10) should become negligible.

Figures 7-6 and 7-7 indicate that both algorithms (7.6.1) and (7.6.4) produce biased coefficient estimates. This is not surprising in view of the coefficient bias present in the SG lattice predictor. This bias is again caused by filter coefficient variance, which alters the statistics driving other filter coefficients,

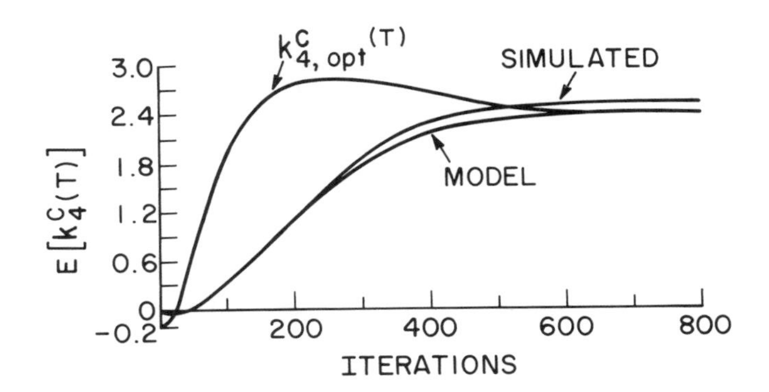

Figure 7-7. Mean value of lattice joint process coefficient $k_4^c(T)$ computed from multiple simulations of the algorithm, and from the adaptation model for the algorithm.

and by correlations between filter coefficients and the input data. In the case of the algorithm (7.6.4), the estimate of k_n^c is unaffected by filter coefficients k_j^c and k_j for $j > n$. This is in contrast to the algorithm (7.6.1) for which $k_j^c(T)$ depends upon *all* other filter coefficients. It therefore seems likely that the statistics which determine the asymptotic values of $k_j^c(T)$, $1 \leqslant j \leqslant N$, when using (7.6.1) are more severely altered by the adaptation of other coefficients than when (7.6.4) is used. This observation is consistent with the fact that the coefficient estimate in figure 7-6 is more biased than the estimate in figure 7-7. This in turn implies that the asymptotic output MSE produced by the first algorithm is typically somewhat larger than that produced by the second.

A first-order estimate of lattice joint process coefficient variance and output MSE produced by (7.6.1) or (7.6.4) can be obtained using the techniques in section 7.2. In particular, squaring both sides of (7.6.4a), and taking expected values of both sides assuming that k_n^c is approximately independent of the inputs to the nth filter stage gives in analogy with (7.2.18),

$$E_\infty[e_c^2(T|n+1)] \approx E_\infty[e_c^2(T|n)] - \left(k_{n+1,opt}^{c^2} - \text{var}_\infty k_{n+1}^c\right) \cdot E_\infty[e_b^2(T|n)] \tag{7.6.21}$$

where $\text{var}_\infty k_n^c(T) \equiv E_\infty[k_n^{c^2}(T)] - \{E_\infty[k_n^c(T)]\}^2$ is the asymptotic variance of k_n^c. Evaluation of $\text{var}_\infty k_n^c$, $1 \leqslant n \leqslant N$, can be accomplished in the same manner as the evaluation of coefficient variance in section 7.2. Assuming

$$E[e_b(T|n-1)e_b(T+j|n-1)] \approx E[e_c(T|n-1)e_b(T+j|n-1)] = 0, \; j \neq 0 \,,$$

then the same procedure used to derive (7.2.26) can be used to obtain (problem 7-14)

$$\text{var}_\infty k_n^c \approx \frac{\beta}{2+\beta}\left[\frac{E[e_c^2(T|n-1)]}{E[e_b^2(T|n-1)]} - k_{n,opt}^{c\,2}\right]. \tag{7.6.22}$$

Assuming that the lattice predictor asymptotically eliminates cross-coupling between the backward error weights, then (7.6.21) and (7.6.22) apply to both algorithms (7.6.1) and (7.6.4). In general, this assumption is not strictly true due to correlations between the filter coefficients and the data, and hence small differences in coefficient variance and output MSE as produced by (7.6.1) and (7.6.4) will be observed.

7.6.3. LS Lattice Joint Process Estimator Adaptation Model

In analogy with the SG lattice, the convergence model for the LS lattice predictor is easily extended to the LS lattice joint process estimator. In addition to the predictor recursions in section 6.3, the (prewindowed) LS lattice joint process estimator uses the additional recursions

$$K_{n+1}^c(T) = wK_{n+1}^c(T-1) + \frac{e_c(T|n)e_b(T|n)}{1-\gamma(T|n)} \tag{7.6.23}$$

and

$$e_c(T|n+1) = e_c(T|n) - k_{n+1}^c(T)e_b(T|n) \tag{7.6.24}$$

where $k_{n+1}^c(T) = \dfrac{K_{n+1}^c(T)}{\epsilon_b(T|n)}$. The residual $e_c(T|n)$ can be represented by the vector $\overline{c}(T|n)$ defined by (7.6.15), and equation (7.6.24) can be rewritten with $k_{n+1}^c(T)$ replaced by its mean value as (7.6.17). The mean values of the joint process coefficients can be computed,

$$E[k_{n+1}^c(T)] \approx \frac{E[K_{n+1}^c(T)]}{E[\epsilon_b(T|n)]} \tag{7.6.25}$$

and

$$E[K_{n+1}^c(T+1)] \approx wE[K_{n+1}^c(T)] + \frac{E[e_c(T|n)e_b(T|n)]}{1-E[\gamma(T|n)]} \tag{7.6.26}$$

where $E[e_c(T|n)e_b(T|n)]$ is given by (7.6.20). Output MSE as a function of time is given by (7.6.19).

The LS lattice predictor model derived in section 7.4.2 combined with the order update (7.6.17), the formulas for second-order statistics (7.6.19) and (7.6.20), and the mean coefficient time updates (7.6.25) and (7.6.26) complete the LS lattice joint process estimator model. The procedure for computing the mean joint process coefficient values and output MSE is essentially the same as that used to generate the same quantities for the SG lattice joint process estimator (7.6.4). The mean coefficient trajectories computed from this convergence model are nearly identical to those obtained from the SG models discussed earlier. This is again due to the fact that the LS gains $\gamma(T|n)$, $1 \leqslant n \leqslant N$, stay relatively close to zero with a Gaussian noise input, thereby eliminating much of the difference between the SG and LS lattice algorithms.

7.7. ALGEBRAIC PROPERTIES OF LS ALGORITHMS

So far we have been concerned with the statistical behavior of LS algorithms. Specifically, approximate statistical averages of coefficients and output squared error have been computed. This type of analysis seems intuitively reasonable when the input to the adaptive filter is Gaussian noise. For this case one would expect that the adaptation of filter coefficients is more "random" than for other types of inputs, so that consideration of individual sample functions (i.e., squared error vs. time) should yield little additional insight. In applications such as channel equalization, where the input to the adaptive filter is derived from something other than a Gaussian noise source, however, the situation is different. This is illustrated in figure 7-8 which shows two adaptive predictors driven by two types of inputs. The input to the top filter is obtained by passing white Gaussian noise samples through an all-pole linear filter. The input to the second adaptive filter is obtained by passing an impulse sequence through the same filter. In both cases the job of the adaptive filter is to estimate the linear filter $H(z)$. Assuming that the power of the impulse sequence is the same as

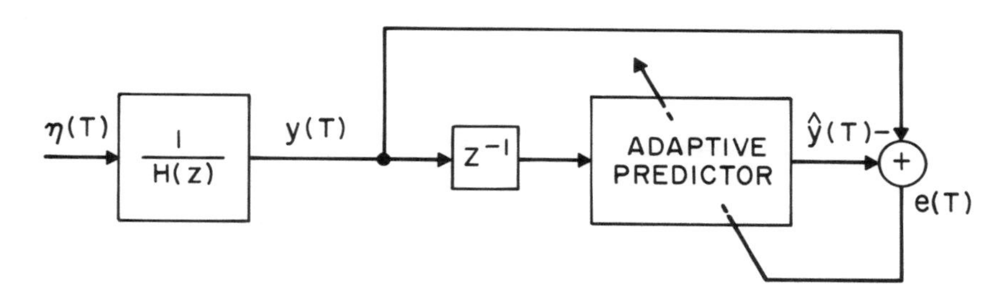

Figure 7-8a. Adaptive predictor with a correlated Gaussian noise input.

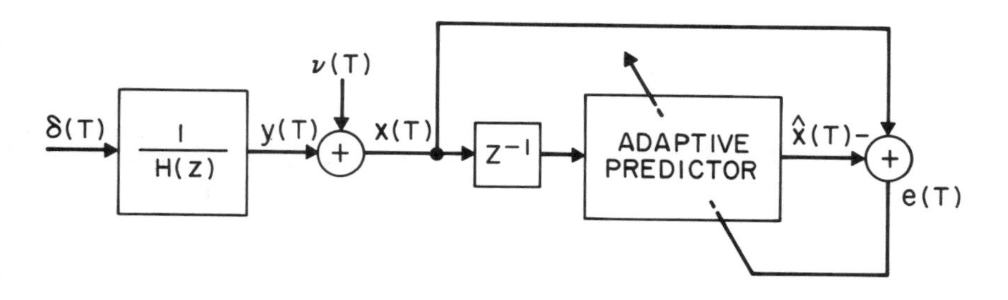

Figure 7-8b. Adaptive predictor with a filtered impulse sequence as the input.

that of the white Gaussian noise sequence and that the additive noise $\nu(T)=0$, the input in figure 7-8b clearly makes the job of the associated adaptive filter much easier than that of the adaptive filter in figure 7-8a. The first N impulse response coefficients of $H(z)$ can be *perfectly* estimated by the bottom filter in N iterations by solving an elementary set of linear equations. In contrast, the top filter must statistically average the input noise and hence can perfectly estimate the first N impulse response coefficients only as the number of iterations becomes infinite.

The previous discussion indicates that the convergence properties of an LS algorithm greatly depend on the type of input sequence considered. In this section LS algorithms are investigated by considering their behavior with *known* input sequences. This is a relatively simple deterministic problem that yields considerable insight. The statistical behavior of an LS algorithm can subsequently be obtained by averaging over the input sequence. Unfortunately, this is an extremely difficult task for almost any type of random input. Nevertheless, by considering individual sample paths it is shown later in this section that for some types of input sequences, such as those that occur in channel equalization and echo cancellation of data signals, whether or not the shifted input data vectors $\mathbf{y}(T)$ are linearly independent plays a crucial role in determining when an LS algorithm converges. Convergence properties of LS algorithms are therefore intimately connected with the "algebraic" properties of the input sequence [10].

7.7.1. Asymptotic Properties

If it is assumed that the filter $H(z)$ in figure 7-8 has a finite impulse response of length N, then the LS estimator, and in particular the asymptotic properties of the LS estimator, can be conveniently characterized. Let $\mathbf{h}$ denote the vector whose elements are the first N impulse response coefficients of $H(z)$. The inputs to the adaptive filters in figures 7-8a and 7-8b are, respectively,

$$y(T) = \mathbf{h}'\mathbf{y}(T-1) + \eta(T) \tag{7.7.1a}$$

and

$$y(T) = \mathbf{h}'\mathbf{y}(T-1) + \delta(T) + \nu(T) , \tag{7.7.1b}$$

where $\delta(T)$ is an impulse sequence, $\eta(T)$ and $\nu(T)$ are white Gaussian noise sequences, and the vector $\mathbf{y}$ has N components and is defined as in previous sections.

Denoting the coefficient vector associated with the adaptive filter as $\mathbf{c}$, the optimum value of $\mathbf{c}$ that minimizes the output MSE in both cases is

$$\mathbf{c}_{opt} = \mathbf{h}, \tag{7.7.2}$$

assuming that the impulse sequence $\delta(T)$ is uncorrelated. The resulting error sequence is denoted as

$$e_{opt}(T) = y(T) - \mathbf{y}'(T-1)\mathbf{c}_{opt} . \tag{7.7.3}$$

In the case of figure 7-8a, $e_{opt}(T) = \eta(T)$, and in figure 7-8b, $e_{opt}(T) = \delta(T) + \nu(T)$. After T iterations the LS estimate of $\mathbf{c}_{opt}$, which is denoted as $\mathbf{c}(T)$, satisfies

$$\left[\sum_{j=0}^{T} w_j \mathbf{y}(j-1)\mathbf{y}'(j-1)\right]\mathbf{c}(T) = \left[\sum_{j=0}^{T} w_j y(j)\ \mathbf{y}(j-1)\right] \tag{7.7.4}$$

where w_j, $j=0, \ldots, T$, is a weighting sequence. If the sample covariance matrix

$$\hat{\mathbf{\Phi}}(T) = \sum_{j=0}^{T} w_j \mathbf{y}(j)\mathbf{y}'(j) \tag{7.7.5}$$

is nonsingular, then (7.7.4) can be rewritten as

$$\mathbf{c}(T) = \hat{\mathbf{\Phi}}^{-1}(T-1)\left[\sum_{j=0}^{T} w_j y(j)\mathbf{y}(j-1)\right]. \tag{7.7.6}$$

Solving (7.7.3) for $y(j)$ and substituting into (7.7.6) gives

$$\begin{aligned}\mathbf{c}(T) &= \hat{\mathbf{\Phi}}^{-1}(T-1)\left[\sum_{j=0}^{T} w_j \left[\mathbf{y}(j-1)\mathbf{y}'(j-1)\right]\mathbf{c}_{opt} + \sum_{j=0}^{T} w_j e_{opt}(j)\ \mathbf{y}(j-1)\right] \\ &= \mathbf{c}_{opt} + \hat{\mathbf{\Phi}}^{-1}(T-1)\left[\sum_{j=0}^{T} w_j e_{opt}(j)\mathbf{y}(j-1)\right].\end{aligned} \tag{7.7.7}$$

The LS estimate of $\mathbf{c}_{opt}$ is therefore corrupted by an additive noise term which depends upon the inverse sample covariance matrix. The matrix $\hat{\mathbf{\Phi}}^{-1}$ can be viewed as an "orthogonalizing" factor since the covariance matrix of the right hand additive noise term in (7.7.7) is diagonal, assuming $e_{opt}(T)$ is uncorrelated from sample to sample.

The performance of a LS algorithm can be conveniently described by using (7.7.7). The covariance matrix of the estimate conditioned on the input data sequence follows directly from (7.7.7) and is given by

$$\begin{aligned}&E\left[\mathbf{q}(T)\mathbf{q}'(T) \mid y(0), y(1), \ldots, y(T)\right] \\ &= \hat{\mathbf{\Phi}}^{-1}(T-1)\left[\sum_{j=0}^{T}\sum_{l=0}^{T} w_j w_l E\left[e_{opt}(j)e_{opt}(l)\right]\mathbf{y}(j-1)\mathbf{y}'(j-1)\right]\hat{\mathbf{\Phi}}^{-1}(T-1) \\ &= \epsilon_{\min}\hat{\mathbf{\Phi}}^{-1}(T-1)\left[\sum_{j=0}^{T} w_j^2\mathbf{y}(j-1)\mathbf{y}'(j-1)\right]\hat{\mathbf{\Phi}}^{-1}(T-1)\end{aligned} \tag{7.7.8}$$

where

$$\mathbf{q}(T) = \mathbf{c}(T) - \mathbf{c}_{opt}, \tag{7.7.9}$$

and the sequence $e_{opt}(T)$ is assumed to be uncorrelated and $E[e_{opt}^2(T)] = \epsilon_{\min}$. Averaging (7.7.8) over the input sequence $\{y(T)\}$ is extremely difficult; however, for large T we can approximate $\hat{\mathbf{\Phi}}$ as

$$\hat{\mathbf{\Phi}}(T) \approx \mathbf{\Phi}\sum_{j=0}^{T} w_j, \tag{7.7.10}$$

where the autocorrelation matrix $\mathbf{\Phi} = E[\mathbf{y}(T)\ \mathbf{y}'(T)]$. Substituting for $\hat{\mathbf{\Phi}}$ in

(7.7.8) gives

$$E\Big[\mathbf{q}(T)\mathbf{q}'(T)\Big] \approx \epsilon_{\min} \frac{\sum_{j=0}^{T} w_j^2}{\left(\sum_{j=0}^{T} w_j\right)^2} \boldsymbol{\Phi}^{-1} . \tag{7.7.11}$$

Using the transformation of variables specified by (7.1.7), it follows that

$$E\Big[\tilde{\mathbf{q}}(T)\tilde{\mathbf{q}}'(T)\Big] \approx \epsilon_{\min} \frac{\sum_{j=0}^{T} w_j^2}{\left(\sum_{j=0}^{T} w_j\right)^2} \mathbf{I} , \tag{7.7.12}$$

and from (7.1.17) and (7.1.19) it follows that the output MSE is

$$\begin{aligned} E\left[e^2(T)\right] &\approx \epsilon_{\min} + \epsilon_{\min} \sum_{j=1}^{N} \lambda_j \frac{\sum_{j=0}^{T} w_j^2}{\left(\sum_{j=0}^{T} w_j\right)^2} \\ &= \left(1 + N\phi_0 \frac{\sum_{j=0}^{T} w_j^2}{\left(\sum_{j=0}^{T} w_j\right)^2}\right) \epsilon_{\min} , \end{aligned} \tag{7.7.13}$$

where λ_j is the jth eigenvalue of $\boldsymbol{\Phi}$. If $w_j=1$ for all j, then

$$E\left[e^2(T)\right] \approx \left(1 + \frac{N\phi_0}{T+1}\right)\epsilon_{\min} , \tag{7.7.14}$$

and if $w_j = w^{T-j}$, then

$$E\left[e^2(T)\right] \approx \left(1 + N\phi_0 \frac{1-w}{1+w} \frac{1-w^{T+1}}{1+w^{T+1}}\right) \epsilon_{\min} . \tag{7.7.15}$$

Letting $T \rightarrow \infty$ gives the following expression for the excess MSE due to adaptation,

$$\epsilon_{ex} = N\phi_0 \epsilon_{\min} \frac{1-w}{1+w} . \tag{7.7.16}$$

Equation (7.7.14) indicates that if the weighting sequence is uniform, then the decay of the output MSE to the minimum fixed-coefficient MSE is asymptotically inversely proportional to the number of iterations T. Assuming an exponential weighting sequence, (7.7.15) indicates that the smaller the exponential weight w, the faster the output MSE converges to the asymptotic MSE, and the larger the asymptotic MSE. It is emphasized that these formulas apply to LS estimators in general, and in particular to lattice realizations.

7.7.2. Equalizer Start-up Properties

The formulas (7.7.13-16) describe asymptotic properties of LS algorithms; however, they provide little information concerning initial start-up properties. An application where start-up properties are of crucial importance is channel equalization, where reliable transmission cannot be presumed until the output MSE of the adaptive equalizer is below some acceptable threshold. The estimation problem encountered in channel equalization is illustrated in figure 7-9. The input to the adaptive filter is in this case derived from a random binary sequence. This type of input sequence is significantly different from both input sequences shown in figure 7-8. Notice also that the adaptive filter is a joint process estimator, rather than a predictor.

Assuming that the order of the equalizer is $N = 2L+1$, where L is an integer, recall from the discussion in chapter 2 that given the channel output data samples $y(T), y(T-1), \ldots, y(T-N+1)$, the adaptive equalizer produces the estimate $\hat{d}(T-L)$. In particular,

$$\hat{d}(T-L) = \mathbf{c}'(T)\,\mathbf{y}(T)\,. \tag{7.7.17}$$

The training sequence $d(T)$ must therefore be appropriately delayed so that the equalizer coefficients multiply both succeeding and preceding channel outputs. It will be convenient in this section to assume that at time T, the coefficient vector $\mathbf{c}(T)$ is estimated from the transmitted symbol sequence

$$d(0), d(1), \ldots, d(T)\,,$$

and the channel outputs

$$y(0), y(1), \ldots, y(T+L)\,.$$

The LS solution for the equalizer coefficient vector at time T satisfies

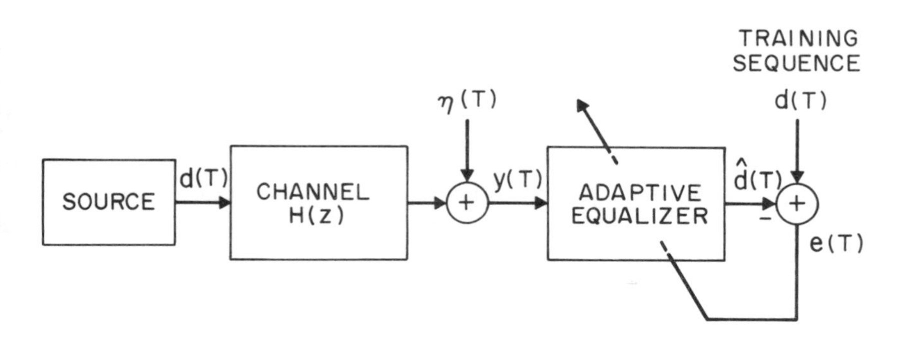

Figure 7-9. Transmission channel with an adaptive equalizer.

$$\left[\sum_{j=0}^{T} \mathbf{y}(j+L)\mathbf{y}'(j+L)\right] \mathbf{c}(T) = \left[\sum_{j=0}^{T} d(j)\,\mathbf{y}(j+L)\right], \qquad \textbf{(7.7.18)}$$

where the output of the channel is

$$y(T) = \sum_{j=-\infty}^{\infty} d(j)h_{T-j} + \eta(T), \qquad (7.7.19)$$

h_T is the channel impulse response, and $\eta(T)$ is an additive white Gaussian noise sequence. At time $T = N-1$, (7.7.18) can be rewritten as

$$\sum_{j=0}^{N-1}\left[d(j) - \mathbf{y}'(j+L)\,\mathbf{c}(N-1)\right]\mathbf{y}(j+L) = 0. \qquad (7.7.20)$$

If the data vectors $\mathbf{y}(j+L)$, $j = 0, \ldots, N-1$, are linearly independent, this implies that

$$d(j) = \mathbf{y}'(j+L)\,\mathbf{c}(N-1), \quad j = 0, 1, \ldots, N-1, \qquad (7.7.21)$$

or equivalently, the prediction error is

$$e(j) = 0, \quad j = 0, 1, \ldots, N-1. \qquad (7.7.22)$$

If perfect equalization is possible (that is, if the channel transfer function is all-pole and of order N), then in the noiseless case (7.7.21) indicates that the optimal coefficient vector will be computed at the Nth iteration. Furthermore, the equalization error will remain zero and the algorithm will cease to adapt. Although this is not the case in general, the algebraic properties, and in particular, linear independence, of the input data vectors play a crucial role in determining the convergence behavior of the LS estimate. This was first pointed out in reference [10], from which this discussion is taken. The condition (7.7.21), where d is replaced by y and $L=-1$, also arises from (7.7.4) if the vectors $\mathbf{y}(j)$, $j=0, \ldots, N-1$, are linearly independent. It is emphasized, however, that this does *not* always imply that the LS estimate $\mathbf{c}(N-1)$ is the optimal MMSE solution. Considering again the predictor in figure 7-8, the condition (7.7.21) becomes

$$y_j = \mathbf{c}_{opt}'\mathbf{y}(j-1) + \eta(j) = \mathbf{y}'(j)\,\mathbf{c}(N-1), \quad j = 0, 1, \ldots, N-1, \qquad (7.7.23)$$

which does not imply that $\mathbf{c}(N-1) = \mathbf{c}_{opt}$ unless $\eta(j) = 0$, $j = 0, 1, \ldots, N-1$. The convergence of the coefficient vector therefore depends not only on the linear independence of the input data vectors, but also on the statistical properties of the input. If the input to the adaptive filter is Gaussian noise, then significantly more than N input vectors are needed in order to gain statistical confidence in the coefficient estimates.

Assuming the input sequence $\{d(T)\}$ is binary, then after T iterations the mean value of the coefficient vector and the output MSE can theoretically be obtained by averaging appropriately modified versions of (7.7.7) and (7.7.8) over 2^T input sequences. Rather than attempt to compute such a complicated average, we now examine the LS solution (7.7.6) given two *specific* start-up sequences. The LS estimate of $\mathbf{c}$ after N iterations has special properties that lend insight into the convergence process for random sequences.

If a single pulse is transmitted at iteration L, then from (7.7.19) it follows that

$$\begin{aligned} \mathbf{y}'(j+L) &= [h_j \quad \cdots \quad h_{j-L} \quad \cdots \quad h_{j-2L}] \\ &= \mathbf{h}(j) \ , \quad j = 0, 1, \ldots, N-1 \ . \end{aligned} \tag{7.7.24}$$

After N iterations the LS solution $\mathbf{c}(N-1)$ satisfies the condition (7.7.21), which can be rewritten as

$$\begin{aligned} \mathbf{h}'(L)\, \mathbf{c}(N-1) &= 1 \\ \mathbf{h}'(j)\, \mathbf{c}(N-1) &= 0, \quad j = 0, 1, \ldots, N-1, j \neq L \ . \end{aligned} \tag{7.7.25}$$

This set of equations uniquely defines the zero-forcing equalizer (see reference [11]), assuming the impulse response $h_j = 0$, $|j| > N-1$. Specifically, it is easily verified from (7.7.25) and (7.7.19) that $\mathbf{c}'(N-1)\, \mathbf{y}(T) = d(T-L)$, assuming $\eta(T) = 0$ for all T. The equalizer $\mathbf{c}(N-1)$ specified by (7.7.25) therefore ensures that the adjacent L transmitted pulses contribute zero intersymbol interference. (If $h_j \neq 0$, $|j| > N-1$, then there will be nonzero intersymbol interference from more distant pulses.)

In the absence of noise, when a single training pulse is transmitted, the LS solution for the equalizer coefficient vector therefore gives the zero-forcing equalizer after N iterations. If noise is present, then the condition (7.7.25) must be rewritten as

$$(\mathbf{H}' + \mathbf{N}')\mathbf{c}(N-1) = \begin{bmatrix} 0 \\ \cdot \\ \cdot \\ \cdot \\ 0 \\ 1 \\ 0 \\ \cdot \\ \cdot \\ \cdot \\ 0 \end{bmatrix}, \tag{7.7.26}$$

where

$$\mathbf{H} = [\mathbf{h}(0)\ \mathbf{h}(1) \cdots \mathbf{h}(N-1)] \tag{7.7.27}$$

and $\mathbf{N}$ is the matrix of noise samples

$$\mathbf{N} = [\boldsymbol{\eta}(L)\ \boldsymbol{\eta}(L+1) \quad \cdots \quad \boldsymbol{\eta}(N-1+L)] \ , \tag{7.7.28}$$

where

$$\boldsymbol{\eta}'(j) = [\eta(j)\ \eta(j-1) \quad \cdots \quad \eta(j-N+1)] \ . \tag{7.7.29}$$

The additive noise may significantly perturb the LS solution after many iterations. Using the previous notation, we can rewrite (7.7.18), which gives the LS solution for $\mathbf{c}$ after T iterations, as

$$\left[\sum_{j=0}^{T}\left(\mathbf{h}(j)\mathbf{h}'(j) + \boldsymbol{\eta}(j+L)\boldsymbol{\eta}'(j+L)\right)\right]\mathbf{c}(T) = \mathbf{h}(L) . \qquad (7.7.30)$$

The left hand term in brackets approaches the channel autocorrelation matrix, i.e.,

$$\lim_{T\to\infty} \sum_{j=0}^{T} \mathbf{h}(j)\ \mathbf{h}'(j) = \mathbf{\Phi} ,$$

where $\mathbf{\Phi}$ has finite elements, and hence (7.7.30) indicates that the influence of noise becomes dominant as T increases.

This distasteful behavior clearly indicates that a single training pulse should not be used to obtain an estimate of the equalizer coefficients. Rather, a training sequence which maintains some constant average power should be used in order to reduce the effects of noise. A commonly used start-up sequence in data transmission which maintains constant average power is a periodic pseudo-random sequence. The binary source symbols $d(T)$ designate a transmitted signal pulse equal to +1 or -1. Defining the vector

$$\mathbf{d}'(j) = [d(j)\ d(j-1)\ \cdots\ d(j-N+1)] , \qquad (7.7.31)$$

a periodic pseudo-random sequence of length N has the properties,

$$d(j+N) = d(j) , \qquad (7.7.32a)$$

$$\mathbf{d}'(j)\mathbf{d}(j) = N \qquad (7.7.32b)$$

$$\mathbf{d}'(i)\mathbf{d}(j) = -1 \quad i \neq j . \qquad (7.7.32c)$$

Assuming that a periodic training sequence is transmitted continuously and that channel noise is negligible, then the channel output at time T is

$$y(T) = \sum_{m=-\infty}^{\infty} h_m d(T-m) = \sum_{m=0}^{N-1} d(T-m) \sum_{l=-\infty}^{\infty} h_{m+lN} . \qquad (7.7.33)$$

Assuming that the vectors $\mathbf{y}(j)$, $\mathbf{y}(j+1)$, . . . , $\mathbf{y}(j+N-1)$ are linearly independent, which is the case for nearly all real channels (see problem 7-16), then the vector $\mathbf{c}(N-1)$ that uniquely satisfies (7.7.21) is guaranteed to exist. For the case considered, where the period of the pseudo-random sequence is equal to N, the length of the equalizer, the set of equations (7.7.21) describes the solution for $\mathbf{c}$ that results from *cyclic* equalization [12]. Cyclic equalization is so named because the input sequence to the equalizer is periodic with period N, and therefore the data vectors $\mathbf{y}(T)$, $\mathbf{y}(T+1)$, . . . , $\mathbf{y}(T+N-1)$ constitute one "cycle", since $\mathbf{y}(T+N) = \mathbf{y}(T)$. The solution for the equalizer coefficient vector given by (7.7.21) has an interesting interpretation in the frequency domain. Denoting the discrete Fourier transform of the cyclic equalizer as

$$C(\omega) = \sum_{m=-L}^{L} c_j e^{-im\omega} , \qquad (7.7.34)$$

and the transfer function of the periodically spaced channel impulse response as

$$H(\omega) = \sum_{m=0}^{N-1} \left(\sum_{k=-\infty}^{\infty} h_{m+kN} \right) e^{-im\omega} \tag{7.7.35}$$

then it can be shown that the magnitude of the combined frequency response $\sum_{k=-\infty}^{\infty} H(\omega+2\pi k) C(\omega)$ is equal to unity at N evenly spaced frequencies in the range $|\omega| < \pi$ [12]. Perfect equalization, where $\mathbf{c}'(T)\, \mathbf{y}(T) = d(T-L)$ in the absence of noise, has the following frequency domain interpretation,

$$\left| \sum_{k=-\infty}^{\infty} H(\omega+2\pi k)\, C(\omega) \right| = 1 \tag{7.7.36}$$

for $|\omega| < \pi$. The cyclic equalizer therefore perfectly equalizes the channel transfer function, which corresponds to the periodically repeated impulse response, at N equidistant points. Although the cyclic equalization solution corresponding to (7.7.21) is not the MMSE solution, it has been found to be very close to the MMSE solution. Nevertheless, this solution is produced by a LS algorithm in only N iterations. SG algorithms in general take significantly longer than N iterations to achieve a similar solution. In contrast to the case of sending a single training pulse, the constant signaling power required to send the periodic pseudo-random sequence prevents additive noise from overwhelming the received samples $y(T)$. It can therefore be shown that the asymptotic solution for the filter coefficients, assuming a periodic pseudo-random sequence is continually transmitted, is the MMSE solution given by (3.1.6) (problem 7-16). Further details are given in references [12] and [10].

The previous discussion considered the start-up properties of a LS adaptive equalizer assuming two specific training sequences. In both cases it was shown that the solution for the equalizer coefficients has special properties after N iterations, where N is the number of equalizer coefficients.[1] The key observation is that if the received data vectors $\mathbf{y}(0), \ldots, \mathbf{y}(N-1)$ are linearly independent, then the equalizer coefficients are uniquely determined and satisfy (7.7.21). This condition states that the LS solution *perfectly equalizes* the first N received data samples. The LS solutions after N iterations for the training sequences considered, namely the zero-forcing equalizer and the cyclic equalizer, are in fact defined by the condition (7.7.21). One is therefore led to believe that if random data is used to train the equalizer, the output MSE converges after approximately N iterations. This has been verified for specific channels by computer simulations (see section 7.8); however, a rigorous theory that predicts this result has not yet appeared. Nevertheless, the previous discussion indicates that the crucial difference between LS algorithms and SG algorithms is that LS algorithms attempt to enforce the perfect equalization criterion (7.7.21) with the received data samples, whereas SG algorithms rely on time averaged estimates of second-order input statistics, which are used to form an estimate of the MMSE solution given by (3.1.6), or in the case of lattice algorithms, (7.1.2). The slower convergence speed exhibited by SG algorithms

---

[1] In the previous discussion no prewindowing was assumed, so that although we refer to N iterations, we assume that $2N$ data samples have been received; i.e., N samples are available at the receiver at time T=0.

is therefore due to the relatively long time it takes to accurately estimate second-order statistics via averaging.

In order to clarify this point, one final example is presented. Rather than estimating the uncorrelated sequence $d(T)$ from the correlated (channel output) sequence $y(T)$, consider the reverse situation, i.e., suppose that the adaptive filter is to synthesize a replica of the channel transfer function. In this case the input to the adaptive filter is the uncorrelated sequence $d(T)$, from which an estimate of $y(T)$ is to be obtained. The MMSE solution for the coefficients of an adaptive transversal filter is the impulse response of the channel, which is assumed to be of finite length N. The previous discussion implies that in the absence of noise, an Nth order LS algorithm will perfectly estimate the impulse response in N iterations. Consequently, the output MSE should be identically zero after N iterations. The output MSE for the SG case, however, is described by (7.1.50), and hence only asymptotically approaches zero. Because in this case the input to the adaptive filter is an uncorrelated sequence, an SG lattice joint process estimator offers no advantage over an SG transversal joint process estimator. There is in fact no reason to use the "lattice section" of the lattice joint process estimator to decorrelate the input sequence. An SG lattice algorithm will therefore not converge faster than a comparable SG transversal algorithm. In fact the output MSE of the SG lattice filter will be greater than that of the SG transversal filter because of the extra variance associated with the lattice PARCOR coefficients, which are being adapted to zero via statistical averaging.

7.8. COMPUTER SIMULATIONS

The purpose of this section is to illustrate the discussion in preceding sections by presenting some specific examples. SG transversal, SG lattice, and prewindowed LS lattice and transversal algorithms were programmed on a general purpose computer in order to present simulated performance comparisons. The simulations shown in this section are by no means comprehensive; however, it is hoped that these examples will provide qualitative insight into the behavior of the simulated adaptive algorithms in many applications of adaptive filtering.

7.8.1. Comparisons of Output MSE

Two sets of simulation results are presented in this section. The first set, shown in figures 7-10 through 7-15, uses the convergence of output MSE as the performance criterion. The second set of simulation results, shown in figures 7-16 through 7-18, uses the accuracy of the autoregressive spectral estimate as the performance criterion. Figures 7-10 through 7-12 show averaged output squared error versus number of iterations obtained from SG transversal, SG lattice, and LS lattice predictors in the context shown in figure 7-8a. The simulation plots in figures 7-10 and 7-11 were obtained by averaging the output squared prediction error vs. time over 200 individual runs of each algorithm. Because the variance of the output squared error relative to the scale of the y-axis is larger in figure 7-12 than in figures 7-10 and 7-11, the simulation plots in figure 7-12 were obtained by averaging 500 runs of each algorithm. In each case the input was generated by passing an uncorrelated Gaussian noise

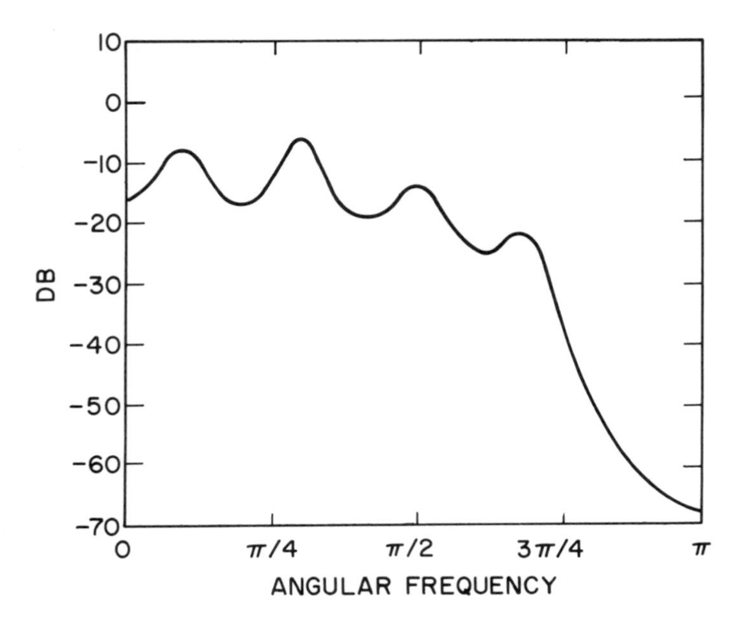

Figure 7-10a. Input Spectrum

sequence with standard deviation 100 through a ten pole recursive filter. The rapid fluctuations of the averaged squared error about its mean value is due to the large statistical variance of the squared error. The order of the adaptive predictors was ten, so that under ideal conditions, the transfer function of the adaptive filter in figure 7-8 should converge to $H(z)$ and the output prediction error should be identical to the input uncorrelated Gaussian sequence $\eta(T)$. The spectra of the inputs to the adaptive predictors corresponding to figures 7-10b, 7-11b, and 7-12b are shown respectively in figures 7-10a, 7-11a, and 7-12a. The spectral peaks shown in figures 7-10, 7-11, and 7-12 are progressively sharper, implying that the associated input processes are progressively more correlated. Specifically, (3.1.30) indicates that the eigenvalue spread, which is the ratio of the maximum eigenvalue to minimum eigenvalue of the autocorrelation matrix, is progressively larger.

The simulated SG lattice algorithm was the one-coefficient algorithm given by (4.2.9), (4.2.10), (4.5.3), (4.5.4), and (4.5.5). The forward and backward coefficients at each stage are the same, i.e., $k_n^f = k_n^b = k_n$, $n = 1, \ldots, N$, hence the reference to "one-coefficient". In order to make a fair comparison with the SG lattice algorithm, the step size of the SG transversal algorithm was also normalized by an estimate of the input variance. The simulated SG transversal algorithm is therefore given by (7.1.1) with the step size β replaced

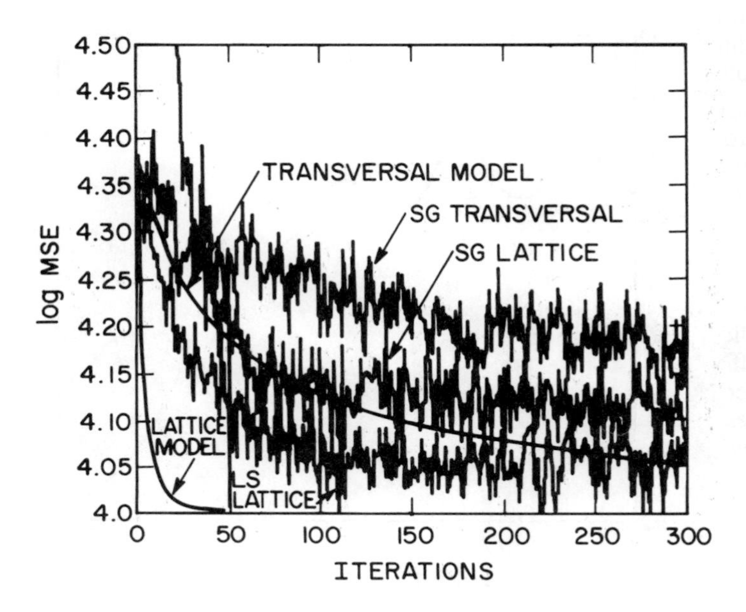

Figure 7-10b. Averaged squared error vs. number of iterations obtained from the SG transversal, SG lattice, and LS lattice algorithms. Also shown is output MSE computed from lattice and transversal adaptation models.

by the normalized step size (3.2.15) and (3.2.17) where $a=1$ and $b=0$. For all three cases the step size and the initial value of the variance estimates were chosen to give relatively fast convergence. The exponential weight w in the LS lattice simulations was 0.97, and the cost functions ϵ_f and ϵ_b were initialized at 50. The algorithm recursions (6.3.64) were computed at each iteration from $n=0$ to $n=N$ starting from $T=0$. Recall that this initialization procedure does not give an *exact* LS solution at each iteration; however, this procedure is less sensitive to finite precision errors than the exact LS start-up procedure described in section 6.3. Besides which, the output MSE that results from using this approximate initialization procedure is nearly identical to the true LS output MSE. Because of the similarity between the SG and LS lattice algorithms, the SG and LS lattice simulations used equivalent parameters. In particular, the SG lattice step size that appears in the recursion (4.5.5) in all three simulations was $\beta = 0.03$ and the initial value of $E(0|n)$ that appears in (4.5.5) was 100. The value of $E(0|n)$ was twice the initial value of the LS cost function ϵ_f because $E(T|n)$ estimates the variance of $E[e_f^2(T|n) + e_b^2(T|n)]$, rather than $E[e_f^2(T|n)]$. Referring to (3.2.15) and (3.2.17), the SG transversal simulations used parameters $\alpha = 0.95$ and $\sigma^2(0) = 10^6$ in figure 7-11 and 7-10, and the parameters $\alpha = 0.97$ and $\sigma^2(0) = 10^8$ in figure 7-12. The initial coefficient values in all cases were zero. In general, decreasing the initial variance estimates for both the lattice and transversal algorithms speeds up convergence; however, reducing these values below those used here either made no noticeable difference or caused instability.

Also shown in figures 7-10 through 7-12 are plots of MSE versus time computed from the adaptation models discussed in section 7.4. For convenience, the transversal model MSE was obtained by assuming that the filter coefficients follow their mean value trajectories, as was assumed for the lattice adaptation models. The SG transversal MSE was therefore computed from the difference equation for the mean coefficient vector (7.1.3) and the expression for output MSE (7.1.17) where $E[\mathbf{c}(T)]$ is substituted for $\mathbf{c}(T)$. These curves are therefore not the same as those obtained from (7.1.21) and (7.1.30), which require the explicit computation of the eigenvalues of the input autocorrelation matrix. (The difference in computed output MSE using these two methods is illustrated for a specific example in problem 7-5.) The error that was squared and averaged in figures 7-10 through 7-12 for each algorithm is the "causal" or oblique forward prediction error

$$e_f^o(T) = y(T) - \mathbf{f}'(T-1)\,\mathbf{y}(T-1)\,, \tag{7.8.1}$$

as opposed to the "current" error that appears in the LS lattice filter,

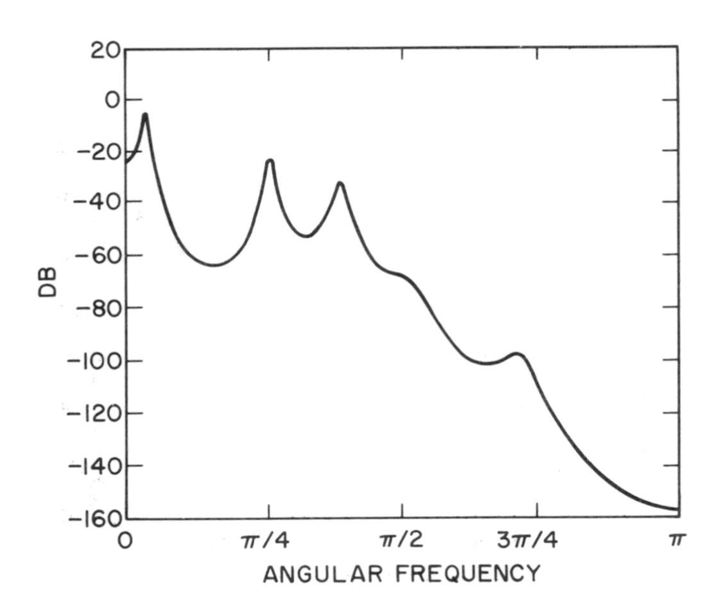

Figure 7-11a. Input Spectrum

$$e_f(T) = y(T) - \mathbf{f}'(T)\, \mathbf{y}(T-1) \,. \tag{7.8.2}$$

In the case of the LS lattice algorithm, the oblique error was computed from the current error by using (6.3.41) (where $\sec^2\theta(T|1,n) = \dfrac{1}{1-\gamma(T-1|n)}$). The PARCOR coefficients in the SG lattice were adapted after the order updates (7.2.1), so that the errors in the lattice at time T were based upon PARCOR coefficients computed at time $T-1$.

Figures 7-10, 7-11, and 7-12 indicate that the more "jagged" the spectrum of the input process, the greater the difference in convergence behavior for all three algorithms simulated. (Observe that the y-axis is scaled differently in each figure, so that the disparity in MSE produced by each algorithm in figure 7-10 is much less than that shown in figures 7-11 and 7-12.) Because of slow modes associated with small eigenvalues of the autocorrelation matrix of the input, the SG transversal algorithm performs considerably worse than the lattice algorithms. The convergence speed of the SG transversal algorithm appears to get progressively slower in figures 7-10 through 7-12, due to progressively larger eigenvalue spreads associated with the inputs. This behavior is predicted by the model curves shown in figures 7-10 through 7-12.

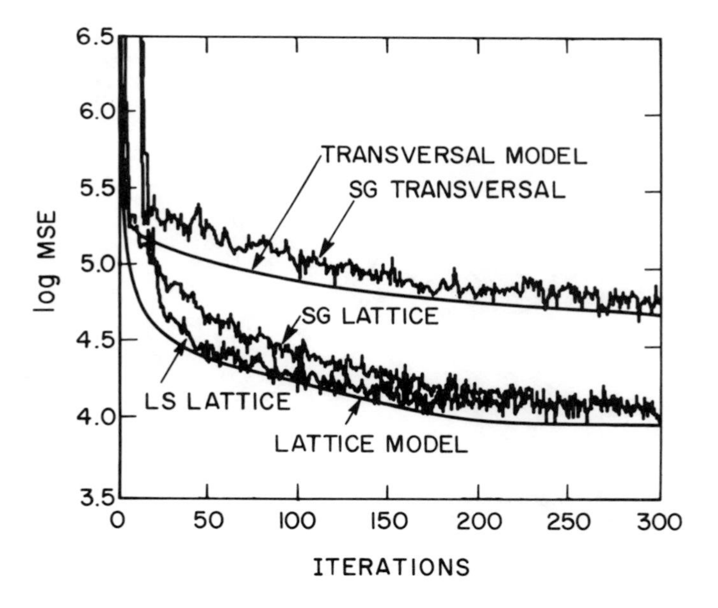

Figure 7-11b. Averaged squared error vs. number of iterations obtained from the SG transversal, SG lattice, and LS lattice algorithms. Also shown is output MSE computed from lattice and transversal adaptation models.

Both the LS and SG lattice algorithms converge with approximately 150 input samples in all three cases, indicating that the convergence speed of these algorithms is insensitive to second-order input statistics. The LS and SG lattice algorithms behave similarly in figures 7-10 and 7-11; however, the SG lattice MSE levels out at a somewhat higher value than the LS MSE in figure 7-12. Also shown in figure 7-11 is averaged squared error vs. time produced by the LS algorithm with $w=1$. The convergence of output MSE is significantly slower in this case. The same type of behavior can be expected of SG lattice algorithms.

The results of two more simulations that use convergence of the output MSE as the performance criterion are shown in figures 7-13 and 7-15. Figure 7-13 shows averaged squared error vs. number of iterations produced by SG transversal, SG lattice, and LS lattice channel equalizers (i.e., joint process estimators) in the context of figure 7-9. This simulation is a modified reproduction of a comparative channel equalization simulation originally presented in reference [13]. The channel shown in figure 7-9 was assumed to have the raised-cosine impulse response defined by:

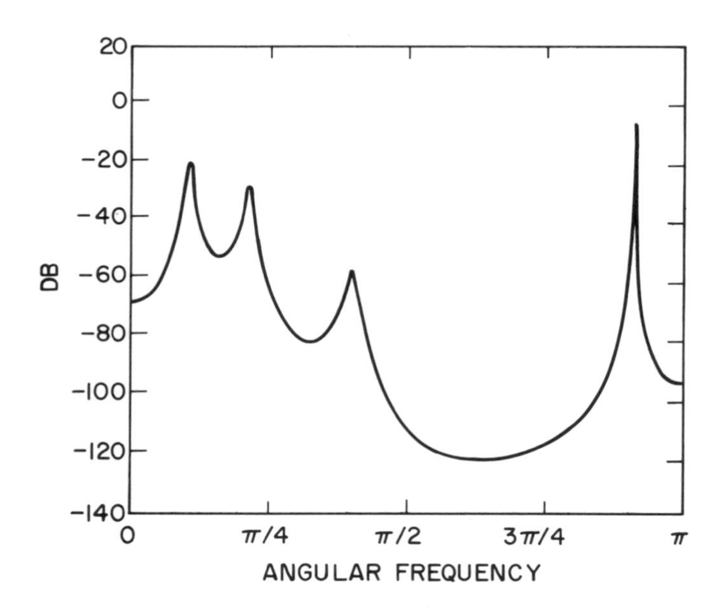

Figure 7-12a. Input Spectrum

$$h_T = \begin{cases} \frac{1}{2}[1+\cos(2\pi(T-2)/W)] & 0 \leqslant T \leqslant 2 \\ 0 & \text{otherwise} \end{cases} \quad . \qquad \textbf{(7.8.3)}$$

The channel output is

$$y(T-1) = \sum_{j=0}^{2} h_j d(T-j) + \eta(T-1) \; , \qquad (7.8.4)$$

where the transmitted symbols $d(T)$ take on values of $+1$ or -1 with equal probability, and $\eta(T)$ is an uncorrelated Gaussian noise sequence with variance 0.001. The parameter W in (7.8.3) can be used to vary the eigenvalue spread of the autocorrelation matrix associated with the channel output $y(T)$. The value W=3.3 was used in this case, which produces an eigenvalue spread of 21. The order of the adaptive filters in each case was 11.

Denoting the transversal equalizer coefficients as $c_j(T)$, $-L \leqslant j \leqslant L$, the simulated SG transversal algorithm is given by

$$e_c^o(T) = d(T) - \sum_{j=-L}^{L} c_j(T-1) y(T-j) \qquad (7.8.5a)$$

$$\sigma^2(T) = \alpha \sigma^2(T-1) + y^2(T) \qquad (7.8.5b)$$

$$c_j(T) = c_j(T-1) + \frac{y(T-j) e_c^o(T)}{\sigma^2(T)}, \quad j = -L, -L+1, \ldots, L \; , \qquad (7.8.5c)$$

where in this case $L = 5$. The variance normalization (7.8.5b) was added in order to make an analogous comparison with the SG lattice equalizer. The lattice equalizer structure is shown in figure 4-12. The one-coefficient SG lattice algorithm was simulated. The PARCOR coefficients were adapted via (4.5.3-5), and the joint process coefficients were adapted via (4.5.15-17).

The parameters selected for the LS algorithm were $w = 1.0$ and $\epsilon_f(0|n) = 0.01$, $0 \leqslant n \leqslant 10$, and for the SG lattice algorithm, $\beta = 0.03$, and $B(0|n) = E(0|n) = 3.0$. Values of $E(0|n)$ smaller than approximately 3 caused the SG lattice coefficients to diverge. For the SG transversal algorithm $w = 0.92$ and $\sigma^2(0) = 3$. The parameters for both the SG lattice and transversal algorithms were selected to make the algorithms converge as fast as possible, while maintaining acceptable asymptotic MSE. The simulation plots were obtained by averaging 200 individual runs of each algorithm. Also shown in figure 7-13 are curves of output MSE vs. time obtained from the adaptation models discussed in section 7.4. Two lattice model curves are shown corresponding to the LS and SG lattice simulations. The difference between the model curves is due to the different initial conditions in each case. The SG lattice algorithm was found to be unstable when used with the same initial conditions as were used for the LS lattice.

The LS error in figure 7-13 is the oblique error given by (7.8.5a) as opposed to the "current" error

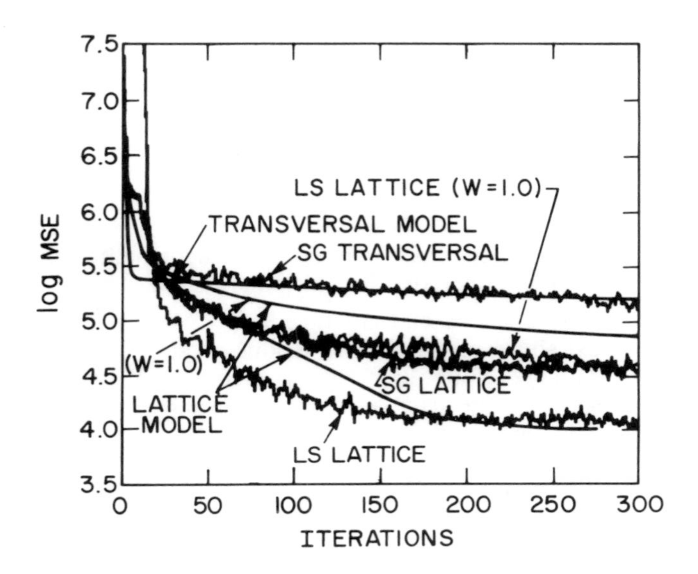

Figure 7-12b. Averaged squared error vs. number of iterations obtained from the SG transversal, SG lattice, and LS lattice algorithms. Also shown is output MSE computed from lattice and transversal adaptation models.

$$e_c(T) = d(T) - \sum_{j=-5}^{5} c_j(T)\, y(T-j) \, . \tag{7.8.6}$$

The oblique error in the LS lattice was computed from the relation (6.5.15). An LS lattice algorithm that computes the oblique error rather than the error (7.8.6) has been derived in reference [13]. (See also problem 6-6). Additional simulations of the LS algorithm with different exponential weighting constants gave approximately the same results as shown in figure 7-13. Although values of β greater than 0.03 cause the SG lattice algorithm to converge somewhat faster (although never faster than the LS algorithm), the asymptotic MSE is greater than that shown in figure 7-13. The SG transversal algorithm again converges somewhat slower than the lattice algorithms due to small eigenvalues of the autocorrelation matrix associated with the channel output. The LS algorithm converges in approximately $2N$ (22) iterations. This is consistent with the discussion of LS convergence properties in section 7.7, which suggests that an LS equalizer requires N linearly independent N-element data vectors $\mathbf{y}(T)$ in order to converge. It takes N iterations to "fill up" the equalizer, and a minimum of N more data samples in order to receive N linearly independent vectors.

Suppose now that the inputs $d(T)$ and $y(T)$ shown in figure 4-12 are reversed; that is, the adaptive filter now estimates the $y(T)$ sequence given the desired sequence $d(T)$. Such a situation, illustrated in figure 7-14, arises in the context of echo cancellation of data signals. The desired sequence in this case is the channel output sequence. Recall from the discussion of echo cancellation in chapter 2 that the purpose of the adaptive filter is to eliminate the linear distortion due to the channel by synthesizing a replica of the channel impulse response, and subtracting the output of the adaptive filter from the output of the channel. Ideally, if the impulse response of the adaptive filter is the same as that of the channel, then the output MSE, $E[e^2(T)]$, will be equal to the channel noise variance $E[\eta^2(T)]$. Because the echo that is to be canceled may experience considerable delay, the order of the adaptive filter may be quite large (i.e., on the order of hundreds of coefficients).

The performance of SG transversal, SG lattice, and LS lattice joint process estimators in the context shown in figure 7-14 is illustrated in figure 7-15. Output squared error vs. number of iterations averaged over 200 individual runs of each algorithm is shown. The data sequence consisted of independent binary samples (± 1), the channel impulse response was

$$h(T) = (0.96)^T, \quad T = 0, 1, \ldots, N, \tag{7.8.7}$$

and the signal to noise ratio was

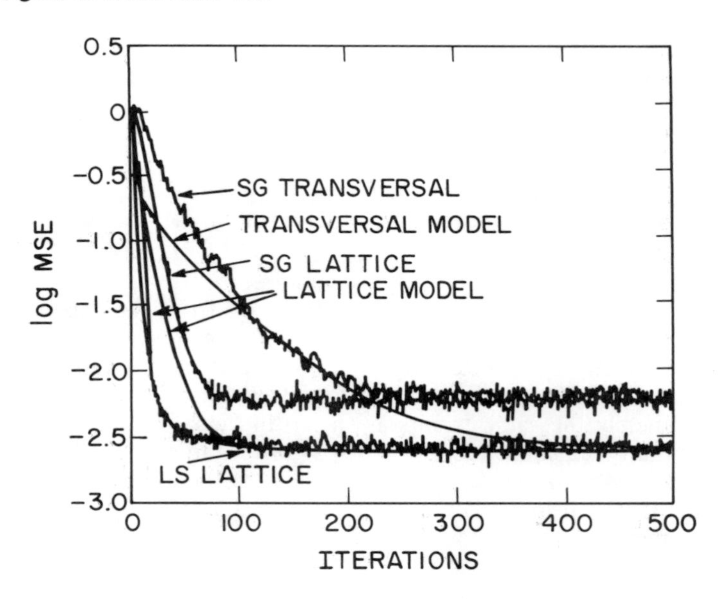

Figure 7-13. Averaged squared error vs. number of iterations produced by SG transversal, SG lattice, and LS lattice channel equalizers. Also shown is MSE computed from transversal and lattice adaptation models.

$$SNR = \frac{E[y^2(T)]}{E[\eta^2(T)]} = 40dB \ . \tag{7.8.8}$$

The order of the adaptive filters was 100. The SG transversal algorithm is again given by (7.8.5), where "y" and "d" are interchanged. Because in this case $d^2(T) = 1$, the step size $1/\sigma^2(T)$ in (7.8.4) converges deterministically to $1/N$, the optimal SG transversal step size given in section 7.1.4. The SG transversal output MSE as a function of time is in fact given approximately by (7.1.51). The oblique error

$$e_c^o(T) = y(T) - \sum_{j=0}^{N-1} c_{j+1}(T-1)\, d(T-j) \tag{7.8.9}$$

was squared and averaged in the LS and SG transversal simulations. The simulated SG lattice algorithm was obtained by setting the LS lattice parameters $\gamma(T|n)$, $0 \leqslant n \leqslant N-1$ in (6.3.64) equal to zero. This is almost equivalent to the gradient updates (4.5.11) and (4.5.12) where the step size β is normalized, the only difference being that (4.5.11) and (4.5.12) indicate that the coefficients k_n^f and k_n^b are updated after the error order updates (7.2.1). In the simulated SG lattice algorithm the filter coefficients are updated before the error order updates, and hence the residuals in the lattice do not correspond to the oblique error e_c^o. The error $e_c^o(T|N)$ can be obtained, however, by freezing the filter coefficients k_n^f, k_n^b, and k_n^c, $1 \leqslant n \leqslant N$, at time $T-1$, and passing the data values $d(T-N+1), \ldots, d(T)$ and $y(T)$ through the fixed coefficient lattice. For computational convenience, however, the following error was computed,

$$\tilde{e}(T) = y(T) - \sum_{j=0}^{N-1} k_{j+1}^c(T-1)e_b(T|j) \ , \tag{7.8.10}$$

and was found to be nearly identical to e_c^o for smaller filter orders. (Notice that $\tilde{e}$ is not identical to e_c^o since $e_b(T|j)$ depends upon the current lattice coefficients $k_n^f(T)$ and $k_n^b(T)$ for $0 \leqslant n \leqslant j$.) The exponential weight w in the LS simulation was 0.99. The step sizes β and α that enter the SG algorithms were 0.01. The initial value of the variance estimators and the LS cost function ϵ_f in all cases was 0.1.

7.8.2. Discussion

In general, the discussion and simulations in this chapter indicate that the SG transversal algorithm performs significantly worse than the SG lattice algorithm, which performs somewhat worse than an LS algorithm. An exception to this rule is shown in figure 7-15. The improved performance of the SG transversal algorithm relative to the SG lattice in figure 7-15 is due to the fact that the (binary) input to the filter ($d(T)$), which is used to adapt the lattice PARCOR coefficients, is uncorrelated. The optimal values of the lattice PARCOR coefficients are therefore zero. Adapting these coefficients to estimate the second-order statistical properties of the input $d(T)$ therefore cannot result in improved performance relative to the SG transversal algorithm, which is obtained by arbitrarily setting these coefficients to zero. To the contrary, adaptation of the coefficients k_n^f and k_n^b, $1 \leqslant n \leqslant N$, causes statistical fluctuations, which increase the variance of the backward errors used in the tapped delay line

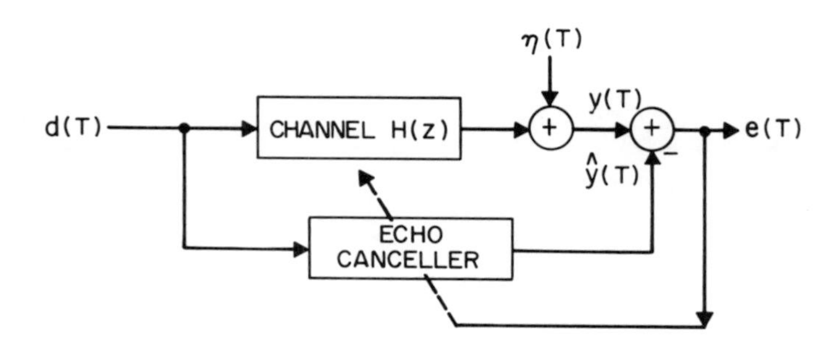

Figure 7-14. Adaptive echo canceler for data signals.

(bottom half of the filter structure) shown in figure 4-12. This in turn increases the output MSE (see problem 7-15). Because the simulations in figures 7-10 through 7-12 use correlated inputs, the coefficients k_n^f and k_n^b are adapted to decorrelate the input sequence and therefore have non-zero optimal values. The SG lattice therefore performs better than the SG transversal algorithm in these cases.

It is emphasized that the previous argument does *not* apply when comparing the LS lattice with the SG transversal algorithm. The LS coefficients k_n^f and k_n^b, $1 \leqslant n \leqslant N$, are *not* adapted to estimate second-order statistics. Instead, they are adapted to orthogonalize the input in an algebraic sense. The statistical property

$$E\,[e_b(T|m)\; e_b(T|n)] \;=\; E\,[e_b^2(T|n)]\;\delta_{mn} \tag{7.8.11}$$

is replaced by the LS property,

$$\sum_{j=0}^{T} e_b(j|m)\; e_b(j|n) \;=\; \epsilon_b(T|n)\;\delta_{mn}\;, \tag{7.8.12}$$

where the residuals $e_b(j|n)$, $0 \leqslant j \leqslant T$, $0 \leqslant n \leqslant N$, result from passing the input through a filter with *fixed* coefficients k_n^f and k_n^b, $1 \leqslant n \leqslant N$, computed at time T. The optimal LS values of k_n^f and k_n^b in this case are therefore *not* equal to zero at any given iteration, although the asymptotic mean values of k_j^f and k_j^b are zero.

The LS MSE shown in figure 7-13 converges in approximately $2N$ iterations, as does the LS MSE in figure 7-15, which is in accordance with the discussion in section 7.7. However, the convergence of the LS MSE in figures 7-10 through 7-12 takes considerably longer than $2N$ iterations. Since the input in these cases is correlated Gaussian noise, it is intuitively reasonable that a relatively large number of samples is needed in order to gain statistical confidence in the estimates of the filter parameters. The difference in performance between the SG and LS lattice is also more pronounced in figures 7-13 and 7-15 as compared with figures 7-10 through 7-12. The algebraic orthogonalization of a correlated Gaussian input that is accomplished by LS algorithms therefore seems to offer little advantage over the statistical orthogonalization of the input accomplished by estimating second-order statistics via time averages.

7.8.3. Numerical Properties

A number of important issues related to the implementation of the adaptive algorithms considered in this section need to be mentioned. Perhaps the most important is finite precision effects. In actual digital implementations of adaptive filters the coefficients and variables that enter the algorithm recursions can never be specified with infinite precision. The performance degradation of adaptive algorithms due to finite precision implementation is difficult to analyze and is a current area of active research. Nevertheless, as discussed in chapter 5, lattice filter coefficients generally require less precision than transversal filter coefficients to achieve a given performance level. In addition, the normalized LS algorithms presented in section 6.6 generally display superior numerical properties relative to the unnormalized versions.

The worst finite precision effect is numerical instability. Depending on the input and start-up conditions, an adaptive algorithm may become unstable, giving arbitrarily large errors and coefficient values. In the case of SG algorithms, this may be due to an excessively large step size; however, it is also possible

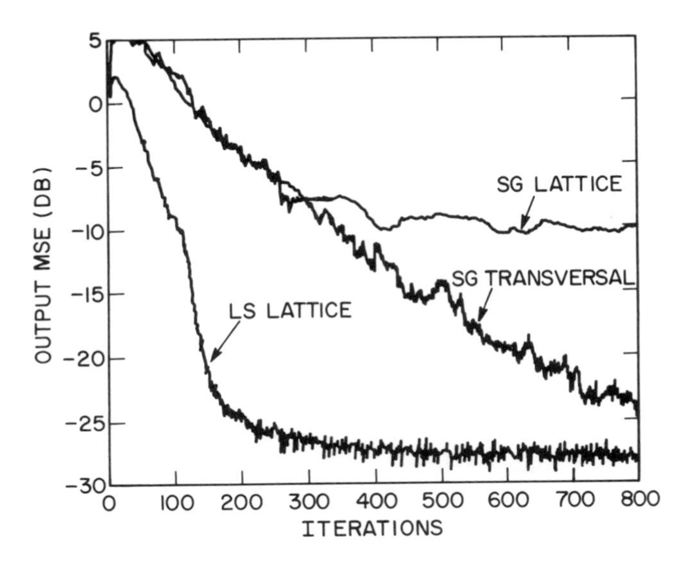

Figure 7-15. Averaged output squared error vs. time obtained from the LS lattice, SG lattice, and SG transversal algorithms in the context of figure 7-14.

that round off errors, even as introduced by a general purpose computer, can propagate and build, eventually overwhelming the intended solution. This problem appears to be most pronounced with the fast Kalman algorithm. The fast Kalman algorithm was also simulated in addition to the other algorithms mentioned in this section. In the case of the correlated Gaussian noise inputs, the performance of the fast Kalman algorithm was the same as that of the LS lattice. In the case of the echo cancellation simulations shown in figure 7-15, however, the fast Kalman algorithm often diverged. Insight into the cause of this problem can be gained by considering the start-up properties of an LS algorithm in the context of figure 7-14.

Given the data samples $d(0), d(1), \ldots, d(n-1)$ and $y(0), y(1), \ldots, y(n-1)$, where $n < N$, the order of the adaptive filter, the discussion of LS start-up properties in section 6.3 stated that a (prewindowed) LS algorithm selects the coefficients $c_1, c_2, \ldots, c_n$ such that

$$\begin{bmatrix} d(0) & 0 & 0 & \cdots & 0 \\ d(1) & d(0) & 0 & & \cdot \\ d(2) & d(1) & d(0) & \cdot & \cdot \\ \cdot & \cdot & \cdot & \cdot & \cdot \\ \cdot & \cdot & \cdot & \cdot & \\ \cdot & \cdot & \cdot & & 0 \\ d(n-1) & d(n-2) & d(n-3) & \cdots & d(0) \end{bmatrix} \begin{bmatrix} c_1 \\ c_2 \\ \cdot \\ \cdot \\ \cdot \\ c_n \end{bmatrix} = \begin{bmatrix} y(0) \\ y(1) \\ \cdot \\ \cdot \\ \cdot \\ y(n-1) \end{bmatrix} . \tag{7.8.13}$$

The error sequence

$$e(T) = y(T) - \sum_{j=1}^{n} c_j(T)\, d(T-j+1) \tag{7.8.14}$$

is zero for $T = 0, 1, \ldots, n-1$, and hence the LS cost function assumes the minimum value of zero. When a new data sample $d(n)$ is received, the same system of equations (7.8.13) is solved where n is replaced by $n+1$. Recall that the first n coefficients remain the same, so that only c_{n+1} must be determined. This procedure continues until N data samples have been received. If the noise term in figure 7-14 is absent, then clearly after N data samples have been received, the elements of the estimated vector $\mathbf{c}_{N-1}$ are the sampled impulse response of the channel, which is the MMSE solution. If, however, the additive noise is nonzero, the estimate $\mathbf{c}(N-1)$ may deviate substantially from the MMSE solution. Examination of (7.8.13) in fact reveals that the deleterious effect of the added noise can increase exponentially with N, and that for moderately large N the last few estimated impulse response values become overwhelmed by channel noise. (An analysis of the numerical properties of the triangular system of equations (7.8.13) appears in chapters 6 and 13 through 18 of references [14, 15]). It has been found empirically that computer round off error alone can cause successive estimated impulse response coefficients to become arbitrarily large, depending upon the input data sequence.

However, if $N+1$, rather than N, data samples are used to obtain a LS estimate for N impulse response values, the triangular system of equations (7.8.13) does not apply since the sequence of errors $e(T)$, $T = 0, 1, \ldots, N$ (with additive channel noise), cannot in general be set equal to zero. The expected value of the sum of the squares of the errors cannot be less than the sum of the noise variance at each iteration, i.e.,

$$E\left[\sum_{T=0}^{N} e^2(T)\right] \geqslant (N+1)\, E[\eta^2(T)]\,. \qquad (7.8.15)$$

The effect of the channel noise upon the estimate of the latter impulse response coefficients is therefore not nearly as severe as when given N data samples.

Typically, this start-up problem is avoided by initializing the diagonal elements of the covariance matrix $\hat{\Phi}(T|n)$, given by (7.7.5) (where "y" is replaced by "d"), to some small value δ such that the LS solution given by (7.7.6) is numerically stable with respect to roundoff errors. This is analogous to setting the LS cost functions ϵ_f and ϵ_b at time zero equal to some small value δ. The simulation results presented here indicate that when this type of initialization is used with the LS lattice algorithm, the performance degradation due to deviation from the true LS solution is hardly noticeable. In the context of figure 7-14, however, the value of δ required to ensure that the fast Kalman algorithm remains stable is quite large (i.e., $\delta = 10$). The resulting convergence of the output MSE in this case is therefore considerably slower than the simulated LS lattice results (which used $\delta = 0.1$) shown in figure 7-15.

To see why the LS lattice algorithm should have superior numerical properties relative to the fast Kalman algorithm during start-up, consider again the case where $T < N$ data samples are used to compute a LS filter. If the transversal structure is used, this implies that the triangular set of equations (7.8.13) must be solved. As T increases, the estimate $\mathbf{c}(T)$ degrades rapidly in the presence of additive noise. This estimate may in fact become numerically very large, in which case a very high degree of numerical precision is required to accurately compute successive coefficient vectors. Also, once $N+1$ data samples are available, (7.8.13) no longer applies and the LS estimate $\mathbf{c}_N$ may be drastically different from $\mathbf{c}_{N-1}$. Because the fast Kalman algorithm consists of a large number of recursions, some of which involve at least N adds and multiplies, it contains numerous sources of roundoff error. This roundoff error can easily lead to instability due to the extreme sensitivity to noise exhibited by the estimate $\mathbf{c}(T)$, for $T < N$, and the large range in values which it may assume.

If $T < N$ data samples are used to compute an LS lattice filter, the first n stages of the lattice, where $n < T$, constitute an nth order LS filter based upon T data samples. Computation of the nth order coefficients is therefore well determined (i.e., the sequence of nth order output errors cannot be zero) and does not exhibit the numerical problems associated with computing $\mathbf{c}_T$ via (7.8.13). Given T data samples, it follows that only the Tth order lattice coefficients may be ill-determined (i.e., the Tth order residuals equal zero), and that when the $(T+1)$st data sample is received, the estimated Tth order coefficients are well-behaved. Because the estimate $\mathbf{c}_T$ obtained via (7.8.13) may be extremely poor, however, given T data samples, it must be expected that the estimated Tth order lattice coefficients will also be poor, and that numerical instability may result in analogy with the fast Kalman algorithm. This is in fact the case, and hence the exact order-recursive lattice start-up procedure described in section 6.3, where the lattice recursions (6.3.64) are computed from $n=0$ to min (T,N), cannot be used when N is very large for the case considered. Because the lattice structure is order recursive, however, latter stages can diverge without affecting preceding stages, and hence the value of δ needed to ensure stability is much less than that needed for the fast Kalman algorithm. The fast LS convergence described in section 7.7 can therefore be obtained by using the lattice structure.

Not only is it likely that an adaptive LS filter will diverge during start-up, but it is also likely in some applications such as channel equalization and echo cancellation, where the adaptive filter must run for hours, that an adaptive LS algorithm may diverge in *steady state*. This is again particularly true of the fast Kalman algorithm. The reason is due to the steady state accumulation of finite precision errors, which eventually drive the coefficients far from the true LS solution. In addition, in some cases the sample covariance matrix of the input sequence may be close to singular, making it especially likely that any LS algorithm will diverge. This phenomena is analogous to the numerical problems encountered with the SG transversal algorithm when the input autocorrelation matrix is close to being singular. The solutions to this problem discussed in section 3.2.2.4 also apply to the LS case. In particular, one solution is to add uncorrelated noise to the input. This has the effect of increasing the diagonal elements of the autocorrelation matrix so that the MMSE solution is not as ill-conditioned. In practice, this is usually not sufficient. In order to prevent the absolute values of the filter coefficients from increasing arbitrarily, an additional constraint can be added to the LS criterion analogous to the leakage term shown in (3.2.24), which can be added to the SG transversal algorithm. An analogous LS criterion to (3.2.24) is to minimize the sum

$$\epsilon_c(T|n) = \mu w^T ||\mathbf{c}(T) - \mathbf{c}_0|| + \sum_{i=0}^{T} w^{T-i} e_c^2(i|n) \tag{7.8.16}$$

where $\mathbf{c}_0$ is a bias reflecting prior knowledge of the input and μ is a small positive constant that determines how much weight is given to the initial estimate $\mathbf{c}_0$. For instance, in speech processing applications $\mathbf{c}_0$ might be the MMSE predictor using long term, rather than short-term, second-order speech statistics. In the case of channel equalization $\mathbf{c}_0$ would most likely be set to zero. If $\mathbf{c}_0 = 0$, then the cost function (7.8.16) presents a tradeoff between minimizing the LS cost function and the norm of the coefficient vector. The cost criterion (7.8.16) has been referred to as LS with a "soft-constraint" [15]. Modification of the LS algorithms in chapter 6 to minimize criterion (7.8.16) has been proposed [15] and is considered in problems 6-5 and 6-12.

If there were a way to detect when an LS algorithm is about to diverge, then criterion (7.8.16), where the last reasonable computed value of $\mathbf{c}$ is substituted for $\mathbf{c}_0$, can be used to restart the algorithm while retaining prior information. In fact an obvious way to monitor the numerical stability of the LS algorithms considered here is to watch the gain $\gamma(T|N) = \mathbf{g}'(T|N)\,\mathbf{y}(T|N)$. If this parameter becomes greater than or equal to one, then the LS solution is ill-determined and the algorithm will diverge. In the case of the fast Kalman algorithm, a better "rescue variable" is [16]

$$r(T|N) \equiv 1 - e_b^o(T|N)\, g_{N+1}(T|N+1) = \frac{1-\gamma(T|N+1)}{1-\gamma(T|N)} . \qquad (7.8.17)$$

Empirical observations have shown that in the case of the fast Kalman algorithm this variable becomes negative just before the algorithm diverges [16, 15]. The fast Kalman algorithm can therefore be "rescued" by monitoring the variable $r(T|N)$, and reinitializing the algorithm using the soft constraint criterion (7.8.16) whenever $r(T|N) < 0$. Similar types of rescue schemes can be envisioned for lattice algorithms. It has been observed that in applications where steady state stability is a problem, the normalized LS algorithms generally remain stable for a much longer period of time than the unnormalized algorithms [15].

One final point concerning the simulation of adaptive algorithms is that extra care must be taken in order to ensure that the algorithm recursions are programmed correctly. In particular, the order updates of the LS lattice recursions (6.3.64) often call for variables computed at the previous time instant rather than the most recently computed values. In order to avoid storage of variables computed at both the previous and present time iterations, it is therefore necessary to compute the recursions (6.3.64a-64g) in a specific order (see problem 6-4). Experience has indicated that making slight errors in implementing the LS lattice algorithm (i.e., using an updated variable rather than the variable computed at the last time interval) has a definite detrimental effect. The algorithm may still converge with some types of mistakes; however, the speed of convergence may be much slower. This is especially true of the joint process estimation examples shown in figures 7-13 and 7-15. It is interesting that SG lattice algorithms are also sensitive to the order in which updates are performed. Specifically, updating the nth PARCOR coefficient *after* the nth order forward residual has been computed from (7.2.1a) gives better performance than the reverse procedure. This is contrary to the LS lattice algorithm, for which the updated coefficient $K_n(T)$ must be computed *before* the order recursions for the forward and backward errors can be computed. Also, the one coefficient lattice algorithm given by (4.5.3-5) appears to perform better than alternative one coefficient algorithms (i.e., (7.2.5)).

7.8.4. Simulated Spectral Estimates

In many applications of adaptive linear prediction, output MSE is an inappropriate performance criterion. In linear predictive coding of speech, for example, the accuracy of the estimated short-term speech spectrum is of paramount importance. For this case the estimated filter coefficients, rather than output MSE, are of interest. This section is completed by a presentation of the performance of LS and SG lattice algorithms using a spectral criterion. Because the SG transversal algorithm converges slowly, and because of the heightened sensitivity of the estimated spectrum to coefficient noise, the SG transversal algorithm is generally not used in spectral estimation applications, and was not simulated.

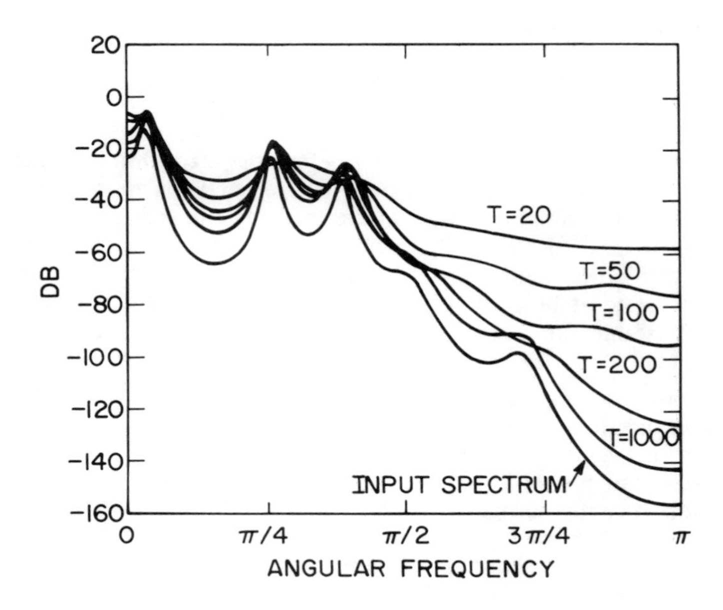

Figure 7-16. Spectral estimates vs. time obtained from the SG lattice algorithm.

Figures 7-16 through 7-18 show spectra obtained from the LS and one coefficient SG lattice algorithms using the same correlated Gaussian noise input that was used to generate figure 7-11. The order of the filters in each case was 10. The spectral estimates were obtained by averaging 200 simulations of each algorithm and sampling the averaged coefficient values at $T=20$, $T=50$, $T=100$, $T=200$, and $T=1000$. In the case of the SG lattice algorithm, the PARCOR coefficients k_n, $1 \leqslant n \leqslant 10$, can be computed from (4.5.3), averaged, and converted to prediction coefficients via the Levinson algorithm in order to compute the estimated spectrum from (2.1.1). The averaged LS PARCOR coefficients used to generate figures 7-17 and 7-18 are similarly given by

$$k_{n+1}(T) = \frac{2K_{n+1}(T)}{\epsilon_f(T|n) + \epsilon_b(T|n)} \; . \tag{7.8.18}$$

Also shown in figures 7-16 through 7-18 is the actual input spectrum. The LS parameters used to generate figure 7-17 were $w = 0.97$ and $\epsilon_f(0|n) = 50$, $0 \leqslant n < 10$. The SG lattice algorithm step size was $\beta = 0.03$ and the initial variance estimate was $E(0|n) = 100$, $0 \leqslant n < 10$. Figure 7-18 shows spectral estimates produced by the LS algorithm with $w = 1.0$.

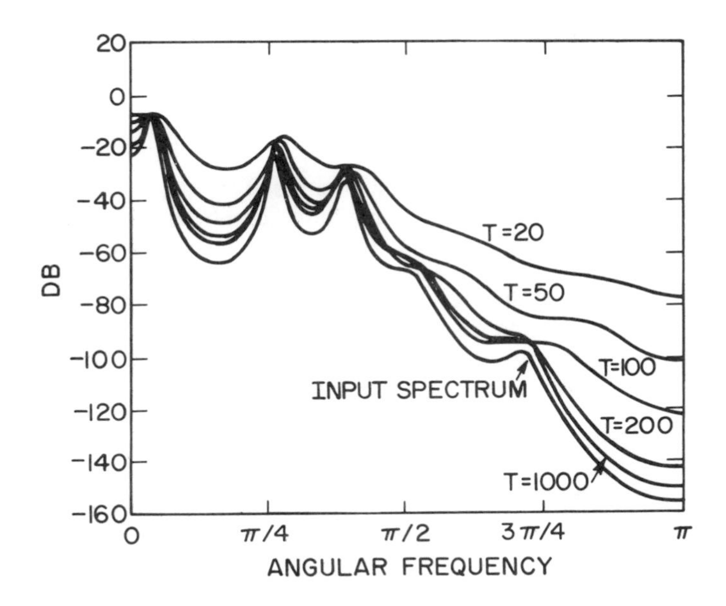

Figure 7-17. Spectral estimates vs. time obtained from the LS lattice algorithm.

Deviation of the estimated spectrum after convergence ($T = 1000$) in figures 7-17 and 7-18 is due to biased coefficient estimates. This effect becomes more pronounced as the exponential weight w decreases. As was mentioned in preceding sections, bias in the coefficient estimates are caused by correlations between the filter coefficients and the inputs to each corresponding stage and by statistical fluctuations in filter coefficients that alter the second order statistics of input signals to successive stages. The asymptotic spectrum generated by the LS algorithm is slightly closer to the actual spectrum than the asymptotic spectrum generated by the SG algorithm. The difference in performance between the SG and LS lattice algorithms is more visible in figures 7-16 and 7-17 than in figure 7-12. The spectral criterion is therefore a more sensitive performance criterion than output MSE. In both cases the algorithms converged after approximately 300 iterations, so that the spectral estimate obtained at $T = 300$ was the same as that shown for $T = 1000$. The output MSE in figure 7-11b converges in approximately 250 iterations. Because exponential weighting was not used in the LS simulation shown in figure 7-18, it can be shown that the estimated coefficients converge in probability to the actual PARCOR coefficients associated with the input (a proof is given in chapter 5 of reference [17]), and the asymptotic estimated spectrum is therefore the same as the input spectrum. However, even at $T = 1000$, which is long after the output MSE is close to its asymptotic value, the estimated spectrum still deviates considerably from the actual input spectrum. Figures 7-17 and 7-18 therefore indicate that in applications of adaptive filtering where accurate estimates of the input spectrum are desired given a moderate number of samples, an exponential forgetting factor $w < 1$ is needed with recursive algorithms in order to speed up convergence.

7.8.5. Summary

This completes the presentation of comparative simulations of adaptive algorithms. To summarize, comparative simulations of LS and SG transversal and lattice algorithms have been presented using two performance criteria, output MSE and the accuracy of the spectral estimate obtained from averaged coefficients. The essential difference between the LS lattice and SG lattice algorithms is the presence of the LS gains $\gamma(T|n)$. In order to measure the effect of these weighting factors most of the comparative simulations used equivalent parameters and initial conditions for both algorithms. In most cases the LS lattice algorithm offers modest performance improvement over the SG lattice algorithms simulated, and substantial performance improvement over the simplest SG transversal algorithm. This difference in performance is more pronounced when the spectral criterion is used. It has been observed that although the fast Kalman algorithm should have identical performance to the LS lattice algorithm in an infinite precision environment, it is much more sensitive to finite precision errors than the LS lattice algorithm.

The channel equalization and echo cancellation simulations presented in this section indicate that when the noise content of the input is very small relative to the binary signal component, the output MSE of an LS algorithm takes approximately $2N$ iterations to converge, which is in agreement with the discussion in section 7.7. In contrast, if the input is correlated Gaussian noise, both LS and SG predictors require considerably more than $2N$ iterations to converge. For many applications of adaptive linear prediction, especially those that involve highly nonstationary environments, the convergence properties of adaptive algorithms are not well understood. Open issues are therefore the characterization of algorithm complexity vs. performance for arbitrary types of random inputs, and a rigorous description of output MSE vs. time for both SG and LS algorithms.

7.9. SUMMARY AND FURTHER READING

In this chapter the convergence properties of SG and prewindowed LS transversal and lattice algorithms have been discussed. In each case by making numerous approximations the time-varying mean values of the filter coefficients, the output MSE, the asymptotic variance of the filter coefficients, and the excess MSE due to the statistical fluctuations of the filter coefficients have been approximated in terms of fixed algorithm parameters and stationary second-order input statistics. The results presented on the convergence properties of the SG transversal algorithm were first presented by Ungerboeck in 1972 [1]. Since then a number of authors have attempted to perform a more rigorous analysis of the SG transversal algorithm (i.e., see references [18, 19, 20, 21, 7, 22]). Nevertheless, the results in section 7.1 have been shown to be quite accurate and are still widely used. In contrast, the first-order analysis of the convergence properties of lattice algorithms presented in sections 7.2-7.5 is more recent (i.e., see references [23] and [24]), and subsequent more rigorous analyses of lattice convergence properties have yet to appear. Despite the necessary approximations made in these analyses, the simulation results in section 7.8 indicate that for the inputs considered the analysis gives a reasonable indication of algorithm performance.

The analysis of recursive LS algorithms in this chapter was included so that a first-order comparison with the performance of SG algorithms could be made. The results in sections 7.3 and 7.4 indicate that the SG and LS lattice algorithms give almost identical performance. While this appears to be true for correlated Gaussian noise inputs, the discussion and simulations in sections 7.7 and 7.8 indicate that in linear system identification applications such as channel equalization and echo cancellation, *algebraic* rather than statistical properties play a crucial role in determining when an LS algorithm converges. In particular, when additive channel noise is relatively small, LS algorithms converge after receiving approximately $2N$ samples, whereas SG algorithms require more samples in order to estimate second-order statistics via time averages. The statistical properties of LS estimators have also been analyzed in the context of classical system identification. See for example chapters 2 and 5 of reference [17] and problems 7-17 to 7-19. A rigorous computation of the mean values of the filter coefficients and output MSE produced by an LS algorithm with the types of random inputs considered in this chapter has not yet appeared, however.

The performance of adaptive algorithms in a finite precision environment is difficult to analyze, but is a very important problem in data transmission applications. An analysis of the performance of the SG transversal algorithm in a finite precision environment has been given [6, 25] as has an analysis of finite precision errors in block LS problems [14] and recursive LS algorithms [26]. (See also [27].) A discussion of numerical accuracy and stability in unnormalized prewindowed LS lattice and transversal algorithms is given in reference [28]. An analysis of the normalized prewindowed lattice algorithm in a finite precision environment is given in reference [29], and references [15] and [30] also discuss finite precision effects in transversal LS algorithms. Results from computer simulations of prewindowed LS algorithms can be found in [28] and [15], and results from simulations of sliding window and growing memory covariance transversal LS algorithms are given in references [30] and [31]. Additional comparative simulations of LS and SG algorithms can be found in references [32, 33], and in references [34, 35] where nonstationary input statistics are considered.

The first example of the fast convergence of a LS equalizer relative to the widely used SG transversal equalizer was presented by Godard [36], who also derived the asymptotic expression for the convergence of the output MSE (7.7.14). The expression for excess asymptotic MSE (7.7.16) first appeared in reference [32]. An extension of the analysis of LS convergence in section 7.7.1 to the case where the input is nonstationary is given in [35] (see also [31]). The performance advantage that the LS covariance algorithms presented in section 6.4 may have relative to the LS prewindowed algorithms in section 6.3 has not yet been reported.

PROBLEMS

7-1 Iterate the SG transversal update equation (7.1.1) and thereby obtain an expression for $\mathbf{c}(T)$ as a function of the data sequence $\{y(i)\}$, $i = 1, 2, \ldots, T$, the desired sequence $\{d(i)\}$, $i = 0, 1, \ldots, T$, the step-size β, and the initial value of the coefficient vector $\mathbf{c}(0)$. What statistical information concerning the sequences $d(i)$ and $y(i)$ is needed in order to exactly compute the mean value of the coefficient vector $E[\mathbf{c}(T)]$?

7-2 Show that $\sum_{j=1}^{N} \lambda_j = N\phi_0$, where λ_j is the jth eigenvalue of the autocorrelation matrix $\mathbf{\Phi} = E[\mathbf{y}(T)\mathbf{y}'(T)]$, and $\phi_0 = E[y^2(T)]$.

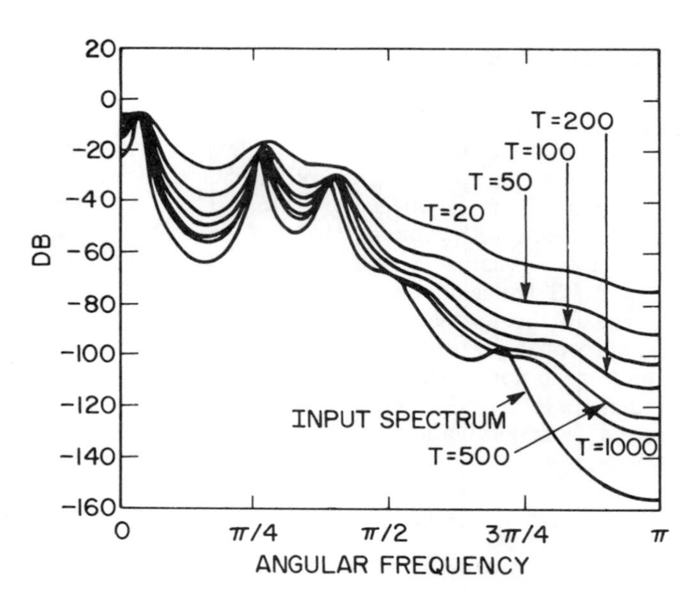

Figure 7-18. Spectral estimates vs. time obtained from the LS lattice algorithm with $w = 1$.

7-3 a. Using (7.1.1) and the coordinate transformation (7.1.6-8), derive (7.1.22).

b. Use (7.1.22) to derive (7.1.24).

7-4 Show that if the row sums of a symmetric matrix with positive elements are less than unity, then all eigenvalues of the matrix must be less than unity. Is the condition (7.1.37) required for the stability of the difference equation (7.1.31)?

7-5 Consider the autoregressive sequence $y(T)$ described by the constant coefficient difference equation

$$y(T) = ay(T-1) + by(T-2) + \eta(T) \tag{7.P.1}$$

where $\eta(T)$ is white noise with unity variance.

a. Compute the second-order MMSE prediction coefficients.

b. Assume that a (unnormalized) second-order SG transversal algorithm is used as an adaptive predictor with $y(T)$ as the input sequence and compute the mean values of the prediction coefficients in terms of a, b, and the initial coefficient values. What are the time constants associated with each mode?

c. Compute the asymptotic coefficient variance and the asymptotic output MSE of the predictor.

d. Derive an expression for the output MSE as a function of T.

e. Compare the expression in part (d) with the expression for output MSE which results from substituting the mean coefficient vector $E[\mathbf{f}(T)]$ for $\mathbf{c}$ in (7.1.17).

7-6 a. Using (7.1.24), (7.2.20), and the independence assumption, show that the covariance matrix for the transformed filter coefficients, $\mathbf{Q}(T)$ defined by (7.1.20), satisfies the following difference equation when the input sequence $y(T)$ to an SG transversal filter is Gaussian:

$$\begin{aligned}\mathbf{Q}(T) &= \mathbf{Q}(T-1)\left[\mathbf{I} - 2\beta\Lambda\right] \\ &\quad + \beta^2\Lambda\left[2\mathbf{Q}(T-1)\Lambda + \text{trace}\,[\Lambda\mathbf{Q}(T-1)] + \epsilon_{\min}\right],\end{aligned} \tag{7.P.2}$$

where Λ is defined by (7.1.5) and $\epsilon_{\min}$ is defined by (7.1.18).

b. Under what conditions do the off diagonal elements of $\mathbf{Q}$ converge to zero?

c. Derive a stability constraint for the step size β. How does this constraint compare with the stability constraint (7.1.37)?

d. Assume that the input to the SG transversal algorithm given by (7.1.1) is Gaussian, and show that the asymptotic excess MSE, defined by (7.1.19), is given by

$$\lim_{T\to\infty} \epsilon_{ex}(T) = \frac{\beta\rho\epsilon_{\min}}{2 - \beta\rho} \tag{7.P.3a}$$

where

$$\rho \equiv \sum_{i=1}^{N} \frac{\lambda_i}{1-\beta\lambda_i} \ . \tag{7.P.3b}$$

7-7 a. Let the nth PARCOR coefficient in an adaptive lattice filter be written as $k_n(T) = E[k_n(T)] + \tilde{k}_n(T)$ where $\tilde{k}_n(T)$ represents the fluctuations of $k_n(T)$ about its mean value. Show that the asymptotic mean value $E_\infty[k_n(T)]$ that exactly minimizes the asymptotic MSE $E_\infty[e_f^2(T|n)]$ can be written as

$$\{E_\infty[k_n(T)]\}_{opt} = k'_{n,opt} + \delta \ , \tag{7.P.4}$$

where $k'_{n,opt} = \dfrac{E_\infty[e_f(T|n-1)e_b(T-1|n-1)]}{E_\infty[e_b^2(T|n-1)]}$ and δ is a constant that depends upon the higher order statistics of the input sequence. If $k_n(T)$ and $e_b(T-1|n-1)$ are assumed to be independent, show that $\delta = 0$.

b. Even if δ in part (a) is approximately zero, deviation of the asymptotic mean of the coefficient estimate, $E_\infty[k_n(T)]$, from the MMSE value $k_{n,opt}$ may be caused by the fluctuations of $k_1(T)$ through $k_{n-1}(T)$ about their mean values, which perturb the second order statistics of $e_f(T|n-1)$ and $e_b(T|n-1)$. (Note that the computation of $k_{n,opt}$ presumes that $k_1, \ldots, k_{n-1}$ are constants.) Estimate the effect of $\tilde{k}_{n-1}(T)$ upon $k'_{n,opt}$ by computing the offset factor $\dfrac{k'_{n,opt}}{k_{n,opt}}$. (Assume that $k_{n,opt}$ is given by (7.2.2) where $k_{n-1}(T) = k_{n-1,opt}$. To compute $k'_{n,opt}$, assume that $k_{n-1}(T) = E[k_{n-1}(T)] + \tilde{k}_{n-1}(T)$. Use the independence assumption and (7.2.3).) Compute this factor for $k_{n-1,opt} = 0.6$, $k_{n,opt} = 0.8$ and $E_\infty[\tilde{k}_{n-1}^2] = 0.0025$.

7-8 a. Derive the coefficient variance formula (7.2.21) from (7.2.19) by using (7.2.20) and assuming that $e_f(T|n-1)$ and $e_b(T|n-1)$ are jointly Gaussian.

b. Derive the coefficient variance formula (7.2.26) from (7.2.25) again assuming that $e_f(T|n-1)$ and $e_b(T|n-1)$ are jointly Gaussian.

7-9 Using (7.2.28) and assuming that the coefficient $k_n(T)$ is adapted via (7.2.4), and that $e_f(T|n-1)$ and $e_b(T|n-1)$ are jointly Gaussian, derive the difference equation (7.2.29) for the output MSE of a single stage of an SG lattice filter.

7-10 In this problem we explore the conditions under which a second-order SG transversal predictor converges faster than a two-stage SG lattice predictor. Specifically, initial conditions and input second-order statistics are derived such that a second-order SG transversal predictor will converge faster than a two-stage lattice predictor. For simplicity, consider only (7.1.1) and (7.2.4) as the adaptive algorithms.

a. Assuming that the same sequence is the input to both the transversal and lattice predictors, and that the step size β is the same in both cases, show that

$$\tau_f^l \leqslant \tau_1^l \leqslant \tau_2^l \leqslant \tau_s^l \tag{7.P.5}$$

where τ_f^l and τ_s^l represent, respectively, the time constants associated with the second-order transversal fast and slow normal modes (problem 7-5) and τ_i^l, $i=1,2$, is the ith stage lattice time constant.

b. Referring to problem 7-5, derive the condition under which the mean transversal coefficients converge as fast as possible.

c. Given the transversal coefficients $f_1(0)$ and $f_2(0)$, compute the PARCOR coefficients $k_1(0)$ and $k_2(0)$ that give the same output MSE at time $T = 0$.

d. To retard the convergence of the two-stage lattice filter, the discussion in section 7.5 indicates that $k_2(0)$ should be chosen greater than $k_{2,opt}$. Assume that the input sequence is described by (7.P.1) and that the autocorrelation coefficient $\phi_1 > 0$, and use the fact that $|k_j| < 1$ to derive the inequalities,

$$0 < a < f_1(0) \leqslant \frac{1 - (b - a)}{2} \tag{7.P.6a}$$

$$-1 < b < f_2(0) \leqslant \frac{1 + (b - a)}{2}. \tag{7.P.6b}$$

e. The initial MSE depends upon how far away the initial coefficient values are from the MMSE coefficients. Show that to maximize the initial MSE,

$$f_1(0) - a = f_2(0) - b \equiv d \tag{7.P.7}$$

and that the maximum value of d is

$$d_{\max} \leqslant 1 - \frac{\phi_1}{\phi_0}. \tag{7.P.8}$$

f. In order to construct an example where the second-order transversal filter converges significantly faster than the two-stage lattice filter, part (a) indicates that $\frac{\phi_1}{\phi_0}$ must be as large as possible. On the other hand, the larger $\frac{\phi_1}{\phi_0}$, the smaller the maximum initial misadjustment $d_{\max}$ will be. Assume that the parameters $a = 0.9$, $b = 0.8$, $f_1(0) = f_2(0) = 0$, and determine whether or not a second-order SG transversal filter will converge faster than an equivalent two-stage SG lattice filter, assuming both filters use the same adaptive step size.

7-11 Derive the single stage LS lattice time constant formula (7.3.8).

7-12 Prove the result (7.4.1).

7-13 The N+1-element vector $\tilde{\mathbf{g}}(T|N)$, which enters the fast Kalman convergence model, is computed at each iteration via (7.4.24). This vector must be replaced by the shifted vector $z^{-1}\tilde{\mathbf{g}}(T|N)$ at the next iteration of the algorithm. Show that the $(N+1)$st component of

$\tilde{\mathbf{g}}(T|N)$ is zero.

7-14 Derive the formula for asymptotic coefficient variance (7.6.22) by using (7.6.4c) and the same type of assumptions that were used to derive (7.2.26).

7-15 a. Assume that a fifth-order adaptive predictor is used to filter a unity variance white Gaussian input sequence. Compute the output MSE of the filter obtained from i) the SG transversal algorithm (7.1.1), ii) the SG lattice algorithm (7.2.4), and iii) the SG lattice algorithm (7.2.5). Assume that $\beta = 0.03$ in all cases. Compute the output MSE of the LS lattice algorithm from (7.7.16) assuming w=0.97.

b. Assume now that the sequence

$$d(T) = 0.9y(T) + 0.8y(T-1) + 0.7y(T-2) \qquad (7.P.9)$$

is the desired input to a fifth-order adaptive joint process estimator. Compute the output MSE of the filter for the four cases listed in part (a).

7-16 a. Assume that a periodic pseudo-random sequence of length N is the input to a channel with transfer function $H(z)$. Assuming that additive channel noise is negligible, derive the condition on $H(z)$ that guarantees that the output vectors $\mathbf{y}(j), \mathbf{y}(j+1), \ldots, \mathbf{y}(j+N-1)$ are linearly independent. How does this condition compare with the condition for the non-uniqueness of the MMSE solution (see section 3.2.2.4)? (Hint: Write the output vector as $\mathbf{y}(j) = \mathbf{B}\mathbf{d}(j)$, where $\mathbf{B}$ is a *circulent* matrix [37], which depends only on the impulse response of the channel, and $\mathbf{d}(j)$ is defined by (7.7.31). Use the facts that $\mathbf{d}(j), \mathbf{d}(j+1), \ldots, \mathbf{d}(j+N-1)$ are linearly independent for all j, and that the eigenvalues of a circulant matrix are given by the discrete Fourier transform of the first row.)

b. Show that the asymptotic LS solution for the equalizer coefficients is the MMSE solution given by (3.1.6). (Do not neglect the channel noise.)

7-17 *Properties of the LS Estimator*

a. Suppose we wish to estimate the parameter vector $\mathbf{c}$ given a vector of noisy observations $\mathbf{Y}$, and suppose that the (stationary) probability distribution $P(\mathbf{Y}|\mathbf{c})$ is known to be Gaussian with mean $m(\mathbf{c})$ and nonsingular covariance matrix Σ. Show that the following weighted LS estimate of $\mathbf{c}$, given by $\hat{\mathbf{c}}$ such that $[\mathbf{Y} - m(\hat{\mathbf{c}})]'\Sigma^{-1}[\mathbf{Y} - m(\hat{\mathbf{c}})]$ is minimized, also maximizes $P(\mathbf{Y}|\hat{\mathbf{c}})$, i.e., $\hat{\mathbf{c}}$ is the *maximum-likelihood estimate* for $\mathbf{c}$.

b. Suppose that $y(T) = \mathbf{x}'(T)\mathbf{c}$, where $\{\mathbf{x}(T)\}$ is a sequence of independent Gaussian random vectors, each of which has mean $\bar{\mathbf{x}} = E[\mathbf{x}(T)]$ and covariance matrix $\Gamma = E[\mathbf{x}(T)\mathbf{x}'(T)]$. Given observations $y(0), y(1), \ldots, y(N)$, what is the maximum-likehood estimate for $\mathbf{c}$?

7-18 Consider a stationary stochastic process $y(T)$, $T = 0, 1, 2, \ldots, N$, for which the mean value is a linear function of the unknown vector $\mathbf{c}$, i.e., $E[y(T)] = \mathbf{x}'(T)\mathbf{c}$, where $\mathbf{x}(T)$, $T = 0, 1, 2, \ldots, N$ is a sequence of known vectors. Also assume that the sequence $y(T)$ is uncorrelated and has variance σ^2. The LS estimate of $\mathbf{c}$ given the observations $y(T)$, $T = 0, 1, 2, \ldots, N$ minimizes the sum

$$S = \sum_{T=0}^{N} [y(T) - \mathbf{x}'(T)\hat{\mathbf{c}}]^2. \tag{7.P.10}$$

Show that the LS estimate of $\mathbf{c}$, denoted by $\hat{\mathbf{c}}$, has the following properties:

a. $\hat{\mathbf{c}}$ is a linear function of the observations $y(T)$,

b. $E[\hat{\mathbf{c}}] = \mathbf{c}$, where the average is with respect to the observations $y(T)$ (i.e., $\hat{\mathbf{c}}$ is unbiased),

c. $E[(\hat{\mathbf{c}} - \mathbf{c})(\hat{\mathbf{c}} - \mathbf{c})'] = \left[\sum_{j=0}^{T} \mathbf{x}(j)\mathbf{x}'(j)\right]^{-1} \sigma^2$, and

d. $E[(\hat{\mathbf{c}} - \mathbf{c})(\hat{\mathbf{c}} - \mathbf{c})'] \leqslant E[(\tilde{\mathbf{c}} - \mathbf{c})(\tilde{\mathbf{c}} - \mathbf{c})']$ where $\tilde{\mathbf{c}}$ is any other linear unbiased estimator, i.e., $\tilde{\mathbf{c}} = \mathbf{B}\mathbf{Y}$, where $\mathbf{B}$ is a constant matrix.

7-19 Suppose, as in problem 7-17, that we are to estimate the parameter vector $\mathbf{c}$ given the observation vector $\mathbf{Y}$. Let $\hat{\mathbf{c}}(\mathbf{Y})$ be an unbiased estimator of $\mathbf{c}$ given $\mathbf{Y}$, that is, $E[\hat{\mathbf{c}}(\mathbf{Y})] = \mathbf{c}$. The probability of observing the data vector $\mathbf{Y}$ given the parameter vector $\mathbf{c}$ is denoted as $P(\mathbf{Y}|\mathbf{c})$. A classical result in estimation theory is the *Cramer–Rao bound*, which states that subject to certain regularity conditions, the covariance of any unbiased estimator $\hat{\mathbf{c}}(\mathbf{Y})$ of $\mathbf{c}$ satisfies the inequality,

$$\operatorname{cov} \hat{\mathbf{c}} \geqslant \mathbf{M}_{\mathbf{c}}^{-1} \tag{7.P.11}$$

where

$$\operatorname{cov} \hat{\mathbf{c}} = E\{[\hat{\mathbf{c}}(\mathbf{Y}) - \mathbf{c}][\hat{\mathbf{c}}(\mathbf{Y}) - \mathbf{c}]'\}, \tag{7.P.12}$$

the average is with respect to $P(\mathbf{Y}|\mathbf{c})$, and $\mathbf{M}_{\mathbf{c}}$ is known as *Fisher's information matrix*. The i,jth element of $\mathbf{M}_{\mathbf{c}}$ is

$$[\mathbf{M}_{\mathbf{c}}]_{i,j} = E\left\{\left(\frac{\partial \log P(\mathbf{Y}|\mathbf{c})}{\partial c_i}\right)\left(\frac{\partial \log P(\mathbf{Y}|\mathbf{c})}{\partial c_j}\right)\right\}. \tag{7.P.13}$$

The proof of this result is in chapter 1 of [17]. Consider a Gaussian stationary stochastic process $y(T)$ for which the mean value is a linear function of the unknown vector $\mathbf{c}$ (see problem 7-17). Show that the LS estimator $\hat{\mathbf{c}}$, which minimizes the sum in (7.P.10), gives the minimum mean square error over all $\mathbf{c}$ in the class of unbiased estimators.

REFERENCES

1. G. Ungerboeck, "Theory on the Speed of Convergence in Adaptive Equalizers for Digital Communication," *IBM J. Res. and Develop.*, pp. 546-555 (Nov. 1972).

2. B. Widrow, "Adaptive Filters I: Fundamentals," *(Tech. Rept. 6764-6)*, (Dec. 1966).

3. B. Widrow, "Adaptive Filters," pp. 563-587 in *Aspects of Network and System Theory*, ed. R.E. Kalman and N. DeClaris,Holt, Rinehart, and Winston, New York (1971).

4. B. Widrow, J. McCool, M. Larimore, and C. Johnson, Jr., "Stationary and Non-Stationary Learning Characteristics of the LMS Adaptive Filter," *Proc. IEEE* **64**(8) pp. 1151-1162 (Aug. 1976).

5. B. Widrow et. al., "Adaptive Noise Canceling: Principles and Applications," *IEEE Proceedings* **63**(12) pp. 1692-1716 (Dec. 1975).

6. R.D. Gitlin, J.E. Mazo, and M.G. Taylor, "On the Design of Gradient Algorithms for Digitally Implemented Adjustment Filters ," *IEEE Trans. on Circuit Theory* **CT-20** pp. 125-136 (March 1973).

7. O. Macchi and E. Eweda, "Second Order Convergence Analysis of Stochastic Adaptive Linear Filtering," *IEEE Trans. on Automatic Control* **AC-28**(1) pp. 76-85 (Jan. 1983).

8. M.L. Honig, *Performance of FIR Adaptive Filters Using Recursive Algorithms,* Ph.D. Dissertation, Univ. of California, Berkeley, CA. (April 1981).

9. E.H. Satorius and S.T. Alexander, "Channel Equalization Using Adaptive Lattice Algorithms," *IEEE Trans. Communications* **COM-27** p. 899 (June 1979).

10. M. S. Mueller, "On the Rapid Initial Convergence of Least-Squares Equalizer Adjustment Algorithms," *Bell System Technical Journal* **60**(10) pp. 2345-2358 (Dec. 1981).

11. R.W. Lucky, J. Salz, and E.J. Weldon, Jr., *Principles of Data Communication,* McGraw-Hill, New York (1968).

12. K.H. Mueller and D. Spaulding, "Cyclic Equalization - A New Rapidly Converging Equalization Technique for Synchronous Data Communication," *Bell System Technical Journal* **54**(7) pp. 369-406 (Feb. 1975).

13. E. Satorius and J. Pack, "Application of Least Squares Lattice Algorithms to Adaptive Equalization," *IEEE Trans. Comm.* **COM-29** pp. 136-142 (Feb. 1981).

14. C. L. Lawson and R. J. Hanson, *Solving Least Squares Problems,* Prentice Hall, Inc., Englewood Cliffs, New Jersey (1974).

15. J. M. Cioffi and T. Kailath, "Fast, Recursive-Least Squares, Transversal Filters for Adaptive Filtering," *IEEE Trans. on Acoustics, Speech, and*

Signal Processing **ASSP-32**(2) (April 1984).

16. D. W. Lin, "On Digital Implementation of the Fast Kalman Algorithm," *IEEE Trans. on Acoustics, Speech, and Signal Processing*, (to appear).

17. G. C. Goodwin and R. L. Payne, *Dynamic System Identification: Experiment Design and Data Analysis,* Academic Press, New York (1977).

18. J. K. Kim and L. D. Davisson, "Adaptive Linear Estimation for Stationary M-dependent Processes," *IEEE Trans. on Information Theory* **IT-21** pp. 23-31 (Jan. 1975).

19. J.E. Mazo, "On the Independence Theory of Equalizer Convergence ," *The Bell System Technical Journal* **58** pp. 963-993 (May-June 1979).

20. D. C. Farden, J. C. Goding , and K. Sayood, "On the 'Desired Behavior' of Adaptive Signal Processing Algorithms," *Proc. Int. Conf. Acoust., Speech, Signal Processing*, (April 1979).

21. S. K. Jones, R. K. Cavin, III, and W. M. Reed, "Analysis of Error-Gradient Adaptive Linear Estimators for a Class of Stationary Dependent Processes," *IEEE Trans. on Information Theory* **IT-28**(2) pp. 318-329 (March 1982).

22. R. R. Bitmead, "Convergence in Distribution of LMS-Type Adaptive Parameter Estimates," *IEEE Trans. on Automatic Control* **AC-28**(1) pp. 54-60 (Jan. 1983).

23. M.L. Honig and D.G. Messerschmitt, "Convergence Properties of an Adaptive Digital Lattice Filter," *IEEE Trans. ASSP* **ASSP-29**(3) pp. 642-653 (June 1981).

24. M.L. Honig, "Convergence Models for Adaptive Gradient and Least Squares Algorithms," *IEEE Trans. on ASSP* **31**(2) pp. 415-425 (April 1983).

25. C. Caraiscos and B. Liu, "A Roundoff Error Analysis of the LMS Adaptive Algorithm," *IEEE Trans. on Acoustics, Speech, and Signal Processing* **ASSP-32**(No. 1) pp. 34-41 (Feb. 1984).

26. S. Ljung and L. Ljung, "Error Propagation Properties of Recursive Least Squares Adaptation Algorithms," Internal Report LITH-ISY-I-0620, Dept. of Electrical Engineering, Lindkoping University Lindkoping, Sweden (Sept. 1983).

27. S. Ljung, "Fast Algorithms for Integral Equations and Least Squares Identification Problems," *Ph.D. Dissertation*, Institutionen for Systemteknik, Linkoping University, (1983.).

28. F. Ling and J. G. Proakis, "Numerical Accuracy and Stability: Two Problems of Adaptive Estimation Algorithms Caused by Round-Off Error," *Proc. 1984 Int. Conf. Acoustics, Speech, and Signal Processing*, (30.0) (March 1984).

29. C. Samson and V. Reddy, "Fixed Point Error Analysis of the Normalized Ladder Algorithm," *IEEE Trans. on ASSP* **ASSP-31**(5) pp. 1177-1191 (Oct. 1983).

30. J. M. Cioffi and T. Kailath, "Windowing Methods and their Efficient Transversal-Filter Implementation for the RLS Adaptive-Filtering Criterion," *IEEE Trans. on ASSP*, (to appear).

31. D. Manolakis, F. Ling, and J. G. Proakis, "Efficient Least-Squares Algorithms for Finite Memory Adaptive Filtering," *1984 Conf. on Information Sciences and Systems*, (March 1984).

32. R.S. Medaugh and L.J. Griffiths, "A Comparison of Two Fast Linear Predictors," *Proc. 1981 IEEE ICASSP*, (April 1981).

33. R. S. Medaugh and L. J. Griffiths, "Further Results of a Least-Squares and Gradient Adaptive-Lattice Algorithm Comparison," *Proc. IEEE ICASSP*, pp. 1412-1415 (May 1982).

34. B. Friedlander, "Lattice Filters for Adaptive Processing," *Proceedings of the IEEE* **70**(8) pp. 829-867 (August 1982).

35. F. Ling and J. G. Proakis, "Nonstationary Learning Characteristics of Least Squares Adaptive Algorithms," *Proc. 1984 Int. Conf. Acoustics, Speech, and Signal Processing*, (3.7) (March 1984).

36. D. Godard, "Channel Equalization Using Kalman Filter for Fast Data Transmission," *IBM J. of Res. and Dev.*, pp. 267-273 (May 1974).

37. Robert M. Gray, "On the Asymptotic Eigenvalue Distribution of Toeplitz Matrices," *IEEE Trans. on Inform. Theory* **IT-18** pp. 725-730 (Nov. 1972).

INDEX

A

B

C

D

E

F

G

H

I

J

K

L

M

N

O

P

T

U

V

W